Springer Series on Environmental Management

John M. Gunn, Editor
Ontario Ministry of Natural Resources
Cooperative Freshwater Ecology Unit
Laurentian University
Sudbury, Ontario, Canada

Restoration and Recovery of an Industrial Region

Progress in Restoring the Smelter-Damaged Landscape Near Sudbury, Canada

With 221 illustrations, 18 in full color

Springer-Verlag
New York Berlin Heidelberg London Paris
Tokyo Hong Kong Barcelona Budapest

John M. Gunn
Ontario Ministry of Natural Resources
Cooperative Freshwater Ecology Unit
Laurentian University
Sudbury, Ontario, Canada

Cover photos: A site at the Copper Cliff tailings area were taken in 1973 (by Tom Peters) and in 1994 (by Ellen Peale).

Library of Congress Cataloging-in-Publication Data
Gunn, John M.
Restoration and recovery of an industrial region: the smelter -damaged landscape near Sudbury, Canada / John M. Gunn.
p. cm.
Includes bibliographical references and index.
ISBN-13:978-1-4612-7568-8 e-ISBN-13:978-1-4612-2520-1
DOI: 10.1007/978-1-4612-2520-1

1. Sulphur dioxide--Environmental aspects. 2. Smelting--Environmental aspects. 3. Acid rain--Ontario--Sudbury (District) 4. Reclamation of land--Ontario--Sudbury (District) I. Title.
TD 196.S95G86 1995
363.73'928'09713133--dc20 95-9909

Printed on acid-free paper.

Softcover reprint of the hardcover 1st edition 1995

Production managed by Princeton Editorial Associates and supervised by Karen Phillips; manufacturing supervised by Rhea Talbert.
Typeset by Princeton Editorial Associates.

9 8 7 6 5 4 3 2 1

ISBN-13:978-1-4612-7568-8

In August, 1888, a roast pile was built and started burning in the middle of a dense growth of spruce and birch trees. Again mind triumphed over matter and the woods fell before the stench. The roast yard was quite a success.

D.H. Browne (circa 1889),
an official with Canadian Copper Company

Last came the prospector and the mining company: but when they came they made the region theirs, and what they found, made all other industries seem of no account. Even the sulphur that blasted all things living, only made nature's grimness grimmer still, substitutued, as it were, deadly purpose for beautiful desolation.

Stephen Leacock (1937),
from *My Discovery of the West*

It was really very encouraging to see the results of all your labours particularly in the quest of trying to reduce the sulphur emissions . . . and the fact that you are setting an example in this area will I'm sure make a great difference to other plants in other parts of the world.

Prince Charles (1991),
during his visit to Sudbury

One of the happiest and most proud moments of the entire Earth Summit, for me as a Canadian, came when the Sudbury community was given the 1992 United Nations Local Government Honours Award for its work to reverse the process of environmental degradation.

Maurice Strong (1994),
UN Secretary General of Earth Summit

Preface

The acid test of our understanding is not whether we can take ecosystems to bits on pieces of paper, however scientifically, but whether we can put them together in practice and make them work.

A.D. Bradshaw, 1983

Home of one of the world's largest metal smelting complexes, Sudbury, Ontario, Canada is well known as a polluted region. Many superlatives and startling statistics describe this area: one of the largest point sources of sulfur dioxide emissions; 17,000 ha of industrial barrens; 7000 acid-damaged lakes; but the picture of "Superstack," the world's tallest smokestack, dispersing pollutants high into the atmosphere, is the powerful image that has brought this area so much unwanted attention. Throughout the much publicized acid rain debates of the 1980s, this 381-m structure became symbolic of the global nature of environmental destruction and the need for international agreements to control these problems.

Sudbury is now beginning to assume a new international reputation, this time a more favourable one. During the debates that preceded the 1990 revisions of the Clean Air Act by the U.S. Congress, the evidence of progressive recovery of acid lakes in the Sudbury area after emission reductions from local smelters was a forceful argument in support of expenditures on control of pollution at the source. Similarly, the recognition of Sudbury's "re-greening program" by the United Nations at the Earth Summit in Brazil has again thrust the region into the international spotlight. With the world now facing the prospect of cleaning up horrendous "Sudburys" in China, India, eastern Europe, and elsewhere, a positive example of improved industrial technology and land and water reclamation techniques is badly needed. In the early acid rain research literature, the word *Sudbury* was occasionally used as a unit or measure of pollution. In the future, it is our hope that *Sudbury* may someday deserve to be used as a measure of restoration.

This book grew out of a series of technical workshops, called the Sudbury Rehabilitation Workshops, that were held at Laurentian University in 1990,

1991, and 1992. These workshops were designed as informal sessions to encourage greater technical exchange and collaboration between researchers from universities, government environmental agencies, and local mining companies. Although most of the participants had conducted research in the Sudbury area for many years, many of them were not familiar with each other's work. However, all shared a common interest in reversing the serious environmental problems of the area, and as a result, several collaborative projects have developed from these meetings. This book is a result of that collaborative spirit. It represents a collective effort to bring an important case history of environmental restoration to as wide an audience as possible.

Another important stimulus for this book was the publication of the proceedings of a very similar workshop dealing with the effects of smelter emissions on the Kola Peninsula in the border area between Russia and Finland (Kozlov, M.V. et al. [eds.]. 1993. Aerial Pollution in Kola Peninsula, Proceedings of the International Workshop, April 14–16, 1992, St. Petersburg, Russia). The descriptions of the ecological damage around the Russian smelters are hauntingly similar to those of Sudbury. Interestingly, the scientific studies in that area have not overlapped significantly with research efforts in the Sudbury area, but rather the studies complement each other. In the Kola area most of the research dealt with terrestrial effects, particularly effects on forest growth. In Sudbury, much of the vegetation damage occurred well before scientific studies began. Here most of the research emphasis was placed on aquatic effects and on remedial studies (aquatic, terrestrial, industrial). I should not overstate the case, but in thinking about the Russian smelter problems, it is also noteworthy that the political changes that allowed Russian and Finnish scientists to meet and exchange information parallel some of the recent changes seen at our workshops. True cooperation between scientists from environmental protection agencies and representatives of the polluting companies to solve problems rather than assign blame is also a relatively recent phenomenon.

In almost every book dealing with the new field of restoration ecology there is a struggle with the term *restoration.* (Should the word be *rehabilitation* or *reclamation* if the chances of returning a system to its original condition are nil?) If I can be so bold as to speak on behalf of the many authors of this volume, we did not worry about this term. For us, *restoration* speaks to the goal that this industrial region is working toward, the re-establishment of attractive self-sustaining "healthy" ecosystems, free of toxic or other deleterious substances. Many will say that this a hopelessly naive dream in such a polluted environment. However, the steady progress that has been made suggests otherwise.

This book was designed to serve a wide international audience, including undergraduate students, environmental resource managers, and the general public. Technical terms and details have been kept to a minimum to allow us to cover a wide range of disciplines. Care has been taken to provide suitable literature sources for people desiring more specific information.

I thank the many distinguished authors and reviewers that contributed to this book. I was fortunate to have Dr. James Kramer (McMaster University), Nels Conroy (Ontario Ministry of Environment and Energy), Dr. Harold Harvey (University of Toronto), Dr. Anthony Bradshaw (University of Liverpool), Dr. John Cairns, Jr. (Virginia Polytechnic Institute), and Maurice Strong (Ontario Hydro) review the section chapters and/or prepare introductory articles. Additional external reviews were provided by Dr. Robert Hedin (U.S. Bureau of

Mines), Dr. Michail Kozlov (University of Turku, Finland), Dr. Robert Benoit (Virginia Polytechnic Institute), and Dr. John Fortesque (Ontario Geological Surveys). Most of the authors also assisted with manuscript reviews, but I must acknowledge the large contributions made by Bill Keller, Ed Snucins, and the members of the synthesis team—Nels Conroy, Bill Lautenbach, Dr. Dave Pearson, Marty Puro, Dr. Joe Shorthouse, and Mark Wiseman. Assistance with typing, graphics, photography, and copy editing was provided by Jim Carbone, Michael Conlon, Kristen Gunn, Jennifer Green-Blair, Jocelyn Heneberry, Cassandra Jacobs, Lisa MacDonald, Pat Smith, and Elizabeth Wright. A special thanks to Chris Blomme for much of the artwork, Léo Larivière for cartography, and Mary Roche and Ed Snucins for photography. Many students helped with the workshops and authors meetings. I thank them all but dare not attempt to name everyone.

This is a royalty-free book for all involved. The principal sponsors of the book were Laurentian University and Ontario Ministry of Natural Resources. Additional support was provided by Ontario Ministry of Environment and Energy and Ontario Ministry of Northern Development and Mines.

John M. Gunn
Sudbury, 23 January 1995

Contents

SECTION E. Planning for the Future

A color insert follows page 182

Contributors

Giuseppe Bagatto, Department of Biology, Laurentian University, Sudbury, Ontario P3E 2C6, Canada

Peter J. Beckett, Department of Biology, Laurentian University, Sudbury, Ontario P3E 2C6, Canada

Nelson Belzile, Department of Chemistry, Laurentian University, Sudbury, Ontario P3E 2C6, Canada

Chris G. Blomme, Department of Biology, Laurentian University, Sudbury, Ontario P3E 2C6, Canada

Dan F. Bouillon, Inco Limited, Copper Cliff, Ontario P0M 1N0, Canada

Anthony D. Bradshaw, Department of Environmental and Evolutionary Biology, University of Liverpool L69 3B8, England

Lise A. Brisebois, Canadian Wildlife Services, Nepean, Ontario K1A 0H3, Canada

John Cairns, Jr., University Center for Environmental and Hazardous Material Studies, Virginia Polytechnic and State University, Blacksburg, Virginia 24061–0415, USA

Nels Conroy, Ontario Ministry of Environment and Energy, Sudbury, Ontario P3E 5P9, Canada

Gerard M. Courtin, Department of Biology, Laurentian University, Sudbury, Ontario P3E 2C6, Canada

Peter J. Dillon, Ontario Ministry of Environment and Energy, Dorset, Ontario P0A 1E0, Canada

Aruna S. Dixit, Department of Biology, Queen's University, Kingston, Ontario K7L 3N6, Canada

Sushil S. Dixit, Department of Biology, Queen's University, Kingston, Ontario K7L 3N6, Canada

Hayla E. Evans, RODA Envir. Research Limited, Lakefield, Ontario K0L 2H0, Canada

E. Ann Gallie, CIMMER/Geography Department, Laurentian University, Sudbury, Ontario P3E 2C6 Canada

John M. Gunn, Ontario Ministry of Natural Resources, Biology Department, Laurentian University, Sudbury, Ontario P3E 2C6, Canada

Harold H. Harvey, Department of Zoology, University of Toronto, Toronto, Ontario M5S 1A1, Canada

Ellen L. Heale, Inco Limited, Copper Cliff, Ontario P0M 1N0, Canada

Wendel (Bill) Keller, Ontario Ministry of Environment and Energy, Sudbury, Ontario P3E 5P9, Canada

James R. Kramer, Department of Geology, McMaster University, Hamilton, Ontario L8S 4M1, Canada

William E. Lautenbach, Regional Municipality of Sudbury, Sudbury, Ontario P3E 5W5, Canada

Karen M. Laws, Ontario Ministry of Natural Resources, Sudbury, Ontario P3A 4S2,Canada

Mark L. Mallory, Canadian Wildlife Services, Nepean, Ontario K1A 0H3, Canada

Christine E. Maxwell, Department of Biology, Trent University, Peterborough, Ontario K9J 7B8, Canada

Don K. McNicol, Canadian Wildlife Services, Nepean, Ontario K1A 0H3, Canada

Robert E. Michelutti, Falconbridge Limited, Falconbridge, Ontario P0M 1S0, Canada

Jim Miller, Regional Municipality of Sudbury, Sudbury, Ontario P3E 5W5, Canada

J. Robert Morris, Department of Biology, Laurentian University, Sudbury, Ontario P3E 2C6, Canada

John J. Negusanti, Ontario Ministry of Environment and Energy, Sudbury, Ontario P3E 5P9, Canada

David Pearson, Department of Geology, Laurentian University, Sudbury, Ontario P3E 2C6, Canada

Tom H. Peters, 31 School Street, Copper Cliff, Ontario P0M 1N0, Canada

J. Roger Pitblado, Department of Geography, Laurentian University, Sudbury, Ontario P3E 2C6, Canada

Raymond R. Potvin, Ontario Ministry of Environment and Energy, Sudbury, Ontario P3E 5P9, Canada

Marty J. Puro, Inco Limited, Copper Cliff, Ontario, P0M 1N0 Canada

R. Kenyon Ross, Canadian Wildlife Services, Nepean, Ontario K1A 0H3, Canada

Joseph D. Shorthouse, Department of Biology, Laurentian University, Sudbury, Ontario P3E 2C6, Canada

John P. Smol, Department of Biology, Queen's University, Kingston, Ontario K7L 3N6 Canada

Ed J. Snucins, Ontario Ministry of Natural Resources, Biology Department Laurentian University, Sudbury, Ontario P3E 2C6, Canada

Maurice F. Strong, Ontario Hydro, Toronto, Ontario M5G 1X6, Canada

E. Keith Winterhalder, Department of Biology, Laurentian University, Sudbury, Ontario P3E 2C6, Canada

Mark E. Wiseman, Falconbridge Limited, Falconbridge, Ontario P0M 1S0, Canada

Tin-Chee Wu, Regional Municipality of Sudbury, Sudbury, Ontario P3E 5W5, Canada

Norman D. Yan, Ontario Ministry of Environment and Energy, Dorset, Ontario P0A 1E0, Canada

Restoration and Recovery of an Industrial Region

SECTION A

History of Geology, Mineral Exploration, and Environmental Damage

Nels Conroy and James R. Kramer

Sudbury, Ontario, is one of at least five "Sudburys" in the world and probably the best known. It was named in 1882 by James Worthington in honor of his wife's birthplace in England. Mr. Worthington was the construction manager for the Canadian Central transcontinental rail line through the area.

The geological and anthropological histories of Sudbury, Ontario, are complex and fascinating, and some aspects are still being researched. In Table A.1, important milestones from earliest of time until the present are documented. Many of the events are discussed in the following chapters.

Ore was discovered in Sudbury in the 1880s. However, attempts to extract the copper was largely unsuccessful because of the high sulfides and the presence of the "devil's metal," nickel. Technology that was developed to recover the metals was not environmentally friendly, and the region suffered deep scars to the landscape from the mining and smelting activity. By the 1950s, Sudbury was famous as the world's largest producer of nickel and infamous as a region where the landscape has been devastated. Soil and vegetation were lost from tens of thousands of hectares of land surrounding the roast beds and the smelters. The scar was readily visible even from outer space (see Plate 1 following page 182). Decades later, the infamy was extended when it was shown that lakes at a considerable distance were acidified and grossly contaminated with metals. Fortunately, the story does not end there. Recent efforts related to reclamation, including emission reductions, have been positive. The latter chapters in this book demonstrate that, given the will and cooperation, environmental degradation can be abated over the short term.

Sudbury is not alone in its history of environmental degradation caused by mineral smelting. In fact, there are several "sister cities" in Russia where nickel smelting has created similar sites of environmental damage related to sulfur dioxide and metal particulate emissions. It is our hope that this book will provide a useful case history to assess progress toward environmental restoration.

TABLE A.1. Chronology of Important Events Influencing the Sudbury Region

	Time (years since present)
Formation/accretion of earth	4,600,000,000
Oldest known rocks of the Canadian Shield	3,920,000,000
Origin of life	3,400,000,000
Formation of Sudbury Basin	1,850,000,000
Beginning of mountain building period causing deformation of Sudbury Basin	1,800,000,000
End of mountain building period	1,630,000,000
Evolution of modern humans	1,000,000
End of last glaciation and formation of glacial deposits	10,000
Ojibways, Hurons, and Ottawas settle in the area	from 10,000
Evidence of first settlement Sheguindah	from 7,000
First European explorer (Champlain) (1615)	380
Chemical and physical properties of nickel recognized	250
Hudson Bay Trading Post established west of Sudbury Basin (1824)	171
First geological reconnaissance of the north shore of Lake Huron (Logan) (1849)	146
First mapping of mineralization in Sudbury area (1856)	139
Chicago fire exerts demands on Sudbury forests for lumber to rebuild (1879)	116
Discovery of mineralization at Sudbury during construction of transcontinental railway (1883)	112
First purchase of mining lands by Murray (1884)	111
First uses of corrosion resistant nickel steel (1885)	110
First smelting of Sudbury ores by roast heap (1888)	107
Thomas Edison stops exploring just short of discovering Falconbridge deposit (1901)	94
First geological map of Sudbury Basin (1905)	90
Ontario Royal Commission on Nickel (1915)	80
Austentite structure for stainless steel determined by Guillet (1915)	80
Damages by Sulphur Fumes Arbitration Act proclaimed (1921)	74
Founding of International Nickel of Canada Ltd. (Inco) (1928)	67
Formation of Falconbridge Nickel Mines Ltd. (1928)	67
Ontario government issues environmental control orders to reduce emissions from area smelters (1969)	26
First published concerns of long-range atmospheric damage from Sudbury (1970)	25
First Earth Day (1970)	25
Completion of Inco's "Superstack" and major emission controls (1972)	23
U.N. Stockholm conference on the environment emphasizes acidification pollution to lakes (1972)	23
First International Conference on Acid Rain at Ohio State University (1976)	19
Regional land reclamation initiated (1978)	17
Ontario government initiates Countdown Acid Rain program (1985)	10
First published evidence of reversal of acidification, Sudbury lakes (1986)	9
U.N. Local Government Honours Award to Sudbury (1992)	3
Legislated reductions in sulfur dioxide emissions achieved (1994)	1

1

Geological and Geographic Setting

David A.B. Pearson and J.Roger Pitblado

The Sudbury area lies near the southern edge of the Precambrian Canadian Shield, just north of Lake Huron (46°00′N–47°30′N; 79°30′N–81° 30′W) (Fig. 1.1) Although the rocks are old, the rugged landscape is young and dominated by rocky hills and ridges, rounded and scoured by glaciers. Lakes and rivers are plentiful and often lie in fractures or weak zones in the bedrock that were more easily eroded by the moving ice. A generally thin veneer of sandy glacial sediment covers the bedrock and supports a thin soil and widespread forest. The focus of geological as well as current economic interest is the Sudbury Basin (Fig. 1.2). It is a puzzling elliptical feature that many geologists believe was produced by a meteorite collision nearly 2 billion years ago. Nickel and copper ore has been mined for more than a century from more than 90 mines around the edge of the structure.

Prevailing winds have carried pollution from smelting over a wide area to the east and northeast, causing the damage to vegetation and lakes described in Chapters 2 and 3. Unfortunately, the bedrock and glacial sediment have contributed very little acid neutralizing capability to the lakes and soils affected by these emissions.

Sudbury Basin

Sudbury is the site of the largest known concentration of nickel on the surface of the planet. Nearly 20 million tons of the metal have either already been extracted or are known to be available in ore reserves. Only the Noril′sk deposits (Box 1.1) north of the Arctic Circle in central Siberia, with a known total of about 15 million tons, are in the same class as Sudbury (Naldrett 1994). Almost all the earth's nickel is concentrated in the core of the planet, and just a very tiny fraction is found in surface rocks. Only about two dozen significant deposits are known (Fig. 1.3).

The Sudbury ore deposits (Dressler et al. 1991) lie in sporadic pockets around the 150-km rim of what is widely known as the Sudbury Basin (see Fig. 1.2). Strictly speaking, geologists reserve this term for the low-lying ground, underlain by sandstone and slate, in the center of the structure. However, in this book, *Sudbury Basin* will be used for the geological structure as a whole. On the surface, the basin is enclosed by the once molten rocks of the Sudbury Igneous Complex. In cross section, the igneous rocks underlie or cradle the rocks of the basin like a spoon. It is an enigmatic feature, and even after 40 years of intensive study, it is not clear how it was formed. Several characteristics make it clear that it was the result of a violent event. Foremost is the enormous volume, estimated to be 1670 km^3 (Stevenson 1972), of breccia or welded broken rock (the Onaping Breccia) that forms a 1.6-km-thick blanket over the igneous rock within the basin. The fragments of the breccia are

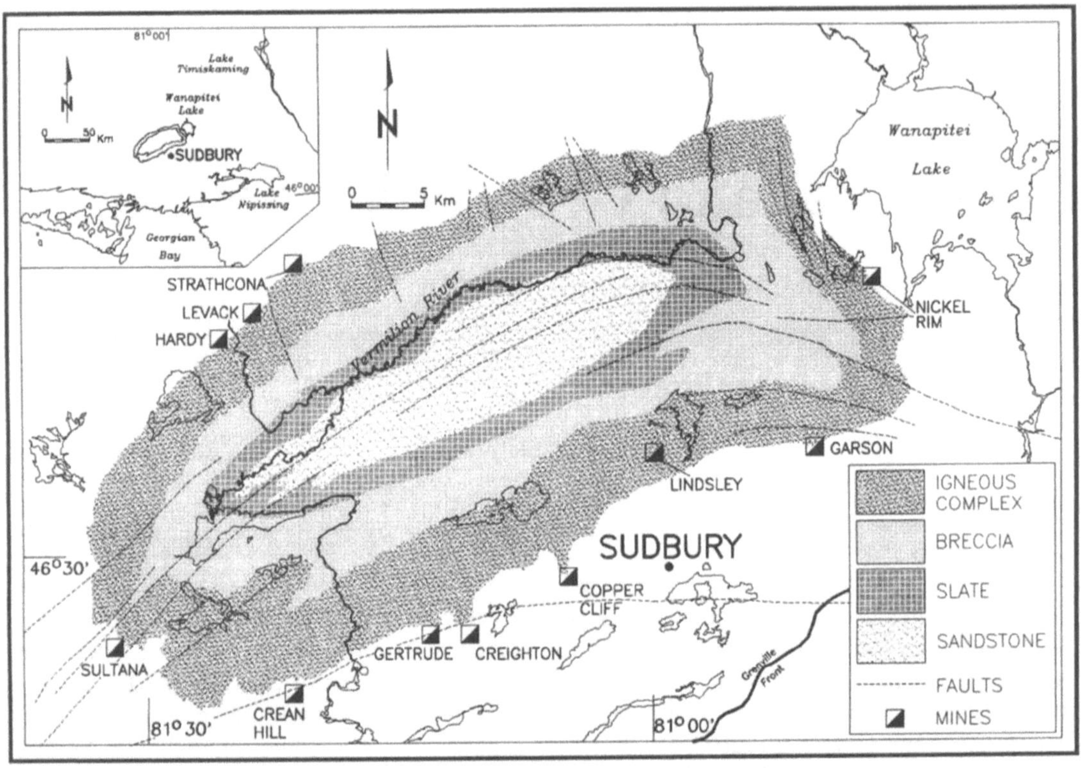

FIGURE 1.1. Generalized location map of the study area and geology of the Sudbury Basin.

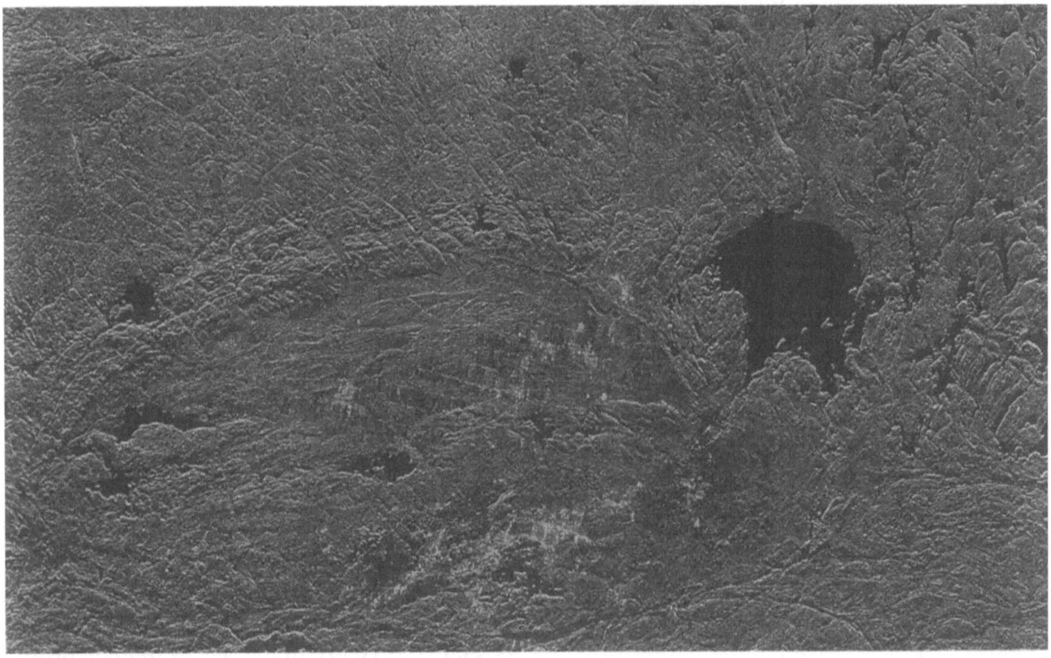

FIGURE 1.2. Radar image of the Sudbury Basin and Wanapitei Lake from 6000 m. (Courtesy of the Canada Centre for Remote Sensing.)

Box 1.1. Noril'sk: The Treasure of Siberia

Open-pit mine near Noril'sk (photo by P. Lightfoot, Ontario Geological Surveys). Noril'sk and Sudbury are the two great nickel sulfide deposits of the world. Also, Noril'sk stands, with deposits in Zimbabwe, as the predominant world resource site for platinum group metals. In relation to Sudbury ore, the Talnakh deposit from the Noril'sk area contains twice the nickel grade, four times the copper grade, and five times the platinum group element grade (Naldrett 1994). The other two areas of Russia where nickel is mined, smelted, and refined are the Kola Peninsula, near the Norwegian and Finnish border, and the middle and southern Urals. Like Noril'sk and Sudbury, the Kola Peninsula deposit is a sulfide ore. The Ural deposits are of the laterite type.

considered to have fallen back to earth into a crater after either one or more volcanic explosions (Muir 1984) or after the impact of a 10-km-diameter meteorite (Grieve 1994). The explosive violence of this catastrophic event, whatever its cause, also shattered the rock around the crater, producing half-cone-shaped fractures called shatter cones (Fig. 1.4, Box 1.2) as well as open fissures that instantly filled with crushed fragments, now referred to as Sudbury Breccia. The igneous rocks forming the hilly rim of the Sudbury Basin are seen either as impact melt generated on the floor of the crater (Golightly 1994) or as a succession of molten intrusions that rose from lower in the crust at roughly the time the crater was produced (Naldrett and Hewins 1984). The ore is thought to have separated as hot sulfide droplets from the molten igneous rock on the crater floor. It is no longer thought, as was once suggested, to have been melted meteorite.

Age dating of the igneous rock has established that it crystallized about 1850 million years ago (Krogh et al. 1984). This is seen as a more or less accurate date for the origin of the basin as a whole.

The elliptical pattern of the outcrops is inherited from compressive forces that built mountains to the southwest of Sudbury after the structure was formed. Dislocation of the south side of the structure may have shoved it several tens of kilometers over what had been the center of the original crater, effectively halving its diameter (Milkereit et al. 1992)

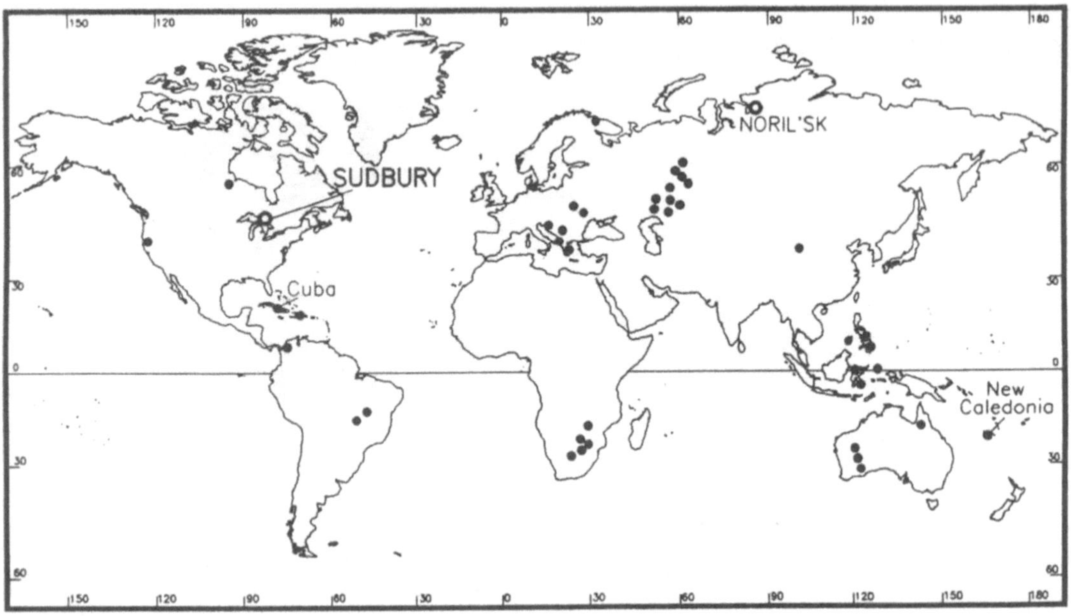

FIGURE 1.3. Nickel mining areas of the world. The two largest producing sites (>100,000 tonnes Ni/yr) are at Sudbury, Canada, and Noril'sk, Russia. At any one indicated site, there may be several individual mines.

FIGURE 1.4. Shatter cones in bedrock outside the southern edge of the Sudbury Basin. Most point toward the center of the basin and are thought to indicate the location of an extremely violent explosion. (Courtesy of Wilf Meyer.)

Box 1.2.

Apollo 16 astronauts Charles Duke and John Young visited Sudbury in July 1971, just a few months before walking on the moon. NASA's purpose was to have the astronauts practice describing the rocks and geological features in preparation for reporting on the geology of the moon. However, this was widely misunderstood by Sudburians still oversensitive to jokes about living in a "moonscape" and deliberately overlooked by outside commentators happy to find an easy target. The Apollo 17 crew, which included Harrison Schmitt, a geologist, also came to Sudbury a year later, but they left their packs behind. Instead of astronauts moonwalking, it was shatter cones (Fig. 1.4) that made the national television news!

Erosion has cut down at least 5 km, but recent seismic reflection work shows the floor of the Sudbury Basin is still between 10 and 15 km below the surface (Milkereit et al. 1992).

Sudbury Ore

Nickel and copper are the main products from the Sudbury ore, but cobalt, platinum, palladium, osmium, iridium, rhodium, ruthenium, gold, silver, selenium, and tellurium are significantly valuable byproducts as well as potential trace contributors to the geochemical environment. Also, zinc, lead, and arsenic are frequently present in trace amounts.

Almost all the minerals in the ore are sulfides, including the waste mineral pyrrhotite, an iron sulfide that dominates the ore, often to the extent of 80–90%. The nickel mineral pentlandite and copper-rich chalcopyrite make up the bulk of the remaining 10–20%. High-grade ore yields between 7 and 10% refined nickel and copper combined, with nickel usually predominating.

Ice and Landscape

A mere 12,000 years ago, the Sudbury area was buried beneath the Laurentide ice sheet, which covered most of Canada with 1–2 km of ice during the last Ice Age. This and several previous advances of ice stripped soil and overburden from the surface, deepened existing rock basins, gouged out many others, and forced the

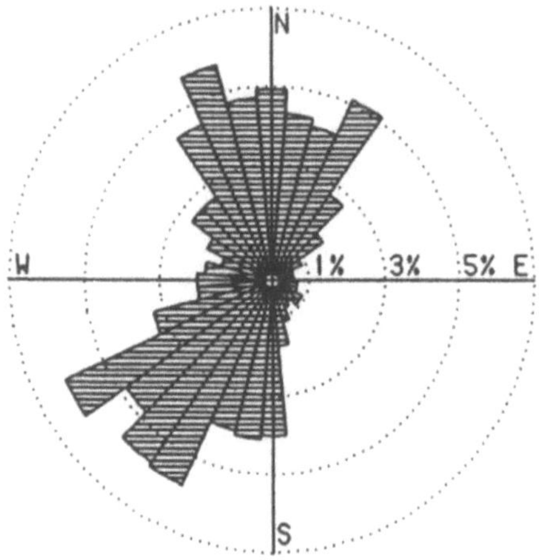

FIGURE 1.5. Frequency distributions shown as a rose diagram highlight the prevailing wind directions.

southward migration of plants and animals (Dredge and Cowan 1989; Trenhaile 1990).

The advancing ice created a bare bedrock surface, in places thinly plastered by a stony and sandy ground moraine (Boissonneau 1968; Burwasser 1979; Barnett 1992). Study of striations scratched into the bedrock and the dispersal of boulders of distinctive lithologies (Shilts 1989) has shown that the ice in Ontario moved toward the southwest or south.

Over the next 2000 years, the landscape was dramatically altered as the ice melted. Meltwater spilled out from the ice front in countless streams and rivers to form massive glacial lakes. One example is Lake Algonquin, which at one time occupied much of what is today Lake Huron and Georgian Bay. All of the area from the north rim of the Sudbury Basin southward was engulfed until the Mattawa-Ottawa river outlet became ice-free to the east. Locally, lakes formed, drained, and re-formed in response to the progressive lowering of post-Algonquin lakes and altering drainage patterns. It was at this time that massive amounts of ice-contact and glaciofluvial materials were deposited to the north and east of the Sudbury Basin. These included the eskers in the vicinity of Falconbridge and sand-gravel deposits to the north and south of Wanapitei Lake. Concurrently, clay and silt were deposited in great quantities in the proglacial lake that formed within the Sudbury Basin itself as well as along the outer margin of the southern basin rim.

Today, the majority of landscapes can be described as rock knobs or ridges of rolling, undulating, sometimes rugged topography of moderate elevation or relief. A wide variety of interesting landforms built from glacial deposits are interspersed. Low relief deposits of silt, silty clay, and organic terrains are scattered throughout the region. The most extensive level areas are the glacial lake and river sediments within the Sudbury Basin itself and, toward the east, parts of the sandy Lake Nipissing lowlands.

Climate

The Sudbury area is alternately buffeted by very cold-dry continental arctic, cool-dry continental polar, hot-dry continental tropical, and warm-moist maritime tropical air masses. The flow of these air masses generates prevailing wind directions from the north and southwest (Fig. 1.5). These prevailing winds have had a profound influence on where airborne pollutants have had an effect on terrestrial and aquatic ecosystems in the region.

The modified continental climate of the area is characterized by relatively long severe winters and short temperate summers. Based on 1951–1980 normals (Environment Canada 1982) for the Sudbury Airport climate station, mean daily minimum and maximum temperatures for the year are 1.6°C and 8.3°C, respectively. The average daily temperature in the month of January is –12.3°C, increasing to 19.8°C in July, the warmest month of the year. Precipitation is uniformly distributed throughout the year, although highest in the summer months. Over the 30-year period from 1951 to 1980, total precipitation has averaged 860 mm/year, with just less than 250 cm of snowfall annually.

Since 1895, the mean annual temperature for Canada has increased by 1.1°C (Gullett and Skinner 1992). Locally, although there have been great variations in monthly total precipitation, the mean monthly temperature shows little variation over the past 100 years. In the past few decades, however, the Sudbury climate has moderated, taking on more of the climatic characteristics associated with parts of southern Ontario.

Soils

With the retreat of the Laurentide ice sheet dated at only 9,000–10,000 years ago, relatively little time has passed for the development of mature mineral soils on the exposed bedrock outcrops. Well-drained to excessively drained shallow, stony, sandy soils, known as regosols, have accumulated in pockets of exposed rock knobs and ridges, in depressions, and along small stream valleys. Similarly, widespread but localized occurrences of poorly drained gleysols (formed when soils are saturated with water either continuously or for long periods during the year) and organic soils are important.

The dominant soil-forming process in the region is podzolization, a process in which organic acids form in the surface horizons, leach basic elements such as calcium, magnesium, iron, and aluminum from the upper layers, and then deposit them in soil horizons immediately below (Canada Soil Survey Committee 1978). Under a coniferous or mixed conifer-deciduous forest cover, this process is enhanced by the cool humid climate and the acidic parent materials produced by the silica-rich Precambrian bedrock. Podzols are very well defined on well-drained sandy tills of the area and are characterized by their dark organic surface layer, a white to ash-gray leached horizon immediately below, followed by yellowish-brown or reddish-brown subsoils.

Vegetation

It is tempting to suggest that today's vegetation (see Chapter 2) represents postglacial communities that followed a simple progression from tundra, to boreal, and finally to the current mixed coniferous-deciduous woodlands. Mounting evidence suggests that is not the case. Pollen analyses (Webb 1985; Gajewski 1988) in the midwest and eastern United States indicate that temperatures 6000 years ago were 1.5°C higher than at present, allowing for increased rates of soil formation and the expansion of temperate vegetation species well beyond their present-day ranges. But over the past 2000 years, there has been a long-term gradual cooling. The latter would permit an expansion of boreal elements at the expense of more-temperate tree species and similarly a decrease in the rates of soil formation.

Similar patterns have been found from pollen studies in Canada (Ritchie 1966, 1988). These indicate a southward extension of the boreal forest in the west and a significant increase in spruce in the transition zone between the boreal and temperate forest of southern Quebec. It is very unlikely that such changes in climate, rates of soil formation, and shifts in species distribution were not experienced as well in the Sudbury area.

Lakes

For both visitors and residents alike, one of the most striking features of the Canadian Shield is the enormous number of lakes that dot the landscape. It has been estimated that within the province of Ontario there are more that 226,000 lakes, and approximately 20,000 of these are within 100 km of Sudbury (Cox 1978). A good impression of the number and complex distribution of these lakes is provided in Figure 1.6. The landscape is strewn with lake basins controlled by bedrock faults and glacial

Figure 1.6. Principal lake patterns derived from a 1:250,000 scale topographic map.

scour, or areas of glacioloacustrine deposits (Sudbury Basin, Nipissing lowlands) virtually unoccupied by lakes. The principal exception is the largest of the lakes, Wanapitei, located in the center of the figure. This is a crater lake created 37 million years ago by meteorite impact (Grieve and Robertson 1987).

Geological Influence on Environmental Geochemistry

The chemical make-up of the bedrock and overlying glacial deposits, as well as processes of soil formation, are critical factors in the ability of aquatic and terrestrial ecosystems to withstand the effects of chemical pollutants. In the Sudbury area, several studies have investigated these relationships in ecosystems affected by acidic precipitation and heavy metals (Conroy and Keller 1976; Semkin and Kramer 1976; Griffith et al. 1984; Jeffries et al. 1984).

With regard to acidification, efforts have been made at a very general level to assess and map "sensitivity" to acidic precipitation (Shilts et al. 1981; Cowell, 1986). Sensitive areas are those in which the bedrock, overlying glacial material, and soil have little ability to neutralize or buffer the acid (Fig. 1.7). Buffering capacity is provided by minerals that accept protons (or hydrogen ions) from acid solutions. The minerals themselves may be dissolved in the process, as occurs with calcite (calcium carbonate), the main mineral of limestone, or they may be altered to another mineral in the case of fine-grained clay.

The most effective buffering is provided by either limestone bedrock or finely ground-up limestone in glacial drift. However, in the hinterland of Sudbury's smelters, limestone is rare. The only extensive outcrops are in the Hudson Bay lowland more than 500 km to the north. Although it is possible for material to be carried that far, Karrow and Geddes (1987) found that the carbonate content of glacial debris fell to less than 10% about 175 km north of Sudbury. Nevertheless, two local sources must be borne in mind: an outlier of limestone around Lake Timiskaming near the Ontario-Quebec border, and small outcrops of Precambrian limestone north of Wanapitei Lake. An example of the influence of the Lake Timiskaming outlier may well be elevated calcium and magnesium levels in near-neutral and alkaline lake

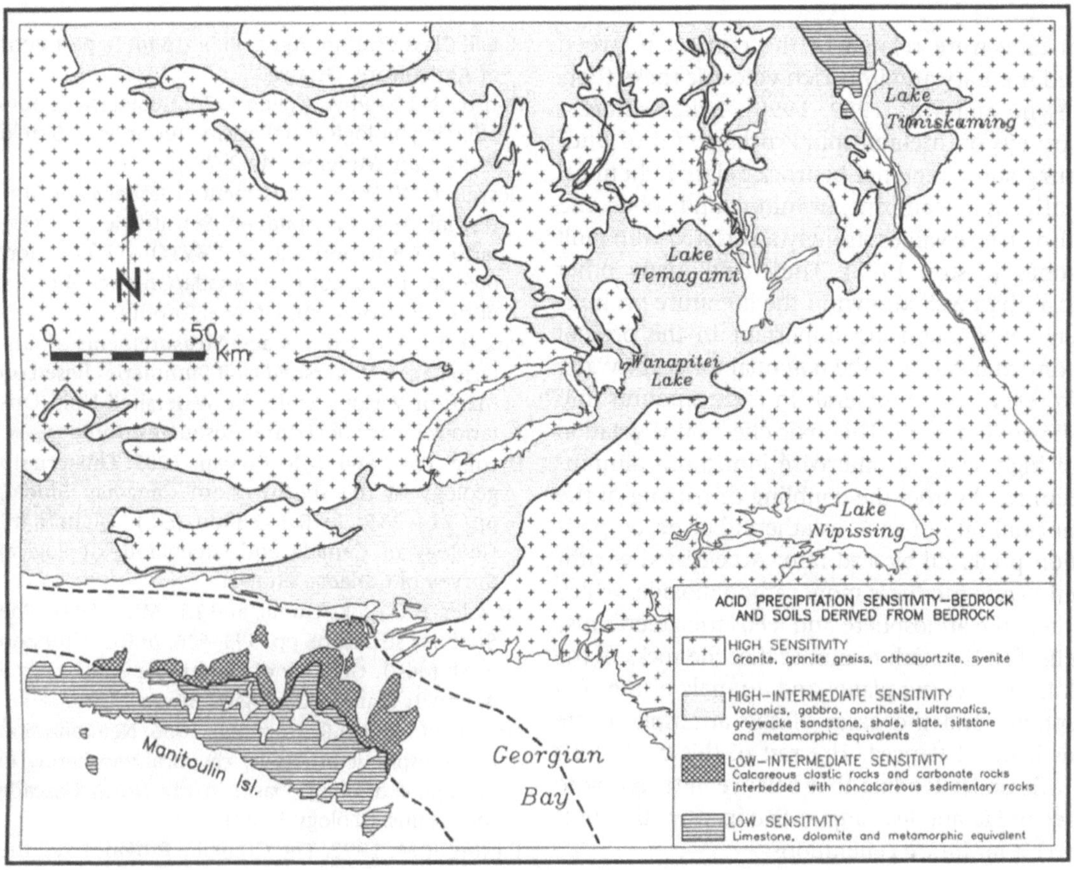

FIGURE 1.7. Sensitivity of bedrock and derived soils to acid precipitation (Shilts et al., 1981).

water in two areas 70 and 115 km north of Sudbury reported by Fortescue (1985). It is also possible that calcite or other carbonate minerals in veins cutting the bedrock may be responsible for surprisingly effective buffering (Drever and Hurcomb 1986).

Acidity is also counteracted by the ability of clay particles to capture hydrogen ions in exchange for other ions such as calcium or potassium. This mechanism may now be significantly effective in areas of former glacial lakes fed by muddy streams as the ice front melted back 9000 years ago. Lake Ramsey, in the shadow of the smelters in downtown Sudbury, may owe much of its remarkable resistance to acidification to the effectiveness of buffering by glacial clay. In small areas with more mature soils, the same mechanism is provided by aluminium and iron hydroxides.

On the whole, transported glacial debris has done little to modify the fundamental pattern of sensitivity produced by the bedrock. To the east, beyond what geologists call the Grenville Front, igneous and metamorphic rocks have produced quartz and feldspar-rich sand and silt. To the northeast, slightly less sensitive terrain is underlain by iron-rich igneous rocks and extensive beds of sandy and silty sedimentary rock. Some slight buffering capacity is provided by clay minerals.

Metal levels in surface environments are also influenced by bedrock. West of Sudbury, near Sault St. Marie, distinct geochemical signatures have been reported for granites, metamorphosed sediments, and two varieties of volcanic rocks. Copper, for example, was closely associated with iron-rich volcanic rocks and uranium with granites, whereas zinc and arsenic were related to the contact between sedimentary and iron-rich volcanic rocks (Fortescue and Vida 1989, 1990). Lake sediment surveys on metamorphic rocks south of Sudbury show regional bedrock-related chemical patterns in lead, zinc, uranium, and cobalt and also arsenic values strongly associated with fault zones (Easton 1992). These and many other examples well known in the literature on mineral exploration are important in the present context because of the potential for acidic water to release these minerals in toxic amounts into the environment. The toxic effect on vegetation of high levels of otherwise innocuous aluminium results from the doubling or tripling of the normal rate of rock weathering or decomposition produced by acid rain. Accelerated weathering also produces more carbon dioxide, which alters the atmosphere and feeds the greenhouse effect. It is a stark reminder that the rocks of the lithosphere, the plants and animals of the biosphere, and the gases of the atmosphere are delicately balanced. The rest of this book deals with understanding that balance, attempting to restore it, and learning to live in ways that protect it for future generations.

Acknowledgments. Thanks are extended to Léo Larivière for drafting the diagrams for this chapter and to Dr. John Fortescue of the Ontario Geological Survey for valuable discussions. Dr. Jim Kramer, Nels Conroy, and Dr. John Gunn made a notable contribution through their comments on an earlier draft.

References

Barnett, P.J. 1992. Quaternary geology of Ontario, pp. 1011–1088. *In* P.C. Thurston et al. (eds.). Geology of Ontario. Special Volume 4. Part 2. Ontario Geological Survey, Toronto.

Boissonneau, A.N. 1968. Glacial history of northeastern Ontario. II. The Timiskaming-Algoma area. Can. J. Earth Sci. 5:97–109.

Burwasser, G.J. 1979. Quaternary Geology of the Sudbury Basin Area, District of Sudbury. Report 181, Ontario Geological Survey, Toronto.

Canada Soil Survey Committee, Subcommittee on Soil Classification. 1978. The Canadian System of Soil Classification. Agric. Publ. 1646. Department of Agriculture, Ottawa.

Conroy, N.I., and W. Keller. 1976. Geological factors affecting biological activity in Precambrian Shield lakes. Can. Mineral. 14:62–72.

Cowell, D.W. 1986. Assessment of Aquatic and Terrestrial Acid Precipitation Sensitivities for Ontario. ARB Report 220-86-PHYTO/APIOS Report 009/86. Environment Canada/Ontario Ministry of the Environment, Ottawa/Toronto.

Cox, E.T. 1978. Counts and Measurements of Ontario Lakes: Watershed Unit Summaries Based on Maps of Various Scales by Watershed Unit. Ontario Ministry of Natural Resources, Toronto.

Dredge, L.A., and W.R. Cowan. 1989. Quaternary geology of the southwestern Canadian Shield, pp. 214–235. *In* R.J. Fulton (ed.). Quaternary Geology of Canada and Greenland. Geological Survey of Canada, Ottawa.

Dressler, B.O., V.K. Gupta, and T.L. Muir. 1991. The Sudbury structure, pp. 593–626. *In* P.C. Thurston et al. (eds.). Geology of Ontario. Special Volume 4, Part 1. Ontario Geological Survey, Toronto.

Drever, J.I., and D.R. Hurcomb. 1986. Neutralization of atmospheric acidity by chemical weathering in an alpine drainage basin in the North Cascade Mountains. Geology 14:221–224.

Easton, R.M. 1992. The Grenville Province and the Proterozoic history of central and southern Ontario, pp. 714–904. *In* P.C. Thurston et al. (eds.). Geology of Ontario. Special Volume 4, Part 2. Ontario Geological Survey, Toronto.

Environment Canada. 1982. Canadian Climate Normals, 1951–1980. Supply and Services Canada, Ottawa.

Fortescue, J.A.C. 1985. Preliminary Studies of Lake Sediment Geochemistry in an Area Northeast of Sudbury, Sudbury and Temiskaming District; Map 80756. Geochemical Series. Ontario Geological Survey, Toronto.

Fortescue, J.A.C., and E.A. Vida. 1989. Geochemical Survey of the Trout Lake Area; Map 80803. Ontario Geological Survey, Toronto.

Fortescue, J.A.C., and E.A. Vida. 1990. Geochemical Survey, Hanes Lake Area; Map 80806. Ontario Geological Survey, Toronto.

Gajewski, K. 1988. Late Holocene climate changes in eastern North America estimated from pollen data. Quat. Res. 29:255–262.

Golightly, J.P. 1994. The Sudbury Igneous Complex as an impact melt; evolution and ore genesis, pp. 105–117. *In* P.C. Lightfoot and A.J. Naldrett (eds.). Proceedings of the Sudbury-Noril'sk Symposium. Special Volume 5. Ontario Geological Survey, Sudbury.

Grieve, R.A.F. 1994. An impact model for the Sudbury structure, pp. 119–132. *In* P.C. Lightfoot and A.J. Naldrett (eds.). Proceedings of the Sudbury-Noril'sk Symposium. Special Volume 5. Ontario Geological Survey, Sudbury.

Grieve, R.A.F., and P.B. Robertson. 1987. Terrestrial Impact Structures; Map 1658A. Geological Survey of Canada, Ottawa; map supplement in Episodes 10:86.

Griffith, M.A., T. Spires, and P. Barclay. 1984. Ontario Soil Baseline Survey—Analytical Data 1980–81. APIOS Report 002/85. Ontario Ministry of the Environment, Toronto.

Gullett, D.W., and W.R. Skinner. 1992. The State of Canada's Climate: Temperature Changes in Canada 1895–1991. SOE Report 92–2. Environment Canada, Ottawa.

Jeffries, D.S., W.A. Scheider, and W.R. Snyder. 1984. Geochemical interactions of watersheds with precipitation in areas affected by smelter emissions near Sudbury, Ontario, pp. 196–241. *In* J. Nriagu (ed.). Environmental Impacts of Smelters. John Wiley & Sons, New York.

Karrow, P.F., and R.S. Geddes. 1987. Drift carbonate on the Canadian Shield. Can. J. Earth Sci. 24: 365–369.

Krogh, T.E., D.W. Davis, and F. Corfer. 1984. Precise U-Pb zircon and baddeleyite ages for the Sudbury area, pp. 431–447. *In* E.G. Pye, A.J. Naldrett, and P.E. Giblin (eds.). The Geology and Ore Deposits of the Sudbury Structure. Special Volume 1. Ontario Geological Survey, Toronto.

Milkereit, B., A. Green, and the Sudbury Working Group. 1992. Deep geometry of the Sudbury Structure from seismic reflection profiling. Geology 20:807–811.

Muir, T.L. 1984. The Sudbury structure; considerations and models for an endogenic origin, pp. 309–325. *In* E.G. Pye, A.J. Naldrett, and P.E. Giblin (eds.). The Geology and Ore Deposits of the Sudbury Basin. Special Volume 1. Ontario Geological Survey, Toronto.

Naldrett, A.J. 1994. The Sudbury-Noril'sk Symposium, an overview, 3–8. *In* P.C. Lightfoot and A.J. Naldrett (eds.). Proceedings of the Sudbury-Noril'sk Symposium. Special Volume 5. Ontario Geological Survey, Sudbury.

Naldrett, A.J., and R.M. Hewins. 1984. The main mass of the Sudbury Igneous Complex, pp. 235–252. *In* E.G. Pye,, A.J. Naldrett, and P.E. Giblin (eds.). The Geology and Ore Deposits of the Sudbury Basin. Special Volume 1. Ontario Geological Survey, Toronto.

Ritchie, J.C. 1966. Aspects of the late-Pleistocene history of the Canadian flora, pp. 66–80. *In* R.L. Taylor and R.A. Ludwig (eds.). The Evolution of Canada's Flora. University of Toronto Press, Toronto.

Ritchie, J.C. 1988. Postglacial Vegetation of Canada. Cambridge University Press, Cambridge.

Semkin, R.G., and J.R. Kramer. 1976. Sediment geochemistry of Sudbury-area lakes. Can. Mineral. 14:73–90.

Shilts, W.W. 1989. Flow patterns in the central North American ice sheet. Nature 386:213–218.

Shilts, W.W., K.D. Card, W.H. Poole, and B.V. Sanford. 1981. Sensitivity of Bedrock to Acid Precipitation: Modification by Glacial Processes. Paper 81–14. Geological Survey of Canada, Ottawa.

Stevenson, J.S. 1972. The Onaping ash-flow sheet, Sudbury, Ontario, pp. 41–48. *In* J.V. Guy-Bray (ed.). New Developments in Sudbury Geology. Special Paper 10. Geological Association of Canada. Toronto, Ontario.

Trenhaile, A.S. 1990. The Geomorphology of Canada. Oxford University Press, Toronto.

Webb III, T. 1985. Holocene palynology and climate, pp. 163–195. *In* A.D. Hecht (ed.). Paleoclimate Analysis and Modeling. John Wiley & Sons, New York.

2

Early History of Human Activities in the Sudbury Area and Ecological Damage to the Landscape

Keith Winterhalder

In contrast to the ancient events that formed the Sudbury Basin and its mineral deposits, its human history spans less than 10,000 years. As the Wisconsin glacier receded, a forest cover developed, and the area was settled by native groups. The events that led to Sudbury becoming one of the largest mining and smelting regions in the world (Fig. 2.1) are from a far briefer period of about 100 years. The environmental damage that occurred during this recent industrial period is the focus of this chapter. Other international examples of mining-related ecosystem damage are described in Box 2.1.

Sudbury lies in a vegetation zone referred to by Rowe (1959) as the Great Lakes-St. Lawrence Forest Region and is located on the margin of the northern Temagami section, once characterized by extensive stands of red and white pine (*Pinus resinosa* and *P. strobus*), and the southern Algonquin section, where white pine formed an admixture with tolerant hardwoods such as sugar maple (*Acer saccharum*) and yellow birch (*Betula alleghaniensis*). Although we do not know the exact nature of the vegetation that existed in the Sudbury area before disturbance, the huge white pine stumps now found on bare stony slopes (Fig. 2.2) and the vestiges of white cedar (*Thuja occidentalis*) that cover many acres of barren peat (Fig. 2.3) hint at a mosaic of pine forests on the slopes and cedar swamps in many of the depressions.

In 1824, the Hudson's Bay Company established a fur-trading post near what was later to become Sudbury, and the hunting and gathering life-style of the local Anishnabe people was modified to include commercial fur trapping. This, however, did not have a permanent effect on the local wildlife population. When the area was opened up to lumbering in 1872, the larger red and white pines were cut and floated down rivers to Georgian Bay and Lake Huron, then rafted to sawmills in the northern United States.

Lumber from the Sudbury area almost certainly played an important role in rebuilding Chicago after that city's devastating fire of 1871. Although the first European settlers in Sudbury would have found a forest that was partly cut-over, with regrowth of birches and poplars, in many areas some fair-sized trees must have remained, because the Sudbury parish came to be known as "Ste. Anne of the Pines."

By the turn of the century, black and white spruce (*Picea mariana* and *P. glauca*), balsam fir (*Abies balsamea*), and later jack pine (*Pinus banksiana*) were also being harvested, and more than 11,000 men were employed in the mills and in the bush around Sudbury. In the broader Sudbury area, lumbering continued to be the dominant industry as late as 1927, despite the emergence of the mining industry after 1886.

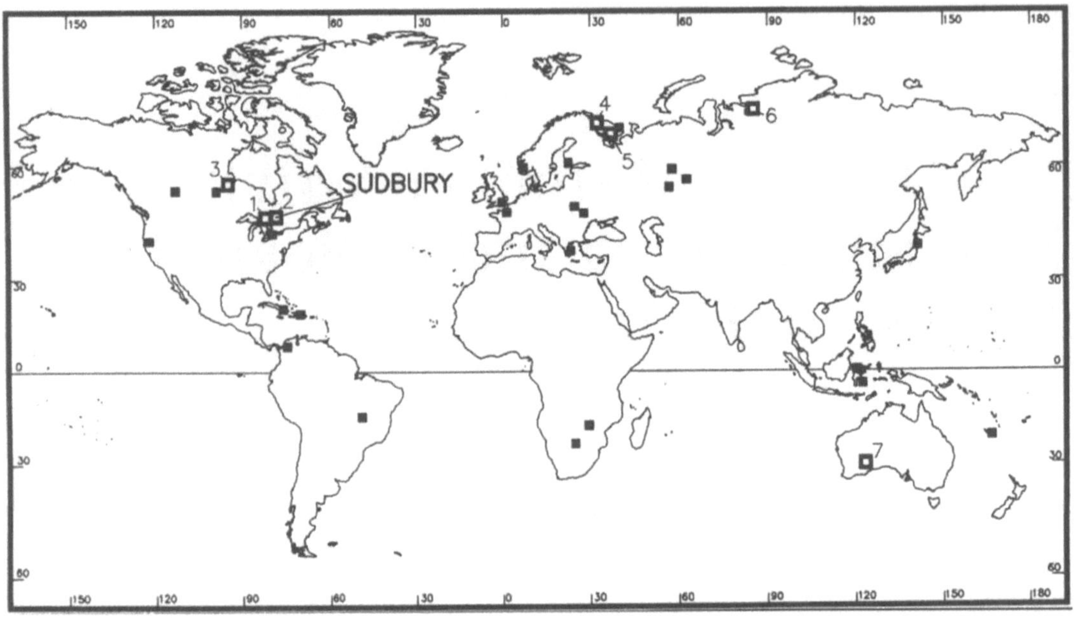

FIGURE 2.1. Worldwide location of nickel smelters and refineries. Numbers indicate the location of relatively large smelters (>30,000 tonnes of nickel per annum): (1) Inco Limited at Sudbury; (2) Falconbridge Limited at Sudbury; (3) Inco Limited at Thompson, Manitoba, Canada; (4) Pechenganikel Company (now Severonikel) at Nikel and Zapolyarny, Russia; (5) Severonikel Complex at Monchegorsk, Russia; (6) Severonikel Complex at Noril'sk, Russia; (7) Western Mining Company at Kalgoorlie, Western Australia.

Early logging practice, involving the selective removal of the larger red and white pine, made a minimal contribution to environmental degradation, because rapidly growing successional species such as white birch (*Betula papyrifera*) and trembling aspen (*Populus tremuloides*) would have colonized quickly. Even the less-selective logging that came later, for railroad ties and locomotive fuel and for pit timbers and pulpwood, should not have led to permanent denudation of the landscape. But events were to take a different direction in 1883 when surveyor William Ramsey accidentally routed the new Canadian Pacific Railway (C.P.R.) to the north of Bimitimigamasing Lake (now Ramsey Lake) rather than to the south. According to legend, it was C.P.R. blacksmith Tom Flanagan's discovery of a rusty stain on rocks near what became Murray Mine that initiated Sudbury's mining history (LeBourdais 1953; DeLestard 1967; Wallace and Thompson 1993).

The arrival of the railroad resulted in increased lumbering and a greater frequency of fire. Early accounts of Sudbury often spoke of blackened stumps and plentiful blueberries (Howey 1938), the blueberry plant being a very fire-tolerant species. The wood-burning locomotives must have contributed to the fire setting, and prospectors were said to have set fires to expose the bedrock under the duff cover. As logging became less selective, a greater proportion of the timber was removed, and the slash left behind created an ideal fuel for forest fires. Later, sulfur dioxide damage led to the death of leaves on the trees, creating tinder. Even today the blueberry patches, maintained to some extent by recurring fire, attract pickers who themselves often accidentally start new fires. As recently as the 1970s, a map of fire frequency for Ontario shows the Sudbury area as a record-setter.

Several different mining companies set up operations in Sudbury, but most of them only survived a few years (LeBourdais 1953). The Canadian Copper Company was the first successful one, beginning operations at Copper

Box 2.1. Selected Examples of Landscapes Denuded by the Effects of Smelter Fumes

Ducktown, Tennessee (Smith 1981)	Open-bed roasting and smelting of copper sulfide ore since 1850 has acidified the soil and created an almost totally barren landscape covering 2700 ha within the richly forested southern Appalachian Mountains.
Monchegorsk, Russia (Kryuchkov 1993)	Smelting of copper-nickel ore has given rise to 21,000 ha of barren land surrounded by 44,000 ha of dwarfed birch forest, in what was originally spruce-pine forest.
Palmerton, Pennsylvania (Jordan 1975)	Smelting of zinc ore in a narrow valley since 1898 has contaminated the naturally acid soil with zinc, cadmium, copper, and lead, giving rise to 485 ha of barren to sparsely vegetated land.
Queenstown, Tasmania (Blainey 1967)	Smelting of sulfide copper ore in this high rainfall area since 1896 has caused the denudation and erosion of the spectacular mountain peaks.
Smelterville, Idaho (Carter et al. 1977)	Smelting of sulfide ore in the Kellogg Valley since 1916 has led to the acidification and zinc, lead, and cadmium contamination of soil, resulting in extensive barren hillsides.
Trail, British Columbia (Archibold 1978)	Sulfide ore smelting since 1926 in the narrow valley of the Columbia River affected vegetation 25 km upstream and 95 km downstream into Washington State. Between Trail and the U.S. border, 17 km to the south, vegetation was almost totally eradicated.
Wawa, Ontario (Gordon and Gorham 1963)	Sulfur dioxide fumes emitted by an iron sintering plant since 1939 have damaged a narrow 37-km strip of vegetation, with complete denudation up to 8 km.

Cliff in 1885 and later being incorporated as part of the U.S.-based International Nickel Company (Inco) in 1902. The Mond Nickel Company, from Britain, began operations in 1900. It later amalgamated with Inco in 1929. Falconbridge Limited was incorporated and became the second largest company in 1928. For the past 65 years, Inco Limited and Falconbridge Limited have remained as the two companies with mining and smelting operations in the Sudbury Basin.

Roast Yards

During the early years of mining, ore was sent to other centers for smelting, but in 1888 the first roast yard and smelter were set up in Copper Cliff (LeBourdais 1953). During the first few years, nickel was looked on as a worthless and problematic contaminant in the copper ore, but in the 1890s, the Orford process was devised to separate the two metals (Boldt 1967). As the century drew to a close, the demand for nickel was boosted by the discovery of nickel-steel, which was used extensively for armorplate and in the Spanish-American War. The history of nickel production at Sudbury is shown in Figure 2.4.

Some of the early smelters used a portion of untreated or "green" ore, but historically, roasting was usually the first step in the processing of Sudbury's sulfide ore. The ore was heated in air to a temperature at which much of the sulfide was oxidized, and the sulfur was burned off as sulfur dioxide. In the early years of the Sudbury industry, crushed ore was piled on beds of cordwood in the open (Fig. 2.5), covered with the finer material to prevent open flames, and ignited. After a period during

FIGURE 2.2. White pine stumps on a barren stony hillside, 3.2 km northeast of the Copper Cliff smelter.

FIGURE 2.3. White cedar stumps on a barren peat area, Happy Valley, close to the Falconbridge smelter.

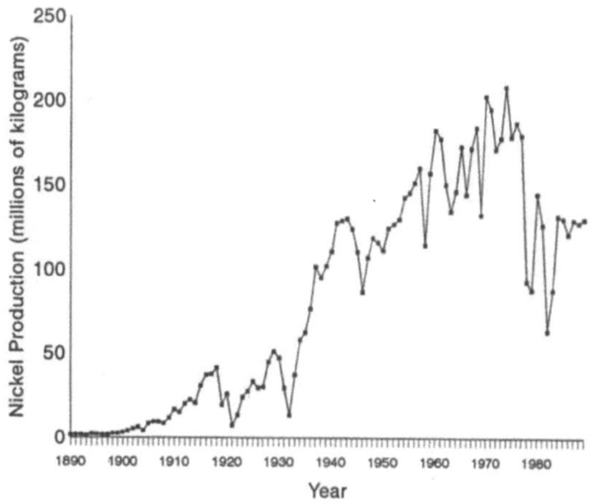

FIGURE 2.4. Annual nickel production from Sudbury smelters. (Data from Ontario Ministry of Natural Resources and the Ontario Ministry of Northern Development and Mines.)

which heat was provided by the burning wood, the ore itself began to burn (Boldt 1967). After burning for 2 months or more, the ore was loaded into rail cars and transported to furnaces for smelting and conversion. In the vicinity of the roast yards, even the smallest timber was removed as fuel, and between 1913 and 1916, the Mond Nickel Company removed nearly all the woody vegetation and tree stumps from the Coniston area to provide fuel for the roasting process (Watson and Richardson 1972). Laroche et al. (1979) estimated that more than 3.3 million m^3 of wood was consumed in the 11 roast yards that were in use during 1888–1929.

Most of the roast yards were located in close proximity to the nine smelters (Fig. 2.6) that operated at various times and locations throughout the area. In 1916, most of the smaller roast yards were replaced by a very large (2286 m long by 52 m wide) and highly mechanized yard, the O'Donnell roast yard, some distance west of Sudbury, to move this source of human discomfort farther from the main population centers. Open-bed roasting was not abandoned until 1929, even though this practice had long since been replaced in Norway because of the damage it had caused there (Peck 1980). In the latter years of heap-roasting at Coniston, the Mond Nickel Company followed the Norwegian example by not operating its roast beds in the summer (Peck 1978), thereby reducing damage to crops and other vegetation (Wallace and Thompson 1993).

Laroche et al. (1979) estimated that the roast beds released an estimated 10 million tonnes of sulfur dioxide. The dense sulfur dioxide fumes, emitted at ground level, killed plants and acidified soils in their path. Indeed, there was once a widely held hypothesis that the extensive destruction of vegetation and the presence of toxic soils in the Sudbury area are mainly the result of early open-bed roasting activities. It seems, however, that the effects of the roast yards were neither as severe nor as permanent as expected; the widespread poisoning of the soil was caused by the smelter fumes, which contained copper and nickel particles as well as sulfur dioxide. The slow-burning roast beds emitted mainly sulfur dioxide, which had a less permanent effect on surrounding soils. Turcotte (1981) showed that the Victoria Mine roast bed, which was distant from smelter activity, had done little to contaminate the surrounding landscape with metals, except in the immediate vicinity of the roast yard, and that vegetation had shown excellent recovery (see Fig. 2.5). The Coniston roast yard, in contrast, is still surrounded by barren metal-contaminated soils (see Fig. 2.5) but only because of its proximity to the Coniston smelter, a source of copper and nickel particulates. Struik (1974) studied a series of air photographs that showed the changes occurring in the vicinity of the O'Donnell roast yard (see Fig. 2.5), which was isolated from smelting activities, after its closure in 1929. By 1946, there was patchy cover by shrubs, whereas by 1959 pioneer trees

FIGURE 2.5. Roast yards in the Sudbury area. (*Upper left*) A 1920 view of the O'Donnell roast yard, showing wood and ore piled before ignition. The O'Donnell (*center left*) and circa 1901 Victoria Mine (*upper right*) roast yards during the roasting process. The Victoria Mine roast yard in 1979 (*center right*) and the O'Donnell yard in 1994 (*lower left*), showing the recovery of vegetation to the edge. (*Lower right*) A 1979 picture of the Coniston roastbed, that operated from 1913 to 1918, set in a landscape that is still almost completely barren. (Photos courtesy of Inco Archives, W. McIlveen and E. Snucins.)

such as poplars and birches had begun to colonize, again in a patchy manner. By 1973, a mosaic of forest cover had developed, but still with intermittent openings. Recent pictures of the site are included in Plate 2 (following page 182) and Figure 2.5.

Extent of Damage

On April 20, 1944, representatives of government and industry met to discuss the smelter emission problem and the forest damages allegedly caused by sulfur dioxide in the Sud-

FIGURE 2.6. Historical pictures of Sudbury smelters. (*Upper left*) A 1900 picture of the Canadian Copper Company's west smelter at Copper Cliff. (*Upper right*) The British-American Nickel Company smelter that operated sporadically at the Murray Mine site between 1917 and 1924. (*Center left*) The Victoria Mine smelter, built by the Mond Nickel Company in 1901, and operated until 1913. (*Center right*) A 1928 view of the Coniston smelter, built by the Mond Company in 1913. This smelter was taken over by Inco Limited in 1928 and operated until 1972. (*Lower left*) A circa 1973 picture of the Falconbridge Limited smelter. (*Lower right*) Inco's Copper Cliff Smelter in 1960.(Photos courtesy of Inco Archives and Falconbridge Limited.)

bury area. Studies were initiated on meteorology, atmospheric sulfur dioxide levels, the sulfur content of conifer foliage, lichen distribution, and forest damage. These studies were conducted during the summers of 1942–1944 when the smelters were in high production because of the war effort and thus may be considered peak conditions for industrial pollutants during the first half-century of smelting operations in Sudbury. The sulfur dioxide study used staff in fire towers and aircraft to determine the dispersal pattern of sulfur dioxide from the smelters. These personal observations demonstrated that smoke from Sudbury smelters could be seen from fire towers at least 120 km away, and the smell of sulfur extended at least 60 km (Murray and Haddow 1945).

The first report on vegetation damage (Murray and Haddow 1945) indicated that "severe burns" of tree foliage had occurred as far away as 35 km to the northeast, 20 km to the north, and 20 km to the south of the smelters. The question of chronic effects was also addressed, and a lichen study indicated that only crustose lichens and *Stereocaulon* were to be found within the most highly polluted area, although *Parmelia physodes* and *P. saxatilis* could be found near the edge of this zone. At a greater distance from the smelters, where effects of atmospheric pollution were less severe, it was possible to find more-sensitive lichen species, such as *Parmelia conspersa* and caribou lichen, *Cladonia rangiferina*. The author of the lichen study, R.F. Cain, suggested that other factors including frequency of fires might have interacted with the fumigation in bringing about this distribution pattern. The use of lichen distribution in tracing change in atmospheric quality is further discussed in Chapter 6.

A similar pattern with respect to higher plants was noted by Linzon (1958, 1971), who made observations on the extreme sensitivity of white pine to sulfur dioxide in the area and detected increased mortality and decreased growth of white pine as far as 40 km northeast of ore smelters. Gorham and Gordon (1960 a,b) reported the absence of white pine and velvetleaf blueberry (*Vaccinium myrtilloides*) within 24 km of the smelter and a sharp drop in species numbers 6.4 km from the smelter. Later, Freedman and Hutchinson (1980b) found that south-southeast of the Copper Cliff smelter, the number of forest floor species peaked at 15–20 km from the smelter, whereas the number of tree species continued to increase up to a distance of 30 km.

H. Struik, a forester with Ontario Ministry of Natural Resources (then Ontario Department of Lands and Forests), produced the first direct measure of the extent of vegetation damage throughout the Sudbury area. He examined air photographs and mapped areas that he referred to as "zones of site and vegetational stability" (Struik 1973). The earliest photographs (1946) revealed a large area of vegetation damage around the three major smelters at Copper Cliff, Coniston, and Falconbridge. This damaged area appeared to expand slightly in subsequent photographs in 1959 and in a composite of pictures for 1970 and 1973. From the maps that Struik produced for the early 1970s, the period just before major pollution reductions began (see Chapter 4), the zones of damage can be measured and summarized into two categories: (1) barren areas around the three smelters that had a total surface area of about 17,000 ha, and (2) a large surrounding semibarren area of approximately 72,000 ha (Fig. 2.7).

Denuded Landscape—The Result of Many Interacting Factors

Soil erosion began with logging and intensified as plant litter was destroyed by fire and as plant cover was killed in the vicinity of roast beds and smelters. Then, as the soil became poisoned and plant cover disappeared completely, erosion continued virtually unchecked. Much of the soil cover from the hillsides was washed into the valleys and often into the creeks and rivers. The only restraining feature, and a valuable one in many areas, was the coarse material in the glacial till. The stony and boulder-strewn slopes that we see today (see Fig. 2.2) result from the removal of fine material from the surface. Fortunately, however, the till was deep enough in some areas

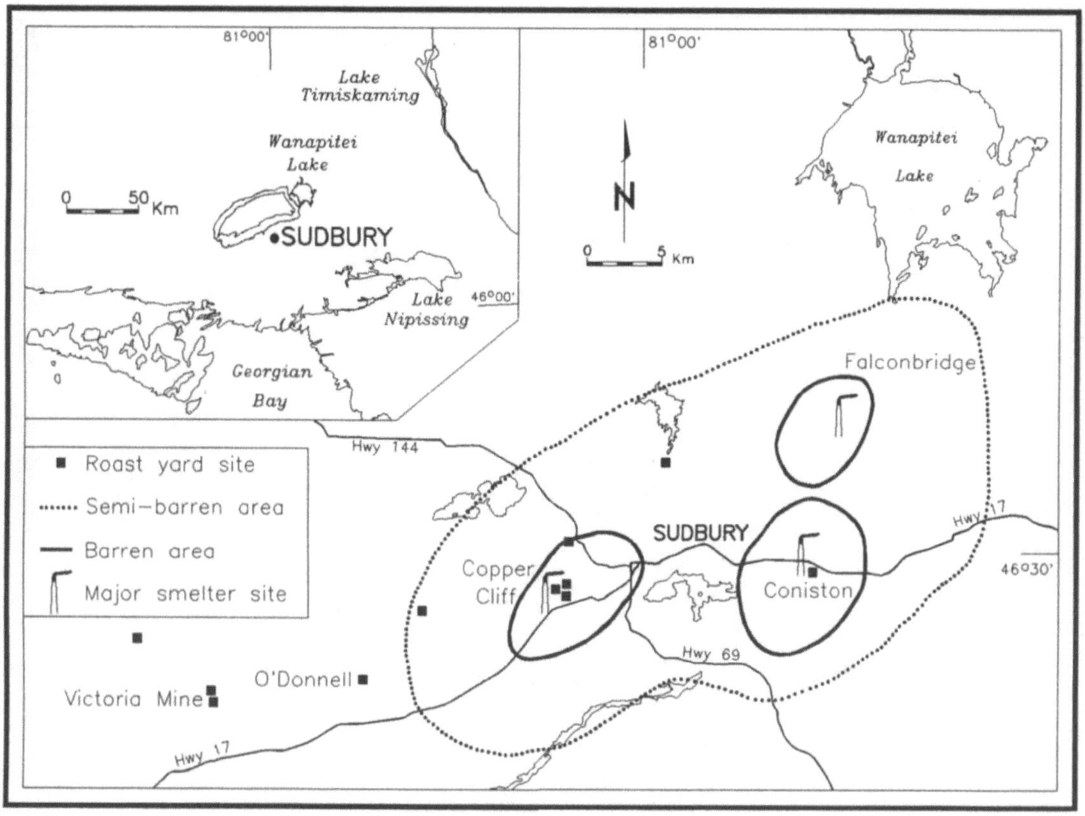

FIGURE 2.7. The location of the Sudbury Basin and the major sites of roasting and smelting activity and Struik's (1973) "zones of site and vegetational stability" based on air photographs.

that a soil base remained under the protection of the stones and boulders. Still, these boulder-covered slopes were not entirely stable, due to the effects of frost action, and each spring, there was some movement of coarse material and the consequent washing-out of fines. As humus-rich surface horizons were lost, so was their microflora.

Until relatively recently, it was assumed that vegetation damage in the Sudbury area was the direct result of sulfur dioxide impact. In the late 1960s, however, interest developed in acidification and metal contamination of the soils. Studies by a local group of ecologists and foresters (Winterhalder 1975) showed that soil acidity and concentrations of copper and nickel were highly elevated in soils near the smelters and could be directly correlated with 1953–1968 air pollution zones (Fig. 2.8), as could soil pH (Table 2.1). Other publications (Hutchinson and Whitby 1974; Freedman and Hutchinson 1980a; Hazlett et al. 1983) have identified the same pattern. In 1974, Whitby and Hutchinson showed that the soil from the Sudbury barrens was inhibitory to plant growth and that the toxic components were water-soluble, apparently including interacting ions of copper, nickel, aluminum (Winterhalder 1983), and at a lower level, cobalt.

The Sudbury landscape of today is the result of several environmental factors acting together over a period of almost a century: sulfur dioxide fumigations; metal deposition; intense logging; wild fires; water and wind erosion; and enhanced frost action (Winterhalder 1984). These environmental factors interacted one with the other. Figure 2.9 attempts to suggest some of the broader interactions that have occurred, including both positive and negative feedbacks.

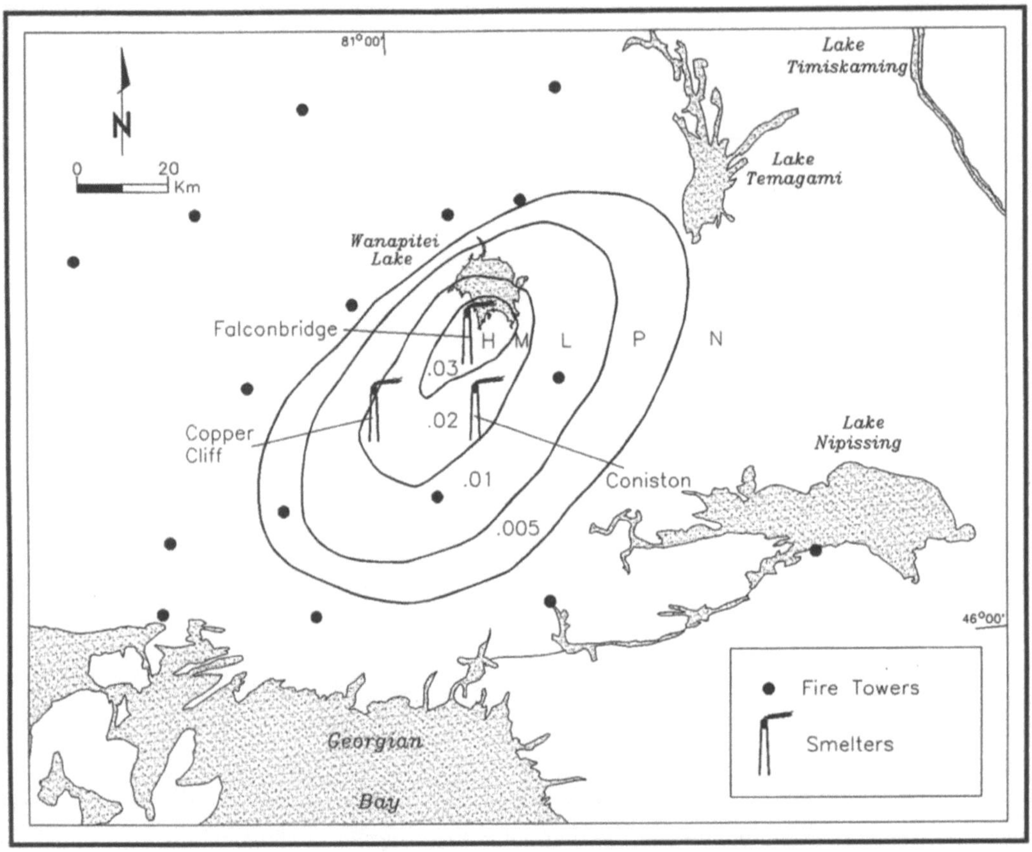

FIGURE 2.8 Mean sulfur dioxide concentrations used to categorize fumigation zones (1953–1968) (Dreisinger and McGovern 1971). The locations of fire towers from which sulfur smoke was frequently seen and smelled during 1939–1944 (Murray and Haddow 1945) are also indicated. See Table 2.1 for definition of zone categories.

In essence, the soil of the barrens is toxic to plants because of the interaction of its low pH (<4.0) with its copper and nickel contaminants, which can exceed 1000 ppm, respectively. At this pH, the contaminating copper and nickel (and possibly cobalt) become toxic. Aluminum is released from the clay minerals, augmenting soil toxicity. The toxic soil solution inhibits root growth, and seedlings are readily killed by drought or by frost heaving.

TABLE 2.1. Chemical characteristics of surface soils from the sulfur dioxide fumigation zones described by Dreisinger and McGovern (1971), as illustrated in Figure 2.8. Sampling was conducted in 1969 with 10 sites sampled in each zone.

Fumigation zone	Mean pH ± SD	Mean total copper (mg/kg ± SD)	Mean total nickel (mg/kg ± SD)
Heavy	3.8 ± 0.3	1250 ± 500	1930 ± 900
Moderate	4.3 ± 0.1	900 ± 300	750 ± 300
Light	4.7 ± 0.1	320 ± 80	400 ± 120
Perceptible	5.0 ± 0.2	200 ± 30	420 ± 120
None	5.0 ± 0.2	100 ± 20	200 ± 30

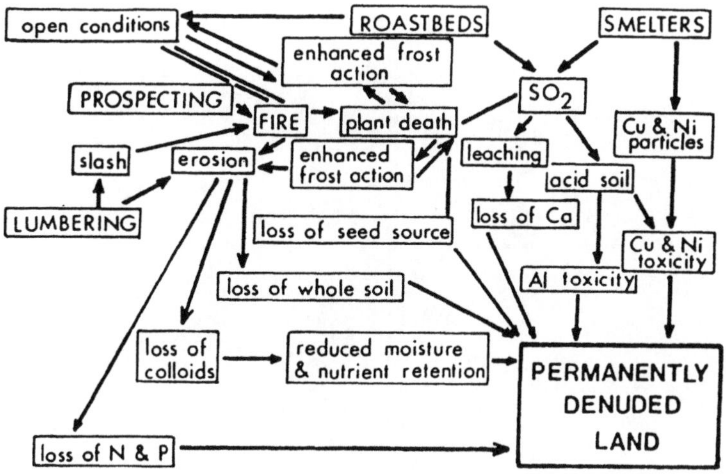

FIGURE 2.9. Simplified representation of some of the major factor interactions leading to the formation of barren land.

Although acidity and metal ions appear to play the major role in the soil degradation, the sulfur itself may also play its part. Some preliminary work (Hutchinson and Whitby 1976) suggests that the fundamental structure of the organic matter fraction of Sudbury soils has been changed by the high sulfur content, giving it a very strong metal-binding capacity.

Plant Communities of Sudbury's Altered Landscape

In 1981, Amiro and Courtin provided the first direct detailed and quantitative description of the vegetation surrounding the smelters today. They described nine different plant community types within the Sudbury area, three of which were confined to the industrially disturbed land. Pitblado and Amiro (1982) showed a similar pattern using remote sensing from a satellite, in which an estimate of the density of the vegetation present was obtained by measuring the ability of the landscape surface to emit radiation of different wavelengths.

The first of the Amiro and Courtin (1981) communities, referred to as "barren," broadly coincides with Struik's (1973) barren areas (see Fig. 2.7) and is characterized by a soil pH of 4.0 and less. In this zone, the degree of denudation was partly dependent on topography, being most severe on hilltops and steep slopes.

Highly depauperate relict trees, mostly red maple (*Acer rubrum*) but occasionally red oak (*Quercus borealis*) or American elm (*Ulmus americana*), are found on some of the barren rocky slopes. Relict shrub species include blueberry, red elderberry (*Sambucus pubens*), and witherod (*Viburnum cassinoides*). The moss *Pohlia nutans* can be found in seepage areas on north-facing slopes, and in moist depressions, tufted hair grass (*Deschampsia caespitosa*) often dominates, sometimes accompanied by patches of hair moss (*Polytrichum commune*). The relict red maples show a phenomenon best described as "regressive dieback," in which the foliage becomes reddened prematurely and the amount of living biomass produced each year gradually decreases, the plant being surrounded by dead limbs of various sizes. Even in low moist meadows with a surface cover of tufted hair grass, the red maples have the same appearance, suggesting that moisture is not the limiting factor.

Three of Amiro and Courtin's plant communities together make up the "semibarren" land more distant from the smelters. This area comprises an open woodland structure, with a sparse understory containing such acid- and fire-tolerant plants as blueberry, sweet fern (*Comptonia peregrina*), and bracken (*Pteridium aquilinum*) alternating with extensive bare areas. It is generally agreed that these stunted and depauperate woodlands, called by Amiro the "Birch Transition," "Maple Transition,"

FIGURE 2.10. Birch Transition community.

FIGURE 2.11. Red maple in a Maple Transition community, showing regressive dieback. The large birches exhibit premature marginal leaf chlorosis.

Box 2.2. The Term Acid Rain

Although we think of the term *acid rain* as modern, devised to describe a modern phenomenon, both the concept and the term are more than a century old. Robert Angus Smith began publishing on the chemistry of rain around the industrial city of Manchester, England, in 1852, and 20 years later he published a book entitled *Air and Rain: The Beginnings of a Chemical Climatology,* based on work in England, Scotland, and Germany. He not only used the term *acid rain,* but he demonstrated that precipitation chemistry is influenced by such factors as the combustion of coal, the decomposition of organic matter, wind trajectories, and the amount and frequency of precipitation. In addition to discussing procedures for the collection and analysis of precipitation, he described damage done to plants and materials by acid precipitation and recognized the atmospheric deposition of copper and other metals in industrial regions.

A century later, Eville Gorham began to further develop and integrate our understanding of the nature of acid precipitation and its effects on vegetation, soils, and surface waters in the United Kingdom and in Canada, but once again the work was largely ignored. A further step forward, which finally stimulated a scientific and public response, was taken by Svante Oden, a Swedish soil scientist. Oden managed to integrate chemical, limnological, and agricultural perspectives on acid precipitation, and he presented to the European scientific community a picture of long-range transport of atmospheric pollutants that had the potential to damage both aquatic and terrestrial environments. After a U.S. lecture tour by Oden in 1971 and his presentation at the 19th International Limnological Congress in Winnipeg, Manitoba, in 1974, scientific and public interest exploded in North America, and in 1975 the U.S. Forest Service sponsored the First International Symposium on Acid Precipitation and the Forest Ecosystem in Ohio.

A full historical perspective on acid precipitation can be found in Cowling (1982).

and "Red Oak" communities, respectively, are relict stands, a fact that is evident from the predominantly coppiced form of the component tree species. Both the Birch Transition (Fig. 2.10) and the Maple Transition contain white birch and red maple, the relative importance of each of the two species defining the community. Red maple shows the same regressive dieback in the woodlands as on the barrens (Fig. 2.11), whereas the white birch is characterized by a premature yellowing of the leaf margins in early summer, followed by the production of normal green leaves in late summer. The three communities tend to form a series related to soil drainage status, with the red oak community clearly at the well-drained end and, according to Amiro and Courtin (1981), Birch Transition at the moist end (a surprising fact in view of the normal preference of red maple for moist sites). The dynamics of these unique plant communities are described in more detail in Chapter 7.

Summary

Vegetation damage began with logging, fire, and roasting beds, but the decades of intensive fumigation from smelters caused most of the damage. The poisoning of soil by the addition of acid and toxic metals from smelter fumes created conditions that were unlikely to allow rapid natural recovery. Damage to the terrestrial ecosystem reached its peak during the 1960s, well before damage to lakes received attention. However, by the early 1970s, this emphasis changed when "acid rain" (Box 2.2) became a household word and the larger zone of ecological damage to surface water, described in the next chapter, became evident.

Ackowledgments. Inco Archives, Falconbridge Limited, and the Ontario Ministry of the Environment and Energy kindly provided the historic pictures of smelters and roast beds. John Gunn, Nels Conroy, James Kramer, and Marty Puro provided review comments. Michael Conlon, Mary Roche, and Léo Larivière assisted with graphics and photographs.

References

Amiro, B.D., and G.M. Courtin. 1981. Patterns of vegetation in the vicinity of an industrially disturbed ecosystem, Sudbury, Ontario. Can. J. Bot. 59(9):1623–1639.

Archibold, O.W. 1978. Vegetation recovery following pollution control at Trail, British Columbia. Can. J. Bot. 56(14):1625–1637.

Blainey, G. 1967. The Peaks of Lyell, 3rd Ed. Melbourne University Press, Melbourne.

Boldt, J.R., Jr. 1967. The Winning of Nickel. Longmans Canada Ltd., Toronto.

Carter, D.B., H. Loewenstein, and F.H. Pitkin. 1977. Amelioration and revegetation of smelter-contaminated soils in the Coeur d'Alene mining district of northern Idaho. Proceedings of the Second Annual Meeting, Canadian Land Reclamation Association, Edmonton, Paper 13. CLRA, Guelph, Ontario.

Cowling, E.B. 1982. An historical perspective on acid precipitation, pp. 15–31. *In* R.E. Johnson (ed.). Proceedings of an International Symposium on Acidic Rain and Fishery Impacts on Northeastern North America. American Fisheries Society Special Publication, Bethesda, MD.

DeLestard, J.P.G. 1967. A History of the Sudbury Forest District. District History Series 21. Department of Lands and Forests, Ontario.

Dreisinger, B.R., and P.C. McGovern. 1971. Sulphur Dioxide Levels and Vegetation Injury in the Sudbury Area during the 1970 Season. Department of Energy and Resources Management, Air Management Branch, Sudbury.

Freedman, B., and T.C. Hutchinson. 1980a. Pollutant inputs from the atmosphere and accumulations in soils and vegetation near a nickel-copper smelter at Sudbury, Ontario, Canada. Can. J. Bot. 58(1):108–132.

Freedman, B., and T.C. Hutchinson. 1980b. Long-term effects of smelter pollution at Sudbury, Ontario, on forest community composition. Can. J. Bot. 58:2123–2140.

Gordon, A.G., and E. Gorham. 1963. Ecological aspects of air pollution from an iron-sintering plant at Wawa, Ontario. Can. J. Bot. 41:1063–1078.

Gorham, E., and A.G. Gordon. 1960a. Some effects of smelter pollution northeast of Falconbridge, Ontario, Canada. Can. J. Bot. 38:307–312.

Gorham, E., and A.G. Gordon. 1960b. The influence of smelter fumes upon the chemical composition of lake waters near Sudbury, Ontario, and upon the surrounding vegetation. Can. J. Bot. 38:477–487.

Hazlett, P.W., G.K. Rutherford, and G.W. Van Loon. 1983. Metal contaminants in surface soils and vegetation as a result of nickel/copper smelting at Coniston, Ontario. Reclamation Revegetation Res. 2(2):123–137.

Howey, F.R. 1938. Pioneering on the C.P.R. Mutual Press Ltd., Ottawa.

Hutchinson, T.C., and L.M. Whitby. 1974. Heavy metal pollution in the Sudbury mining and smelting region of Canada, I. Soil and vegetation contamination by nickel, copper and other metals. Environ. Conservation 1:123–132.

Hutchinson, T.C., and L.M. Whitby. 1976. The effects of acid rainfall and heavy metal particulates on a boreal forest ecosystem near the Sudbury smelting region of Canada, pp. 745–765. *In* L. S. Dochingen and T. A. Seliga (eds.). Proceedings of the First International Symposium on Acid Precipitation and the Forest Ecosystem, Columbus, Ohio, May 1975. USDA Forest Service General Technical Report NE-23. U.S. Department of Agriculture, Upper Darby, PA

Jordan, M.J. 1975. Effects of zinc smelter emissions and fire on a chestnut-oak woodland. Ecology 56:78–91.

Kryuchkov, V.V. 1993. Degradation of ecosystems around the "Severonikel" smelter complex, pp. 35–46. *In* M.V. Kozlov, E. Haukioja, and V.T. Yarmishko (eds.). Aerial Pollution in Kola Peninsula: Proceedings of the International Workshop, April 14–16, 1992, St. Petersburg, Russia. Kola Scientific Center, Apatity, Russia.

Laroche, C., G. Sirois, and W.D. McIlveen. 1979. Early roasting and smelting operations in the Sudbury area—an historical outline. Unpublished report. Ontario Ministry of Environment and Energy, Sudbury, Ontario.

LeBourdais, D.M. 1953. Sudbury Basin. Ryerson Press, Toronto.

Linzon, S.N. 1958. Influence of Smelter Fumes on the Growth of White Pine in the Sudbury Region. Department of Lands and Forests, Department of Mines, Ontario.

Linzon, S.N. 1971. Economic effects of SO_2 on forest growth. J. Air Pollut. Control Assoc. 21: 81–86.

Murray, R.H., and W.R. Haddow. 1945. First report of the subcommittee on the investigation of sulphur smoke conditions and alleged forest damage in the Sudbury region, February 1945. Unpublished report.

Peck, G.R. 1978. The not-so-distant past. Sudbury Star, September 30, 1978.

Peck, G.R. 1980. The not-so-distant past. Sudbury Star, April 5, 1980.

Pitblado, J.R., and B.D. Amiro. 1982. Landsat mapping of the industrially disturbed vegetation communities of Sudbury, Canada. Can. J. Remote Sensing 8(1):17–28.

Rowe, J.S. 1959. Forest Regions of Canada. Bulletin 123. Canada Department of Northern Affairs and National Resources, Forestry Branch, Ottawa.

Smith, W.H. 1981. Air Pollution and Forests. Springer-Verlag, New York.

Struik, H. 1973. Photo interpretive study to assess and evaluate the vegetational and physical state of the Sudbury area subject to industrial emissions.Unpublished report. Ontario Ministry of Natural Resources, Sudbury, Ontario.

Struik, H. 1974. Photo interpretive study to assess and evaluate vegetational changes in the Sudbury area. Internal report. Department of Lands and Forests, Ontario.

Turcotte, C.K. 1981. A comparative study of soils and vegetation in the vicinity of two roast yards in Sudbury, Ontario. Hons. B.Sc. Thesis, Laurentian University.

Wallace, C.M., and A. Thompson (eds.). 1993. Sudbury: Rail Town to Regional Capital. Dundurn Press, Toronto.

Watson, W.Y., and D.H.S. Richardson. 1972. Appreciating the potential of a devastated land. Forestry Chron. 48:312–315.

Whitby, L.M., and T.C. Hutchinson. 1974. Heavy metal pollution in the Sudbury mining and smelting region of Canada. II. Soil toxicity tests. Environ. Conservation 1(3):191–200.

Winterhalder, K. 1975. Reclamation of industrial barrens in the Sudbury area, pp. 64–72. *In* Transactions: Annual Meeting, Ontario Chapter, Canadian Society of Environmental Biologists, Sudbury, February 1975.

Winterhalder, K. 1983. Limestone application as a trigger factor in the revegetation of acid, metal-contaminated soils of the Sudbury area, pp. 201–212. *In* Proceedings of the Annual Meeting of the Canadian Land Reclamation Association, University of Waterloo, August 1983. Unpublished proceedings. CLRA, Guelph, Ontario.

Winterhalder, K. 1984. Environmental degradation and rehabilitation in the Sudbury area. Laurentian Univ. Rev. 16(2):15–47.

3

Reading the Records Stored in the Lake Sediments: A Method of Examining the History and Extent of Industrial Damage to Lakes

Sushil S. Dixit, Aruna S. Dixit, John P. Smol, and W. (Bill) Keller

In this chapter, the effects of air pollutants from the roasting beds and smelters on Sudbury area lakes are examined. A rather novel approach has been used to track the lake water quality changes that occurred in the past century. This approach uses the rapidly expanding science of paleolimnology, the study of the fossil record in the lake sediments. In the absence of long-term data, paleolimnological techniques using biological remains in lake sediment cores are being used extensively to provide quantitative assessments of past water quality in North America (Charles et al. 1990; Dixit et al. 1987, 1992c) and Europe (Battarbee et al. 1990).

A large area around Sudbury is characterized by a geological environment that is highly resistant to chemical weathering (see Chapter 1). As a result, many lakes have low acid-neutralizing (buffering) capacity, making them vulnerable to inputs of strong acids. The widespread acidification and metal contamination of area lakes are major environmental problems resulting from metal mining and smelting activities near Sudbury. Acidification is also a global concern (Fig. 3.1).

The very acidic nature of some lakes in the Sudbury area was observed as early as the late 1950s (Gorham and Gordon 1960). Evidence of fish population disappearance (Beamish and Harvey 1972; Keller 1978; Kelso and Gunn 1984) indicates that the acidification of many Sudbury area lakes was severe by the 1950s and 1960s. However, long-term water quality data are lacking, because the actual monitoring of water chemistry of some Sudbury lakes only began in the 1970s. Recently, it has been estimated that lakes in a 17,000-km^2 area (Fig. 3.2) have been measurably affected by the Sudbury emissions. If pH 6.0 is considered the acidity level at which significant damage begins to occur to the most acid-sensitive components of lake communities (e.g., some aquatic insects, crustaceans, and small fish species; see Fig. 5.2), then more than 7000 lakes within this zone have likely suffered biological damage (Neary et al. 1990).

Paleolimnology and Environmental Assessment

Paleolimnology uses the physical, chemical, and biological information contained in lake sediments to assess past environmental characteristics (Smol and Glew 1992). This multidisciplinary science has made a unique contribution to environmental assessment studies by making data available that would otherwise be unattainable. It has provided answers to questions such as (1) Has there been a change in the lake? (2) If so, what was the magnitude and rate of change? (3) Is the observed change greater than the natural variability? and (4) What caused the change?

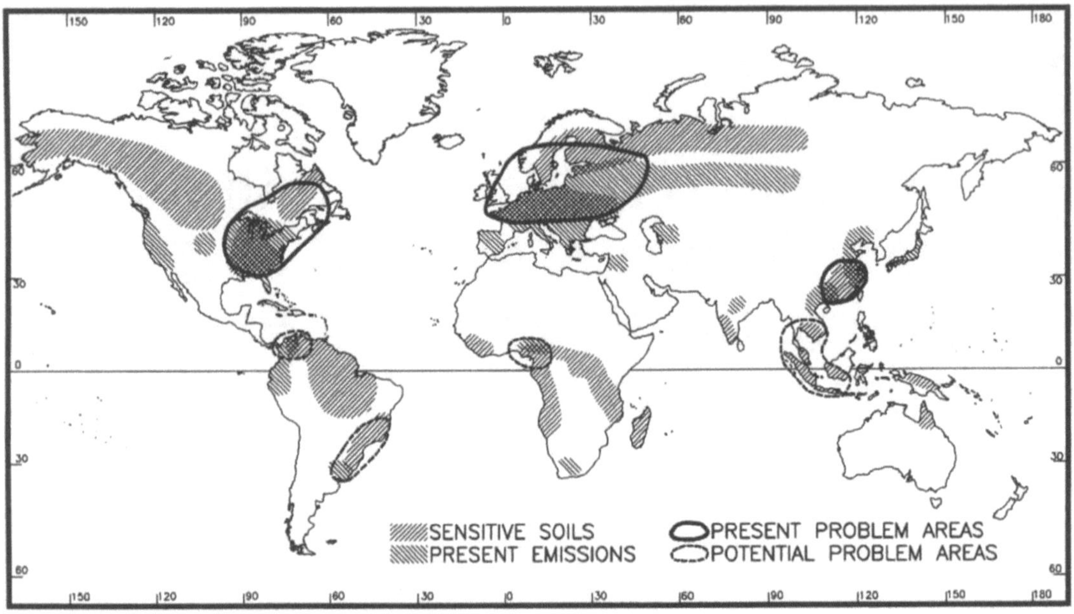

FIGURE 3.1. The combination of acid-sensitive terrain and high levels of acid deposition has resulted in widespread acidification, with effects documented in areas of North America, Europe, and Asia. Other areas of the world may develop acidification problems in the future if industrialization increases (from Rodhe and Herrera 1988).

Lake sediments contain many indicators of environmental conditions, including geochemical markers, plant pigments, and the fossil remains of aquatic organisms (Smol and Glew 1992). The remains of diatoms (Fig. 3.3) are the most widely used biological indicators of past lake water characteristics (Dixit et al. 1992c), although the fossil remains of other organisms such as *Chaoborus* (Box 3.1) have also provided much useful information. Within Canada, sedimentary diatoms have been used most extensively in Sudbury lakes to assess the impacts of industrial activities (Dixit et al. 1987, 1990, 1991).

Diatoms are single-celled microscopic plants belonging to the algal class Bacillariophyceae. They are made up of a highly ornamented cell wall composed primarily of glass (SiO_2). Each cell wall is made up of two pieces called valves and beltlike elements called girdle bands that hold them together. The size, shape, and sculpturing of the valve are species-specific (Fig. 3.3) and provide the basis for identification. Because diatom valves are made of silica, they are generally well preserved in the lake sediments (see Plate 3, following page 182).

Diatoms are key components of nearly all fresh and saline environments. They are ecologically diverse and colonize virtually every microhabitat in lakes and rivers. Diatom species have narrow optima and tolerances for many environmental variables and respond quickly to changes because of their ability to immigrate and replicate rapidly. Also, changes in diatom assemblages correspond closely to shifts in other biotic communities such as other algae, zooplankton, aquatic macrophytes, and fish. Although paleolimnology and study of sedimentary diatoms have become highly sophisticated in recent years, refinements to paleolimnological approaches are constantly being sought and implemented as techniques and protocols are evaluated.

Field and Laboratory Methods

The approach commonly used for paleolimnological monitoring and assessments is summa-

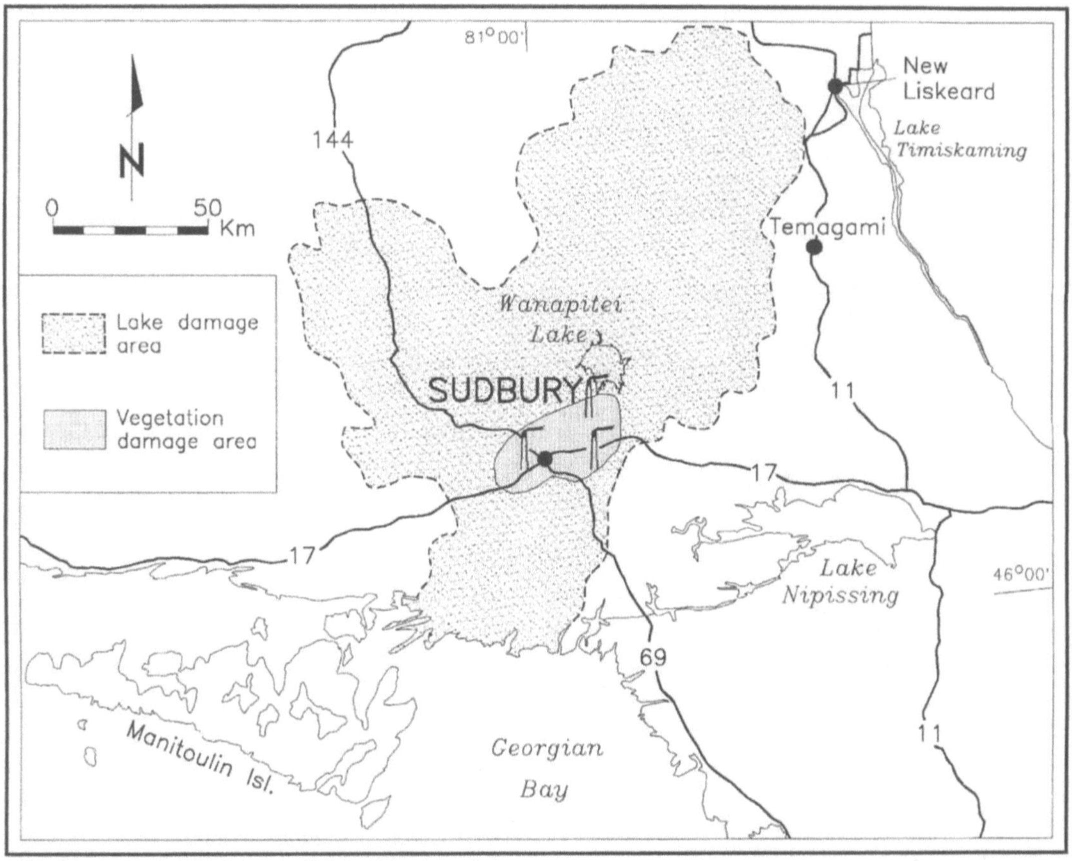

FIGURE 3.2. Approximate zone of influence of the Sudbury smelters on lake chemistry, based on water quality surveys (from Neary et al. 1990).

rized in Figure 3.4. Sediment cores are obtained from the deep basin of the lake using gravity or piston corers. Sediment cores of 30–40 cm generally cover the past two centuries of sediment accumulation in Sudbury and so are of sufficient length to study the post-industrial environmental changes. The cores are slowly pushed out of the top of the core tube and sectioned at specified intervals. The thickness of the sections is set by the investigators to meet the goals of the project (i.e., reconstruct environmental conditions for the past 3 years, 10 years, a century, or thousands of years). The upper portions of the cores from the Sudbury study were sectioned at fine intervals (i.e., 0.25 cm) so that recent (<10 years) as well as long-term environmental trends could be established. Techniques are also available to obtain information at annual or seasonal levels in some lakes (Simola 1977; Renberg 1981).

The sections of the sediment cores are dated to determine the time when the sediment was deposited at any particular depth (a depth–time profile must be established for each sediment core). Although a variety of techniques are available, the lead-210 radioisotope method (Appleby and Oldfield 1978) is the most commonly used dating method for recent (<150 years) lake sediments.

Diatoms are first cleaned and separated from the rest of the sediment by using strong acids, followed by repeated washes in distilled water. The resulting siliceous slurries are then mounted on glass slides for microscopic examination. Identifications are made to the lowest possible level (e.g., variety), because it is not uncommon

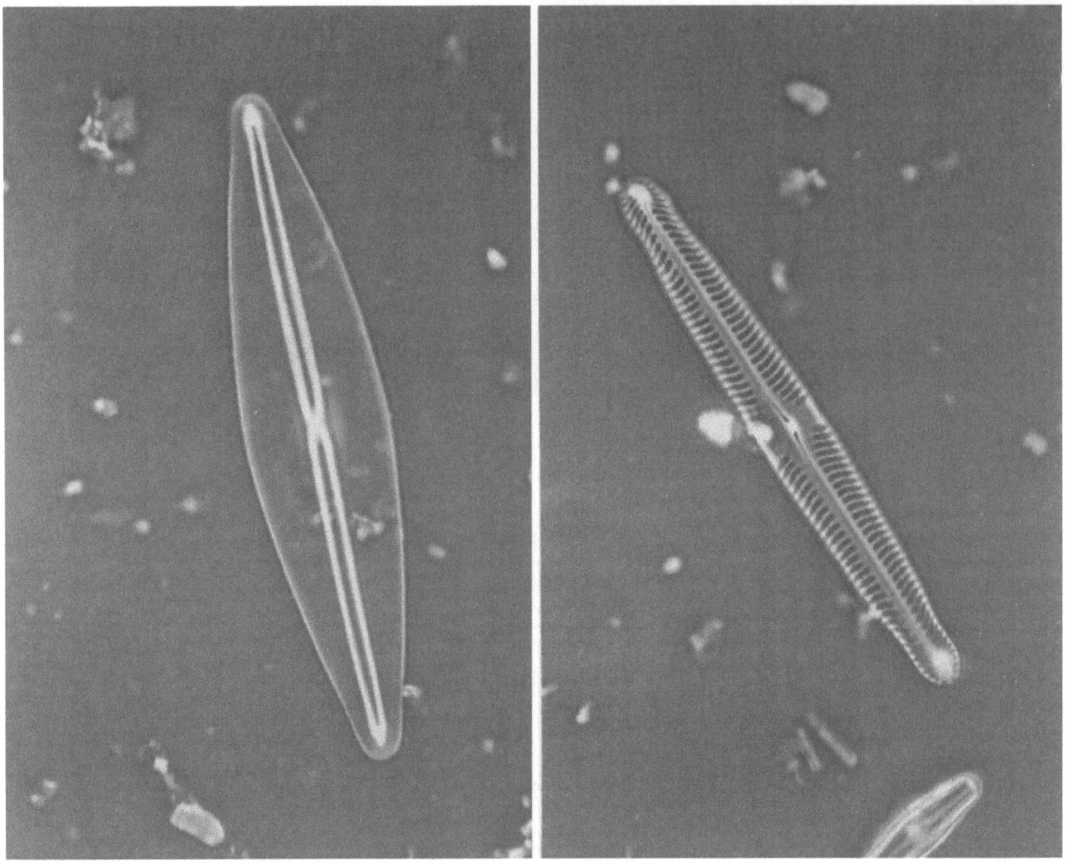

Figure 3.3. Light micrograph of microfossils of two diatom species.

for species belonging to the same genus to have different environmental tolerances and optima.

Diatom Calibration for Sudbury Lakes

In many lake regions of North America and Europe, surface (recent) sediment studies (calibration or training sets) have shown that the distribution of most diatoms is significantly correlated with variables such as lake water pH, metals, nutrients, conductivity, and many morphometric (e.g., lake size and depth) characteristics. Using a variety of statistical treatments (Charles et al. 1993), mathematical relationships (transfer functions) can be developed that quantitatively relate species distributions to environmental variables, such as lake water pH. Fortunately, a single sediment sample can provide a realistic picture of conditions in the whole lake because the sediments integrate diatoms both in space and time from different habitats within the lake.

Diatom assemblages present in the top 0.25-cm sediment of 72 lakes located within a 100-km radius of Sudbury have been calibrated (Dixit et al. 1991). Diatom-based models were then developed to infer past lake water pH and conductivity and concentrations of aluminum, nickel, and calcium. These models have been used to reconstruct past lake water characteristics of many Sudbury lakes (Dixit et al. 1991, 1992a,b).

Environmental Shifts in Clearwater Lake: A Case Study

Clearwater Lake is located 13 km southwest of Sudbury. It is a moderate-sized lake (76.5 ha), with maximum and mean depths of 21.5 and

Box 3.1.

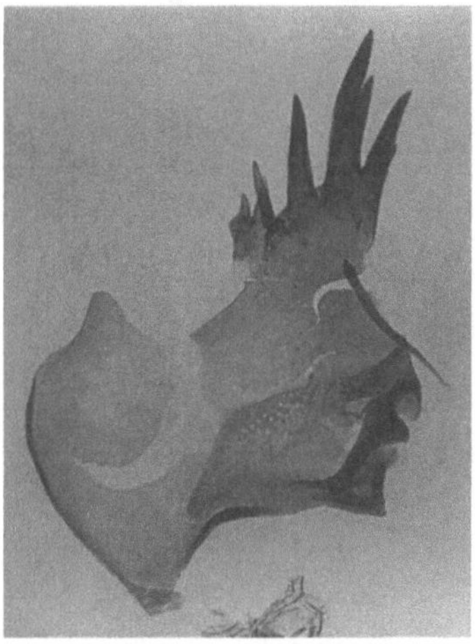

Paleolimnological methods can be used to infer historical biological conditions as well as the chemical and physical characteristics of lakes. In particular, the remains of hard mouthparts (mandibles; shown in the photograph) of larvae of the phantom midge, *Chaoborus* (see Plate 4, following page 182), in lake sediments can be used to reconstruct general patterns in historical fish populations (Uutala 1990; Uutala et al. 1994). Some *Chaoborus* species can coexist with fish, whereas others, notably *Chaoborus americanus,* cannot withstand fish predation. Thus, the species of *Chaoborus* found at different levels in the sediments indicate the presence or absence of fish at that time. This technique has been used to show that the current absences of fish from lakes in Ontario and the Adirondack Mountains of New York are not natural conditions but resulted from lake acidification.

8.3 m, respectively. Since 1973, this lake has been monitored as a reference site for lakes manipulated by liming (see Chapter 15) and to study seasonal and long-term changes in water chemistry and biota of an acidic lake (Yan and Miller 1984). Since the monitoring began, the lake has remained fishless, and no records are available to indicate whether the lake supported a viable fish population in the past.

Diatoms deposited over the past 200 years were analyzed from a sediment core retrieved in 1984. The core was dated using the lead-210 chronology method. The distribution of common diatom taxa in the past (Fig. 3.5) indicates that until about 1920, the diatom community remained relatively unchanged; however, since then major shifts have occurred. The relative abundance of acid and metal-tolerant species (e.g., *Tabellaria quadriseptata, Eunotia exigua,* and *Frustulia rhomboides* var. *saxonica*) increased, while *Cyclotella stelligera* (a species characteristic of nonacidic waters) declined sharply. These species shifts provide a strong indication that the lake has experienced marked water quality changes during this century.

Mathematical relationships (Dixit et al. 1991) for estimating past conditions of pH, aluminum, and nickel were then applied to the above diatom data. The resulting diatom-inferred pH profile (Fig. 3.6) indicates that the lake's preindustrial pH ranged between 5.7 and 6.1, whereas since 1920, the lake acidified very rapidly to a pH of 4.7 in 1980.

The acidification of Clearwater Lake appears to be directly related to sulfur dioxide emissions from the smelters in Sudbury. The study also shows that acidification began about 20 years after metal smelting commenced in Sudbury; this delay was likely due to the natural buffering capacity of this lake and its watershed. It is also possible that the open-pit roasting of ore, which started in the 1880s, did not adversely affect the lake and that the acidification only started after tall stacks were installed in the 1920s. Between 1980 and 1984, the inferred pH profile indicates that lake water pH has recovered slightly. Recent pH recoveries have also been observed in many other Sudbury lakes (see Chapter 5).

The Sudbury region is one of the few regions in which fossil diatoms found in lake sediments have been used to establish past trends in lake water metal concentrations. Inferred aluminum and nickel profiles indicate that lake water metal concentrations increased greatly in Clearwater Lake after the smelters

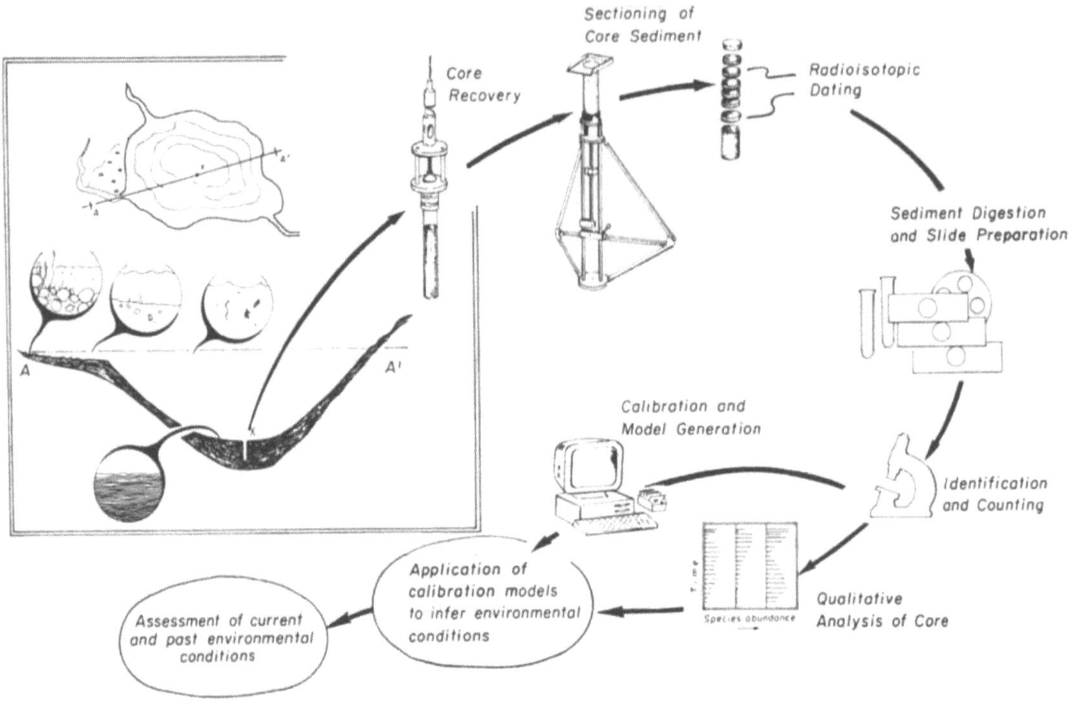

FIGURE 3.4. Microfossil inputs from various habitats to the lake bottom and the steps used in paleolimnological investigation of aquatic ecosystems (from Dixit et al. 1992c).

started operating (Fig. 3.6). Nickel and aluminum increases started about the same time that the lake water began to acidify. In the 1970s, lake water metal concentrations stabilized, and coincident with the pH recovery between 1980 and 1984, nickel and aluminum declines occurred in Clearwater Lake. The recovery is undoubtedly a reflection of the recent decline in smelter emissions. Paleolimnological inferences of the chemical history of Clearwater Lake agree well with actual chemistry monitoring data collected since 1973 (see Chapter 20).

Similar reconstructions of past lake water acidity have also been completed for 13 other Sudbury lakes. From these studies, it is possible to generalize that in acidic lakes located close to the smelters (<15 km), acidity-related changes started early in this century (i.e., 1920s–1940s), whereas in lakes located farther away from the emission sources, acidification began in the late 1950s and 1960s. In some of the study lakes, recent lake water pH increases and declines in metal concentrations have also been identified.

Changes in Lake Water Chemistry since Preindustrial Times

The analysis of complete sediment cores provides continuous assessment (e.g., timing, rate, and magnitude) of postindustrial changes in lake water quality. However, this is very time-consuming work, and it is not logistically feasible to analyze complete sediment cores from a large number of lakes to provide regional assessments of lake water quality change. The most effective approach for such a study is to analyze only the top (recent) and bottom (preindustrial) sediments of cores from a large number of lakes (Charles and Smol 1990). The difference between preindustrial and recent inferences provides an estimate of the change in

Figure 3.5. Relative percentage abundance of common diatoms in a sediment core from Clearwater Lake (modified from Dixit et al. 1987).

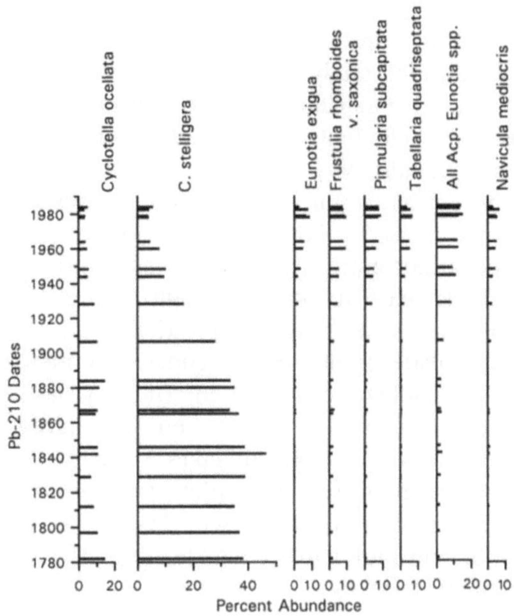

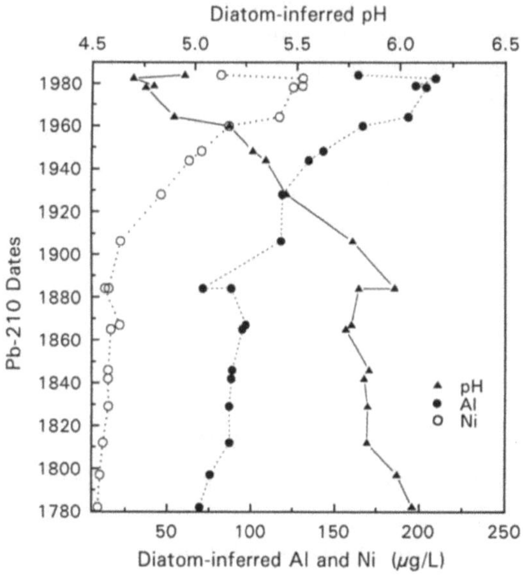

Figure 3.6. Diatom inferences of past changes in lake water pH, aluminum, and nickel in Clearwater Lake.

lake water chemistry as a result of industrial activity. Moreover, inferences for the bottom samples also provide information on background or reference limnological conditions.

The recent and preindustrial diatom-inferred water chemistry for 72 lakes within a 100-km radius of Sudbury (Dixit et al. 1992a) showed that extensive acidification has occurred in presently acidic (pH <6.0) lakes. The region also contains a few naturally acidic lakes; however, these lakes have acidified further as a result of industrial activity. Lakes that have current measured pH between 6.0 and 7.0 have either declined or increased in pH in the past, whereas high pH lakes (pH >7.0) have become more alkaline. Aluminum has increased in acidified lakes, whereas nickel increased in all study lakes. In general, in both acidic and nonacidic lakes, concentrations of dissolved elements (identified through inferred conductivity values, a measure of the total concentration of dissolved materials) have increased (Dixit et al. 1992a) because of direct atmospheric deposition to lakes and watersheds (e.g., sulfate) and high leaching of watershed soils by acid deposition (e.g., calcium). Patterns of elevated concentrations of some dissolved elements in Sudbury lakes have been shown by water chemistry surveys (Jeffries et al. 1984). In lakes close to the smelters, the loss of vegetation cover and resulting erosion of soils (see Chapter 2) no doubt contributed to the high loadings of some elements to lakes.

In addition to preindustrial and recent diatom assemblages, sediments deposited during about 1900, 1930, 1950, and 1970 were also studied for 22 Sudbury lakes to assess the rate of change in lake water chemistry since preindustrial time. Diatom-inferred pH, aluminum, and nickel were computed for these six sediment levels of lead-210 dated cores. Differences (increases or decreases) between preindustrial conditions (pre-1880) and the above time periods are plotted in Figure 3.7.

The inferred pH differences show that between pre-1880 and 1900, lake water pH remained relatively unchanged (Fig. 3.7). The small declines or increases that occurred in some lakes likely reflect natural variation, especially because there was no pattern of change with respect to current measured pH. By 1930, the lakes had started to show some pH decline, and in five of the 22 lakes, pH had declined 0.3 of a pH unit or more since pre-1880. In three lakes, pH increased 0.3 of a pH unit or more. Lakes continued to acidify during 1950 and 1970. Although further acidification occurred in some lakes between 1970 and recent time, a distinct pH recovery occurred in others. Marked pH recovery in Hannah and Middle lakes is a response to the liming of these lakes in the 1970s (see Chapter 15). The post-1970 lake water pH recovery in Baby and Clearwater lakes corresponds with the reductions in sulfur dioxide emissions by almost 80% within the past two decades (see Chapter 4).

The inferred aluminum data indicate that since pre-1880, aluminum has consistently increased in lakes that have current measured pH of 5.6 or lower (Fig. 3.7). In other lakes, aluminum has either increased or decreased. Although by 1930 aluminum had increased in some lakes, most of the increases occurred after 1950, the period of maximum acidification. Between 1970 and recent time, aluminum has declined in those lakes where pH has increased, whereas in other low pH lakes, aluminum has increased further.

The increase in nickel in almost all lakes (Fig. 3.7) suggests that these increases were largely independent of pH changes. With the exception of Hannah Lake, increases in nickel generally occurred after 1930, and by 1970 the maximum increase had occurred in most lakes. Generally the highest inferred nickel concentrations are for lakes located close to smelters and/or for mine tailing ponds in a few cases. The decline of lake water nickel in these lakes since 1970 follows the post-1970 reductions in smelter emissions. The absence of a close relationship between aluminum and nickel increases was expected, because nickel inputs were mainly atmospheric, whereas aluminum inputs were from the mobilization of aluminum from watersheds and possibly lake sediments.

Temporal and Spatial Patterns in Lake Water pH

Temporal and spatial patterns in lake water pH were further examined by drawing distribution maps for five inferred pH categories for preindustrial time, about 1930 and 1970, and recent time (Fig. 3.8). These maps provide a graphic display of lake water pH changes over space and time.

In the 22-lake data set (Fig. 3.8) during preindustrial time and 1930, none of the lakes had pH less than 5.0, whereas 13.6% of the lakes were in this category by 1970. In the pH range 5.0–5.6, the percentage of lakes contin-

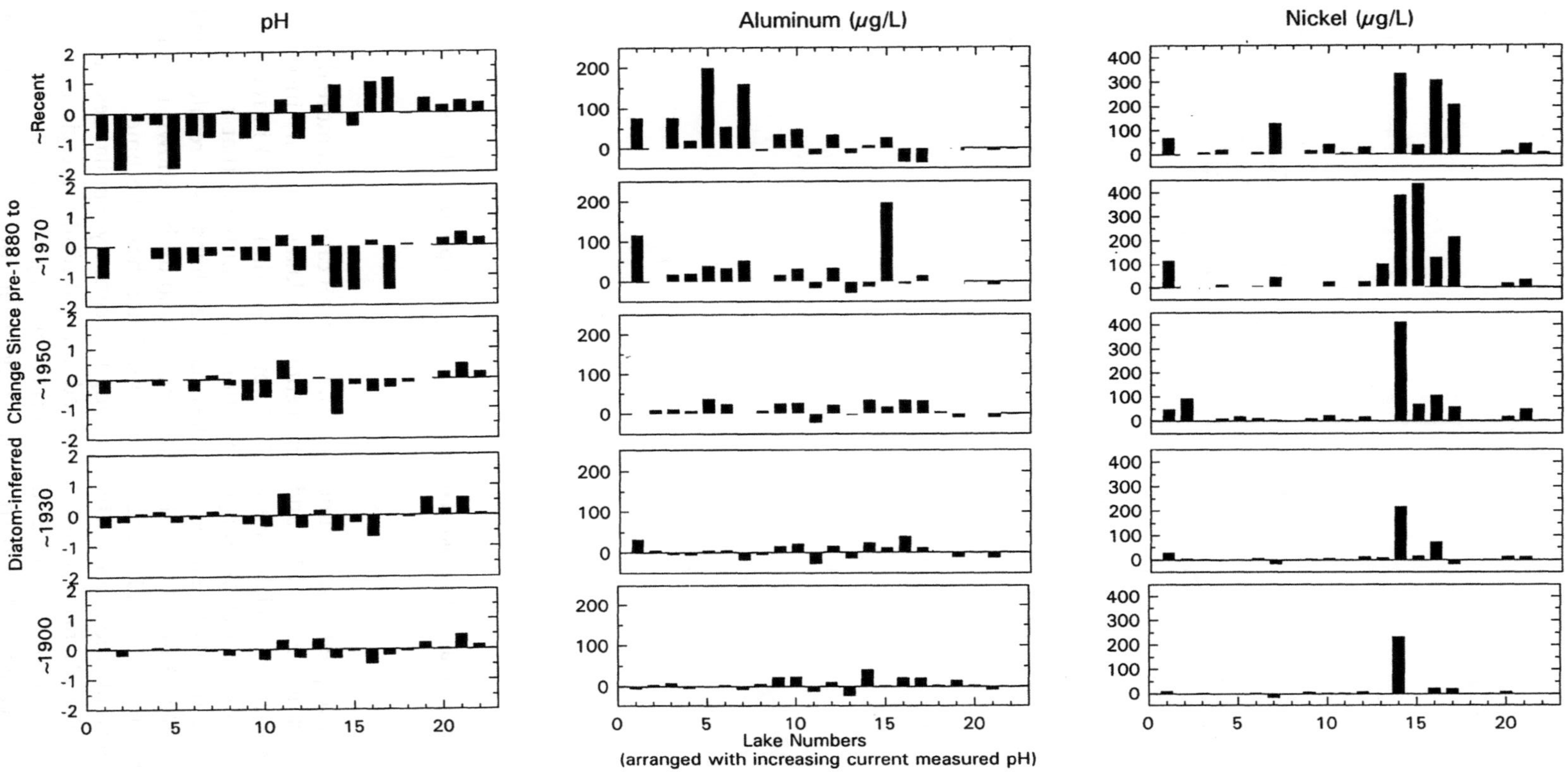

FIGURE 3.7. Inferred changes in lake water pH, aluminum, and nickel in 22 lakes between preindustrial time, about 1900, 1930, 1950, and 1970, and recent time. The lakes and their recent measured pH are (1) Clearwater, 4.5; (2) Daisy, 4.7; (3) Mountaintop, 4.8; (4) Wavy, 4.8; (5) Chiniguchi, 4.9; (6) Telfer, 5.0; (7) Swan, 5.6; (8) Alphretta, 6.2; (9) Laura, 6.3; (10) Whitson, 6.4; (11) Horseshoe, 6.7; (12) Southeast Baby, 6.7; (13) Labelle, 6.8; (14) Hannah, 6.8; (15) Baby, 6.9; (16) Clarabelle, 6.0; (17) Middle, 7.1; (18) Emerald, 7.3; (19) Fairbank, 7.5; (20) Ramsey, 7.5; (21) Round, 7.5; (22) Little Panache, 7.8.

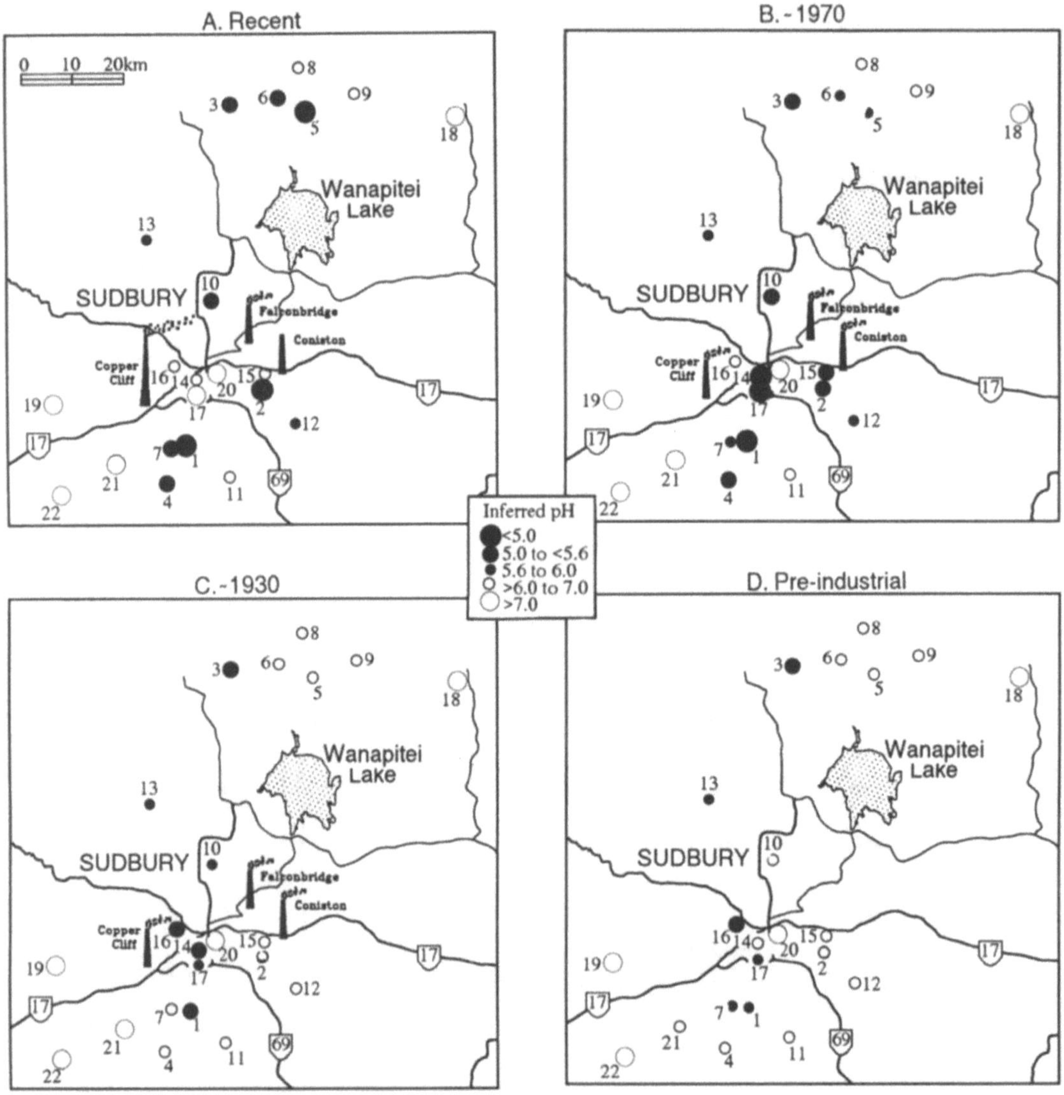

FIGURE 3.8. Spatial and temporal distribution of inferred pH for 22 lakes.

ued to increase from 9.1 in preindustrial time to 18.2, 22.7, and 22.7% in 1930, 1970, and recent time, respectively. Only two lakes were naturally acidic (pH <5.6). In the pH range 5.6–6.0, the percentage of lakes reached a maximum of 22.7 in 1970. Only 9.1% of the study lakes are presently in this pH range, most likely due to emission reductions. The percentage of lakes in the pH range 6.0–7.0 declined from 54.5 (preindustrial) to 18.2 in 1970 but since then increased to 27.3% (recent). In the pH category of more than 7.0, the percentage of lakes gradually increased from 18.2 (preindustrial) to 27.3 in recent time. In addition to identifying that many Sudbury lakes have acidified as a result of industrial activity, this study has shown that almost all high-pH lakes have become more alkaline during this century. Similar results were obtained for Adirondack region lakes (New York) (Cumming et al. 1992). Various factors are responsible for the alkalization of high-pH lakes (Cumming et al. 1992; Dixit et al. 1992a).

The pH distribution maps show that, in general, maximum acidification occurred by 1970, and the lakes that experienced most acidification are located close to smelters and/or lie within the northeast–southwest zone of impact. However, not all lakes were affected. Alkaline lakes are often located in close proximity to recently acidified lakes. This is explained by the presence of acid-neutralizing minerals in some watersheds, which allowed lakes to maintain a high pH despite high atmospheric acid loadings. Inferred pH has even increased in these lakes since preindustrial times.

Summary and Conclusions

Lake sediments have revealed the history of damages to Sudbury lakes from smelter emissions. Inferred patterns of increasing lake water acidity and metal concentrations began about 1920 for many lakes close to the smelters and later, in the 1950s and 1960s, in many lakes farther from the emission sources. Only a few Sudbury lakes appear to be naturally acidic, and these acidified further after industrialization of the region. With high-acid deposition, some well-buffered lakes became even more alkaline over time. Recent increases in pH and decreases in metal concentrations, related to reduced smelter emissions, are evident in the sediment record for several lakes.

Because paleolimnological techniques provide long-term data on ecosystem condition and changes, they are being used as an integral part of environmental monitoring and assessment programs. The approach can be used to detect and quantify lake water quality changes, provide data on preindustrial (predisturbance) conditions, establish natural variability, identify long-term trends, and monitor the effects of remedial action plans on aquatic environments. Although this chapter mainly deals with lake acidification, this approach has been used to address a wide variety of environmental issues, such as lake eutrophication and climate change. The research in the Sudbury region has not only provided answers to questions that could not be answered in any other way, but it has also helped to refine paleolimnological techniques currently being used in other lake regions.

Acknowledgments. Our paleolimnological work in Sudbury was mainly funded by the Natural Sciences and Engineering Research Council of Canada. Lead-210 dating was done at Chalk River Nuclear Laboratory. We thank Dr. A. Uutala for the photograph of the *Chaoborus* mandible.

References

Appleby, P.G., and F. Oldfield. 1978. The calculation of ^{210}Pb dates assuming constant rate of supply of unsupported ^{210}Pb to the sediment. Catena 5:1–8.

Battarbee, R.W., J. Mason, I. Renberg, and J.F. Talling (eds.). 1990. Paleolimnology and lake acidification. Phil. Trans. R. Soc. Lond.

Beamish, R. J., and H.H. Harvey. 1972. Acidification of the La Cloche Mountain lakes, Ontario, and resulting fish mortalities. J. Fish. Res. Board Can. 29:1131–1143.

Charles, D.F., M.W. Binford, B.D. Fry, E. Furlong, R.A. Hites, M.J. Mitchell, S.A. Norton, M.J. Patterson, J.P. Smol, A.J. Uutala, J.R.White, D.R. Whitehead, and R.J. Wise. 1990. Paleoecological investigation of recent lake acidification in the Adirondack Mountains, N.Y. J. Paleolimnol. 3: 195–241.

Charles, D.F., and J.P. Smol. 1990. The PIRLA II Project: regional assessment of lake acidification trends. Verh. Int. Verein. Limnol. 24:474–480.

Charles, D.F., J.P. Smol, and D.R. Engstrom. 1993. Paleolimnological approaches to biomonitoring. *In* S. Loeb and A. Spacie (eds.). Biomonitoring of Freshwater Ecosystems. Lewis, Ann Arbor, MI.

Cumming, B.F., J.P. Smol, J.C. Kingston, D.F. Charles, H.J.B. Birks, K.E. Camburn, S.S. Dixit, A.J. Uutala, and A.R. Selle. 1992. How many lakes in the Adirondack Mountain region of New York (U.S.A) have acidified since preindustrial times? Can. J. Fish. Aquat. Sci. 49:128–141.

Dixit, S.S., A.S. Dixit, and R.D. Evans. 1987. Paleolimnological evidence of recent acidification in two Sudbury (Canada) lakes. Sci. Total Environ. 67:53–67.

Dixit, S.S., A.S. Dixit, and J.P. Smol. 1990. Paleolimnological investigation of three manipulated lakes from Sudbury, Canada. Hydrobiologia 214: 245–252.

Dixit, S.S., A.S. Dixit, and J.P. Smol. 1991. Multivariable environmental inferences based on diatom assemblages from Sudbury (Canada) lakes. Freshwater Biol. 26:251–266.

Dixit, S.S., A.S. Dixit, and J.P. Smol. 1992a. Assessment of changes in lake water chemistry in Sudbury area lakes since preindustrial times. Can. J. Fish. Aquat. Sci. 49 (Suppl. 1):8–16.

Dixit, A.S., S.S. Dixit, and J.P. Smol. 1992b. Long-term changes in lakewater pH and metal concentrations in 3 Killarney Provincial Park lakes, near Sudbury, Ontario (Canada). Can. J. Fish. Aquat. Sci. 49 (Suppl. 1):17–24.

Dixit, S.S., J.P. Smol, J.C. Kingston, and D.F. Charles. 1992c. Diatoms: Powerful indicators of environmental change. Environ. Sci. Technology 26:22–33.

Gorham, E., and A.G. Gordon. 1960. The influence of smelter fumes upon the chemical composition of lake waters near Sudbury, Ontario, and upon the surrounding vegetation. Can. J. Bot. 38:477–487.

Jeffries, D.S., W.A. Scheider, and W.R. Snyder. 1984. Geochemical interactions of watersheds with precipitation in areas affected by smelter emissions near Sudbury, Ontario, pp. 196–241. *In* J. Nriagu (ed.). Environmental Impacts of Smelters. John Wiley & Sons, New York.

Keller, W. 1978. Limnological Observations on the Aurora Trout Lakes. Ontario Ministry of the Environment Report, Sudbury, Ontario.

Kelso, J.R.M., and J.M. Gunn. 1984. Responses of fish communities to acidic waters in Ontario. *In* G.R. Hendrey (ed.). Early Biotic Responses to Advancing Lake Acidification. Butterworth Publishers, Woburn, MA.

Neary, B.P., P.J. Dillon, J.R. Munro, and B.J. Clark. 1990. The Acidification of Ontario Lakes: An Assessment of Their Sensitivity and Current Status with Respect to Biological Damage. Technical Report. Ontario Ministry of the Environment Report, Dorset, Ontario.

Renberg, I.G. 1981. Improved methods for sampling, photographing and varve-counting of varved lake sediments. Boreas 10:255–258.

Rodhe, H., and R. Herrera (eds.). 1988. Acidification in Tropical Countries. John Wiley & Sons, New York.

Simola, H. 1977. Diatom succession in the formation of annually laminated sediment in Lovojarvi, a small eutrophicated lake. Ann. Bot. Fennici 14:143–148.

Smol, J.P., and J.R. Glew. 1992. Paleolimnology, pp. 551–564. *In* W.A. Nierenberg (ed.). Encyclopedia of Earth System Science, vol. 3. Academic Press, San Diego, CA.

Uutala, A.J. 1990. Chironomidae (Diptera) as paleolimnological indicators of acidification in some Adirondack Mountain Lakes (New York, USA). J. Paleolimnol. 4:139–151.

Uutala, A.J., N.D. Yan, A.S. Dixit, and J.P. Smol. 1994. Paleolimnological assessment of damage to fish communities in three acidic, Canadian Shield lakes. Fish. Res. 19:157–177.

Yan, N.D., and G.E. Miller. 1984. Effects of deposition of acids and metals on chemistry and biology of lakes near Sudbury, Ontario, pp. 244–282. *In* J. Nriagu (ed.). Environmental Impacts of Smelters. John Wiley & Sons, New York.

Section B

Trends in Natural Recovery after Emission Reductions

Harold H. Harvey and John M. Gunn

In the background of the picture on the previous page the scene is one of desolation—the two smokestacks of the abandoned Coniston nickel smelter, the bare hills blackened from decades of intensive fumigation, and the dried remains of a fallen pine forest. These are the results of destructive smelting activities described in detail in earlier chapters of this book. However, this section deals not with the past but with recent evidence that natural processes are beginning to repair the damage to terrestrial and aquatic ecosystems in the Sudbury Basin. The focus is, therefore, on recovery, represented in the foreground of the picture by a young birch tree, one of the first tree species to recolonize the harsh landscape of the Sudbury barrens.

Resilience or the ability to *re-create* are terms often used when describing healthy ecosystems. Ecologists argue strenuously about the appropriateness of such terms, but whether it is a seedling pushing its way up through urban pavement or wind-blown insects landing on a newly formed volcanic island, disturbed and even severely damaged land does not remain empty of plants and animals indefinitely. The Sudbury landscape was repeatedly scoured of soil and vegetation by the passage of mile-high ice during six glaciations in the past 190,000 years. This fact does not make the current damage any more acceptable, but it does offer hope through example that natural recovery can occur despite what may appear to be irreparable damage. However, recovery cannot happen unless something else happens—the cause of the damage ceases. The Sudbury situation illustrates this point.

In recent decades, emissions of sulfur dioxide and metal particulates from Sudbury smelters have been reduced dramatically. As the authors of the following chapters show, controlling pollution at the source not only protects sensitive ecosystems from being damaged but also permits damaged systems to begin to recover. Preventing damage from occurring is the ultimate solution to pollution problems, but seeing evidence of recovery also provides powerful support for the legislated effort to reduce present and future emissions. Too often the argument has been made that, although tragic, "the damage has been done." Evidence of natural recovery after abatement programs at an internationally known site such as Sudbury indicates clearly that damage can be undone and that environmental degradation is not a license to continue to pollute. Extensive remedial work may be needed to speed the process, but the evidence of natural recovery indicates that we are on the right track—that the corrective measures being taken currently are effective indeed.

Studies in the Sudbury area have contributed to international awareness of environmental problems. This was the case in the mid-1960s when one of us began working on some of the many remote, seemingly pristine lakes in the LaCloche Mountains, approximately 60 km southwest of the Sudbury smelters. During the course of these studies, which initially had nothing to do with air pollution, Beamish and Harvey discovered that the lakes were acidifying and fishes of many species were dying and disappearing from these lakes. These results opened the eyes of many people in North America to the problem of "acid rain." This finding was soon confirmed for other lakes in Ontario near point sources of sulfur dioxide pollution such as the Wawa area and for lakes in south-central Ontario: Parry Sound, Muskoka, and Haliburton, which were far removed from any point source. These findings, plus a rapidly accumulating list of similar results from many places, especially Scandinavia and northeastern United States, instigated an enormous public campaign that propelled governments into restricting the use of the atmosphere as a dumping ground for contaminants. Several important pieces of provincial and state legislation have been passed since, and more important, Canada and the United States have signed an accord requiring a large reduction in sulfur dioxide emissions within 10 years.

Chance also played a role in the observations of natural biological recovery of damaged lakes in the Sudbury area when, in the early 1980s, Gunn and Keller observed that a remnant population of lake trout was starting to reproduce again. This time the observation came from a study site far to the north of the smelters, but again links to Sudbury were made—water quality was improving, which allowed reproduction to occur, and all this was related directly to declining emissions. Unfortunately, on the global scale, observations of recovery are, as of yet, far less common than were the earlier reports of damage.

We thank the authors of the following chapters for their descriptions of air quality improvements and for the evidence they present of the beginnings of natural recovery of terrestrial and aquatic biological communities. We do not exaggerate the progress made, given the severe problems that still exist near Sudbury, but the evidence presented does provide encouragement and strong support for continuing efforts to reduce atmospheric emissions. The people who live and work in the Sudbury Basin are the greatest benefactors of this progressive improvement in the quality of the natural environment. All Canadians will benefit from the changing image of Sudbury from an international center of pollution to that of a green and pleasant land.

4

Declining Industrial Emissions, Improving Air Quality, and Reduced Damage to Vegetation

Raymond R. Potvin and John J. Negusanti

During the century of mining and smelting activity in the Sudbury basin, more than 100 million tonnes of sulfur dioxide and tens of thousands of tonnes of copper, nickel, and iron have been released into the atmosphere (Ontario/Canada Task Force 1982). In the early 1960s, Sudbury's copper/nickel smelting complex represented one of the largest point sources of sulfur dioxide in the world (Summers and Whelpdale 1976), contributing approximately 4% of the global emissions (Freedman 1989). During this peak period, Sudbury emissions of sulfur dioxide approached current-day emissions from the whole of the United Kingdom (Table 4.1).

Environmental improvements in the Sudbury area during recent decades illustrate the importance of controlling the release of pollutants into the atmosphere. In the middle part of this century, when the regional and global nature of air pollution was not well understood, dispersing pollutants through tall stacks was considered an acceptable solution to a local environmental problem. In later years, it was recognized that discharging pollutants high into the atmosphere was simply creating problems elsewhere.

This chapter describes progress in reducing air pollutants from the Sudbury smelters. Attention is focused on two main pollutants: sulfur dioxide and metal particulates. Oxides of nitrogen, the other major precursors of acidic deposition, are released in relatively small amounts from the Sudbury smelters.

Recognition of the Problem

In Ontario, the Canadian province with the largest population, more than 95% of the sulfur dioxide emissions are from large point sources, predominantly smelters and electrical power plants. Of these, the nickel and copper smelters of Inco Limited and Falconbridge Limited in the Sudbury area contribute slightly more than half of the provincial and approximately 20% of the Canadian emissions (Table 4.1).

The damaging effects of smelter fumes on gardens, forests, and other vegetation were well known at the turn of the century (Barlow 1907; Haywood 1907), but early government response to public complaints appears to have been designed to protect the industry rather than people and ecosystems. Government legislation passed in 1915 provided that all patents issued to settlers of land within a defined area include a clause exempting mining companies from liability due to smoke damage. In 1921, The Damages by Sulphur Fumes Arbitration Act was passed to facilitate the settlement of claims of damage to agricultural crops and other vegetation. This act was repealed and replaced with a similar one in 1924, which led to the hiring of a claims arbitrator in 1925.

Table 4.1. Sulfur dioxide emissions from Sudbury smelters relative to total emissions from selected countries

	Total emissions (1000s of tonnes)	Year
Sudbury		
present	216	1994[a]
recent	700	1988[a]
historical	2560	1960[a]
(Comparison of 1986–1988)		
United States	20,700	1988[b]
China	20,000	1987[c]
Soviet Union[d]	9270	1988[e]
Eastern Germany	4365	1988[e]
Canada	3800	1988[f]
Poland	3760	1988[e]
United Kingdom	3400	1988[e]
India	3070	1987[c]
Spain	2925	1983[e]
Czechoslovakia	2520	1988[e]
Italy	2133	1986[e]
France	1368	1987[e]
Western Germany	1350	1988[e]
Netherlands	260	1988[e]
Sweden	198	1987[e]
Switzerland	67	1988[e]
Norway	67	1988[e]

[a]Unpublished data for Inco Limited plus Falconbridge Ltd.
[b]U.S. EPA 1990.
[c]Kato and Akimoto 1992.
[d]European part of the former Soviet Union.
[e]French 1990.
[f]OECD 1991.

As production of nickel increased, spurred on by increased demand during the two world wars, so, too, did the air pollution problems (Katz 1954; Dreisinger 1955; see Chapter 2). The mining companies took some important initiatives to reduce sulfur dioxide emissions through process improvements (e.g., rejection of high sulfur pyrrhotite components of ore before smelting, development of oxygen flash-furnace smelting, and the capture of sulfur for marketable byproducts such as sulfuric acid and liquid sulfur dioxide) (see Chapter 21). However, continuing fumigation problems within nearby urban and rural areas and the mounting scientific evidence of extensive damage to area forests (Dreisinger and McGovern 1964; Linzon 1971) demanded more stringent controls.

First Control Orders

The provincial government of Ontario finally responded to these concerns by imposing annual limits on smelter sulfur dioxide emissions, with the first control orders issued against Sudbury mining companies in 1969 and 1970. This control program was solely directed at improving local air quality. It did not address the contribution of Sudbury sulfur dioxide emissions to worldwide acid precipitation. Thus, the 381-m "Superstack," the world's tallest smokestack, was constructed at the Copper Cliff smelter of Inco Limited and began operating on August 21, 1972 (Fig. 4.1). Inco's Coniston smelter and Falconbridge's iron ore sintering plant (Fig. 4.2) were closed that same year, with the result that total emissions also declined after 1972.

The combination of increased dispersal of pollutants, reduced emissions, and the closure of obsolete plants led to dramatic improvements in air quality (Fig. 4.3). The concentration of sulfur dioxide in the Sudbury area dropped immediately by at least 50% after 1972. During the 1980s, the annual average concentration of sulfur dioxide remained fairly constant and well below the provincial objective of 0.020 ppm. There were also readily observed visual signs of the improving air quality. The occurrence of smoke and haze at the local airport, located 16 km northeast of Sudbury, declined sharply after the early 1970s (Fig. 4.4).

However, despite the reductions in annual average sulfur dioxide concentrations, severe short-term ground-level fumigations still occurred under certain climatic conditions. This continuing fumigation problem, together with mounting concern about the impacts of acidic deposition, led to further reductions in annual allowable emissions, starting in 1978, and required that the smelters curtail daily production/emissions under weather conditions that restricted the dispersal of the plumes. Under

FIGURE 4.1. Inco Limited Superstack (Sudbury Star Files) constructed in 1972 to reduce local impacts of industrial emissions. Technologies at Inco to reduce the quantity of pollutants released into the atmosphere are described in Chapter 21.

FIGURE 4.2. Falconbridge smelter (circa 1955). As part of the emissions abatement program, the pyrrhotite plant (center, rear) was closed in 1972. The sintering plant on the right was closed in 1978. Details of the abatement technology at Falconbridge Limited are provided in Appendix 4.1.

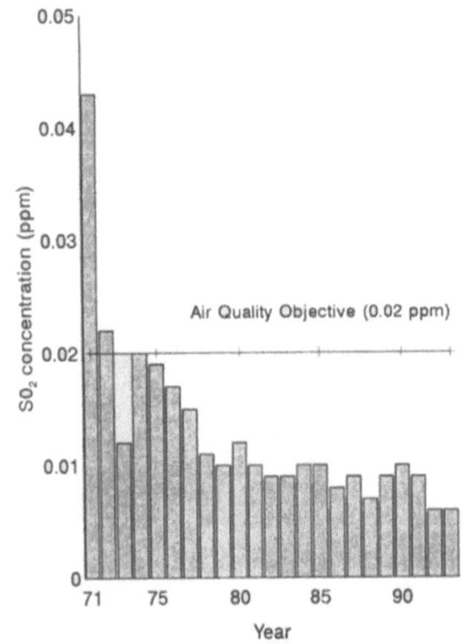

FIGURE 4.3. Annual average sulfur dioxide concentrations recorded at three sites in the Sudbury area over the period 1971–1993.

these requirements, production at the smelters is significantly reduced to avoid hourly sulfur dioxide ground-level concentrations greater than 0.50 ppm. These more-stringent requirements reduced the occurrence of ground-level fumigations but have still not eliminated them at all sites (Fig. 4.5).

A Regional and International Problem

The commissioning of the Superstack and the other changes to deal with what was perceived to be a local air pollution problem occurred in the same year that Sweden brought the transboundary problem of air pollution to the attention of the world. Sweden presented its case history of the impacts of atmospheric pollutants at the UN environment conference at Stockholm in 1972 (Anonymous 1972). By the mid- to late 1970s, North American scien-

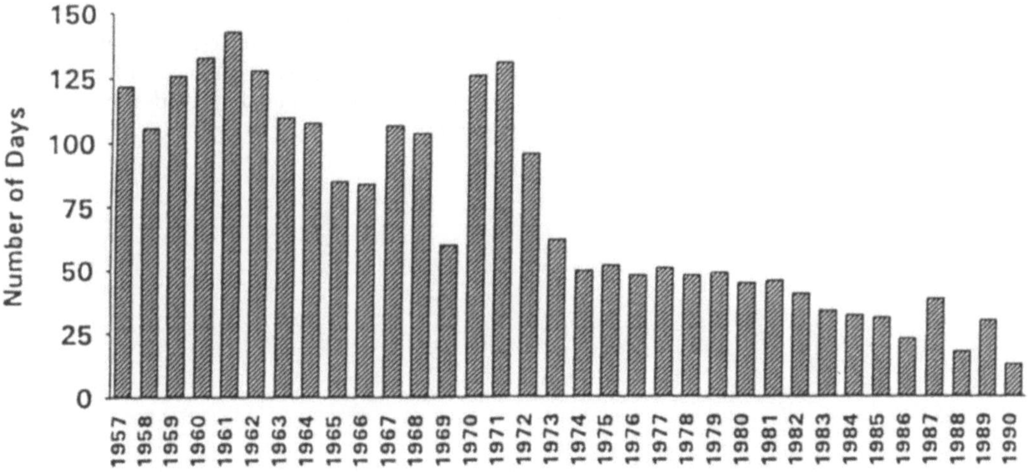

FIGURE 4.4. Changes in the recorded instance of smoke or haze at the Sudbury airport weather station, 1957–1990 (courtesy of R. Pitblado).

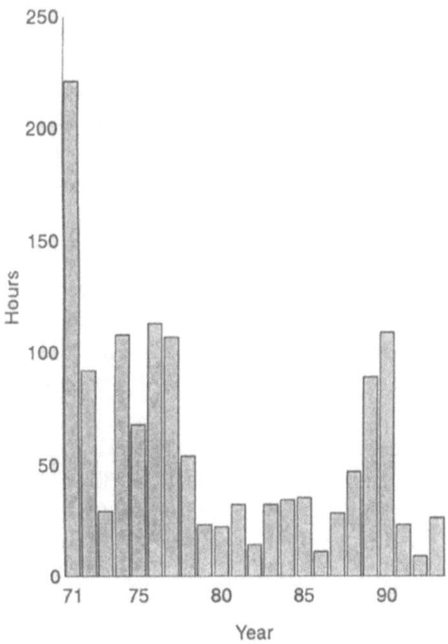

FIGURE 4.5. Frequency of short-term sulfur dioxide fumigations with an hourly average in excess of 0.50 ppm at Ontario Ministry of the Environment monitoring stations, 1971–1993.

tists were rapidly accumulating evidence that air pollution was a regional rather than a local problem. The words *acid rain* and the chemical reactions that were involved in the transformation of emissions from automobiles and factories to acid precipitation quickly became "common knowledge."

$$\underset{\text{(sulfur dioxide)}}{SO_2} + \underset{\text{(water)}}{H_2O} \rightarrow \underset{\text{(sulfuric acid)}}{H_2SO_4}$$

$$\underset{\text{(nitrogen oxide)}}{NO_x} + \underset{\text{(water)}}{H_2O} \rightarrow \underset{\text{(nitric acid)}}{HNO_3}$$

From extensive monitoring studies conducted during the 1970s, large point sources of sulfur dioxide, such as the smelters of Sudbury, were shown to be contributing to air pollution problems (e.g., acidified lakes) far from the sources of the pollution. For example, computer models indicated that 19% and 5.5%, respectively, of the total sulfur deposition in the Muskoka (central Ontario) and Nova Scotia areas originated from Sudbury (Ontario/Canada Task Force 1982). Within the Sudbury area itself, dry deposition and fallout of metal particulate were primarily of local origins, but even the Sudbury area was subject to substantial inputs of acid pollutants from far away (Jeffries 1984).

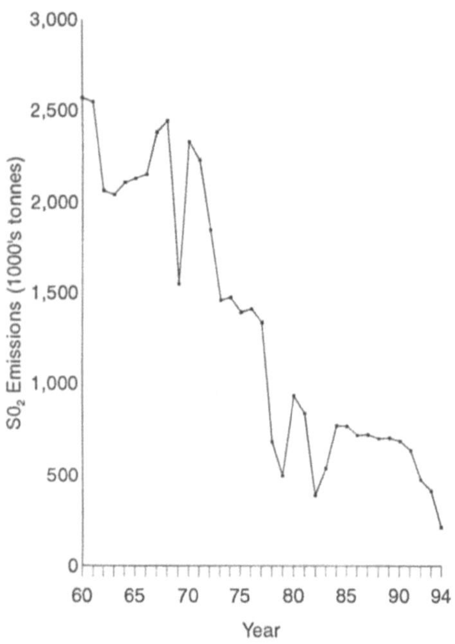

FIGURE 4.6. Sulfur dioxide emissions for the Sudbury area (1960–1994). In addition to legislated changes, labor strikes (1969, 1978, 1979) and periods of reduced nickel production (1982, 1983, 1994) contributed to the decline in sulfur dioxide emissions.

From modeling studies, it has been estimated that approximately 80% of the wet (snow and rain) deposition of acid in the Sudbury area originated from industrial areas in the United States and southern Ontario (Chan et al. 1984; Lusis et al. 1986).

New provincial and national emission control legislation was needed to address this international problem. In 1985, the Ontario government established a control program called the Countdown Acid Rain Program to meet nationally and provincially negotiated reduction objectives for eastern Canada (Ontario Ministry of the Environment 1985). Under this program, the annual legal limit for Sudbury area sulfur dioxide emissions was set at 365,000 tonnes, to be achieved by 1994. This would bring the annual emission of sulfur dioxide down to about 14% of the highest emission year: 2.56 million tons in 1960 (Fig. 4.6).

Declining Industrial Emissions

Historical trends in emissions of sulfur dioxide in Fig. 4.6 illustrate the magnitude of the Sudbury area sources and the success of control programs. Both mining companies were able to achieve the 1994 limit. In fact, because of additional reductions due to extended shutdowns and some initial problems with one of the new furnaces at Inco Limited, the 1994 total of 216,000 tonnes was well below the required level.

In addition to regulating sulfur dioxide, the control orders have resulted in lower emissions of suspended particulates and trace metals in the Sudbury area. During 1973–1981, approximately 15,000 tonnes of particulate matter were released into the atmosphere each year (Ontario Ministry of the Environment 1982; Ozvacic 1982). These emissions included approximately 1800 tonnes of iron, 700 tonnes of copper, 500 tonnes of nickel, 200 tonnes of lead, and 100 tonnes of arsenic on an annual basis (Ozvacic 1982). Average levels of suspended particulates at two sites in the city of Sudbury ranged from 50 to more than 80 $\mu g/m^3$ during the early 1970s (Potvin and Balsillie 1976, 1978). By the late 1980s, these annual averages had dropped to less than about 30 $\mu g/m^3$, well below the provincial objective of 60 $\mu g/m^3$ (Dobrin and Potvin 1992). Among trace ele-

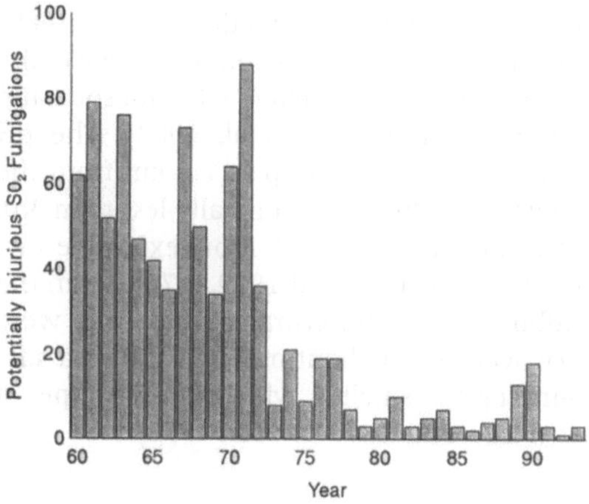

FIGURE 4.7. Frequency of potentially injurious sulfur dioxide fumigations in the Sudbury area, 1960–1993. The occurrence of potentially injurious fumigation events is calculated from the measured duration and concentration of sulfur dioxide at ground-level monitoring stations. Some international examples of the damaging effects of air pollutin on buildings, statues, and rock surfaces are provided in Box 4.1

ments, nickel is probably the most suitable indicator of temporal trends. In 1971, the mean concentration of nickel in air samples at a site 4 km east of the Copper Cliff smelter was 0.37 μg/m³; after abatement efforts of 1972, concentrations immediately dropped by at least 50%. Additional reductions followed ongoing abatement activities and by the mid- to late 1980s, annual average concentrations of nickel at the monitoring station were in the order of 0.05 μg/m³ (Dobrin and Potvin 1992). By 1995, it is projected that total emission of particulate matter will be approximately 1500 tonnes/year, equal to approximately 10% of the levels in the early 1970s.

Terrestrial Effects

Direct Damage to Vegetation and Acidification of Soils

Sulfur dioxide causes direct damage to vegetation (Katz et al. 1939; Linzon 1978). The severity of damage depends on the fumigation dosage and frequency, as well as the species tolerance. Damage can range from discoloration of leaves and needles to reduced growth rate and mortality of plants (Linzon 1978). Some observable damage to plants includes terminal necrosis on white pine (*Pinus strobus*) foliage and interveinal necrosis on leaves of white birch (*Betula papyrifera*) (see Plates 5 and 6, following page 182). Excellent color photographs of sulfur dioxide symptoms are also available in Malhotra and Blauel (1980) and Skelly et al. (1987).

During the period before the major emission reductions, many cases of vegetation injury associated with severe fumigations, sometimes up to 80 km from the Sudbury smelters, were documented (McCallum 1944–1964; Dreisinger 1952–1971). The size of area where sensitive species such as white pine were affected with severe-to-moderate injury covered more than 1800 km², with an additional 4000 km² of sporadic injury (Linzon 1978). These direct effects of air pollution, common in the 1960s, were largely eliminated by the control programs of the 1970s and 1980s.

The dramatic decline since 1971 in the number of potentially injurious fumigation events (Tebbins and Hutchinson 1961; Dreisinger and McGovern 1970) provides a quantitative measure of the improving conditions for vegetation in the Sudbury area (Fig. 4.7). Another valuable indicator of the improving conditions for plants are the lichens, which have recolonized much of the Sudbury area in recent decades. Lichens are useful monitors because they are particularly sensitive to air pollution (see Chapter 6).

In addition to the direct damage to exposed vegetation, the high emissions of sulfur from smelters also leads to the severe acidification of soils (reviewed in Chaudhry et al. 1982). Extremely acid soils were found in the vicinity of the Coniston smelter (Hutchinson and Whitby 1974; Hazlett et al. 1983). The pH values of surface soils up to 7.4 km from the Coniston smelter were generally less than 3.0, with values as low as 2.2. More extensive surveys between 1970 and 1979 at 70 sites in the Sudbury area demonstrated that soils were very acidic throughout much of the area surrounding the smelter, with an average pH of 4.1 (Negusanti and McIlveen 1990).

Trace Metal Accumulation in Soil and Vegetation

Although sulfur dioxide is generally recognized as the most important cause of vegetation damage in the Sudbury area, many researchers have studied and demonstrated the significance of the terrestrial effects of metals (Hutchinson and Whitby 1974; McGovern and Balsillie 1975; McIlveen and Balsillie 1978; Freedman and Hutchinson 1980; Negusanti and McIlveen 1990). These studies have shown that metal levels in soil (see Chapter 2) and vegetation decline with increasing distance from the smelters. Extremely high metal levels have been recorded in soil and vegetation in close proximity to Sudbury smelters. Nickel and copper at localized sites in surface soil near the Coniston smelter were determined to be as high as 12,300 μg/g and 9700 μg/g, respectively (Hazlett et al. 1983), but generally heavily contaminated sites have metal concentrations in the 200–500 μg/g range (Negusanti and McIlveen 1990).

There is very little conclusive information available regarding the changes in soil conditions that may have occurred after the major industrial abatement programs of the early 1970s. Gunderman and Hutchinson (1993) recently reported that the concentration of nickel and copper and the acidity of soils near the abandoned smelter at Coniston have declined substantially during the past 20 years. This pattern has also been observed from soil surveys conducted in the Sudbury area by Ontario Ministry of Environment scientists (Negusanti and McIlveen 1990). However, despite the apparent changes, the soils remain toxic to sensitive plant species (Gunderman and Hutchinson 1993), and it is difficult to determine what factors are responsible for the chemical changes. For example, the organic content of the soil also declined during the same period, suggesting that the surface soils are continuing to be eroded by wind and rain. Support for these findings was obtained from a recent extensive survey of Sudbury soils conducted by Dudka et al. (1994).

Monitoring of soil conditions and assessing the process of natural recovery represent important information needs. However, the structural instability of the soil and the extreme temporal and spatial variability in soil chemistry (Negusanti and McIlveen 1990) make this a challenging research area.

Other Air Pollution Problems

Particulate fallout, especially during smelter startups and shutdowns, has occasionally caused significant direct damage to vegetation and property (vehicles, patio decks, house sidings, etc.) at sites near the smelters (Ontario Ministry of Environment 1978). The injury symptom, referred to as "black spotting," is caused by particulate fallout consisting principally of nickel, copper, and iron sulfates. Leafy garden crops such as lettuce and cabbage, due to their large surface areas, have been observed to be very susceptible to fallout injury. Normal washing of affected garden produce was found to reduce metal levels well below those considered safe for normal consumption. Fallout damage to vegetation still occurs but appears to have been greatly reduced since the early 1980s.

Other air pollutants that have caused injury to vegetation on an episodic basis are sulfur trioxide and ground-level ozone (Negusanti and McIlveen 1990). Isolated atmospheric releases of sulfur trioxide resulting from malfunctions at the sulfuric acid plants have caused injury to sensitive species such as tomato, bean, and cucumber at downwind distances greater than 20 km.

Ground-level ozone, largely formed through the long-range transport of its precursors (oxides

Box 4.1. Fading Statues and Black Rocks

The corrosive effects of acid-forming air pollutants have created significant damage to buildings and statues throughout the world. The picture of the stone carvings in the front of the Confucian Temple in a suburb of Anshun, Guizhou Province, China, illustrates this damage. (Photo by Xiong Jilin.) Damage to the famous Taj Mahal, in India, is another example of this large-scale problem. Fortunately, recent pollution abatement programs in India appear to be effective at reducing or preventing further damage to the Taj.

In Sudbury, the blackened rocks also bear witness to the severity of air pollution problems. The relative contribution of roast bed fumes, smelter emissions, or smoke from forest fires in creating this distinctive feature of the barren landscape is unknown, but the combined effects have created a substantial challenge for restoration efforts. The blackened surfaces create extremely hot and dry conditions in summer, microclimate conditions that exclude all but the hardiest of plant species.

of nitrogen and reactive hydrocarbons) from southern industrial areas, has been reported to exceed the 1-hour provincial criterion of 80 ppb in the Sudbury area during the growing seasons (Potvin and Balsillie 1976; Dobrin and Potvin 1992). From 1975 to 1990, the criterion was exceeded during slightly more than 1000 hours, or on average about 63 hours per growing season. The injury noted in the Sudbury area has been minor and limited to species such as sweet corn, grape, potato, and Manitoba maple (Negusanti and McIlveen 1990).

Conclusions and Future Projections

As illustrated in later chapters, the reductions in industrial emissions have allowed the denuded Sudbury landscape to begin slowly to recover. Natural processes have begun to repair the damage inflicted since before the turn of the century. Improvements in air quality have also allowed highly successful artificial reclamation programs to speed up the recovery process (see Chapter 8). This is particularly evident in the successful re-establishment of white pine, which is among the most sensitive tree species to sulfur dioxide injury. With the influx of vegetation into the denuded areas from natural growth and colonization, as well as from the reclamation efforts, the organic content and pH of surface soil should increase; soil metals should then become less bioavailable and hence less toxic to plants.

The prospects for additional reductions in smelter emissions and further ecosystem recovery are promising. If anticipated progress in ore processing and smelting technology is realized (see Chapter 21), it will be possible to reduce annual sulfur dioxide emissions in the Sudbury area to less than 250,000 tonnes. Additional reductions in particulate and trace metal emissions would also occur, possibly resulting in an annual particulate emission level of about 1000 tonnes. However, there are still concerns about the frequency and intensity of local site-specific fumigations. Further improvements will be required in the daily produc-

Box 4.2. Noril'sk, the World's Largest Point Source of Sulfur Dioxide

Noril'sk, a city of 260,000 built during the Stalin era to harvest the vast mineral resources in central Siberia, is now considered one of the most polluted cities on earth (Saunders 1990; Peterson 1993). The current emissions of sulfur dioxide from the nickel, copper, and cobalt smelters and refineries are estimated to exceed 2.5 million tonnes/year (Saunders 1990). Located at approximately 300 km above the Arctic Circle, Noril'sk supplies roughly two-thirds of the total nickel production of Russia. At present, there is little information available on the effects of industrial emissions on the tundra ecosystems near Noril'sk. It has been estimated that air pollution has severely affected 10,000 km^2 of forests in Russia and other parts of the former Soviet Union (cited in Peterson 1993). (Photo of the nickel smelting complex in Noril'sk by D.J. Peterson in 1992.)

tion/emission reduction program to reap the full benefits of continuing smelter sulfur dioxide abatement initiatives.

Sudbury is not alone in its efforts to control emissions. Significant reductions in sulfur emissions and deposition have been achieved in several areas in North America and in parts of Europe (Hedin et al. 1987; Dillon et al. 1988; Wright and Hauhs 1991). However, in contrast, the conditions at the giant Noril'sk smelting complex (Box 4.2) and at other large Russian smelters are still very similar to historical conditions at Sudbury. Also, significant increases in sulfur dioxide emissions are expected in China and India where rapid industrial development is increasing the use of high sulfur coal (Galloway 1989).

Acknowledgments. We thank D. Bouillon, A. Bradshaw, J. Gunn, H. Harvey, and W. Keller for providing review comments. M. Conlon and M. Courtin assisted with graphics.

References

Anonymous. 1972. Sweden's case study for the United Nations Conference on the Human Environment, 1972: Air Pollution across National Boundaries. The Impact on the Environment of Sulfur in Air and Precipitation. Stockholm, Sweden.

Barlow, A.E. 1907. Origin, geology, relations and composition of the nickel and copper deposits of the Sudbury mining district. Geol. Surv. Can. 961:1–244.

Chan, W.H., R.H. Vet, C.U. Ro, A.J.S. Tang, and M.A. Lusis. 1984. Impact of Inco smelter emissions on wet and dry deposition in the Sudbury area. Atmos. Environ. 18:1001–1008.

Chaudhry, M., M. Nyborg, M. Molina-Ayala, and R.W. Parker. 1982. Reactions of SO_2 emissions with soils. Proceedings of the Symposium on Acid Forming Emissions in Alberta and Their Ecological Effects, Edmonton, Alberta.

Dillon, P.J., M. Lusis, R.A. Reidand, and D. Yap. 1988. Ten year trends in sulphate, nitrate, and hydrogen deposition in central Ontario. Atmos. Environ. 22:901–905.

Dobrin, D.J., and R. Potvin. 1992. Air Quality Monitoring Studies in the Sudbury Area. Ontario Ministry of the Environment, Northeastern Region, Sudbury, Ontario.

Dreisinger, B.R. 1952–1971. Sulphur Dioxide Levels in the Sudbury Area and the Effects of the Gas on Vegetation. Yearly reports 1952–1971. Ontario Department of Mines, Sudbury, Ontario.

Dreisinger, B.R. 1955. Atmospheric Sulphur Dioxide Observations and the Effects of the Gas upon Foliage in the Sudbury Area. Ontario Department of Mines, Sudbury, Ontario.

Dreisinger, B.R., and P.C. McGovern. 1964. Sulphur Dioxide Levels in the Sudbury Area and Some Effects of the Gas on Vegetation in 1963. Ontario Department of Mines, Sudbury, Ontario.

Dreisinger, B.R., and P.C. McGovern. 1970. Monitoring atmospheric sulphur dioxide and correlating its effects on crops and forests in the Sudbury area. Proceedings of Conference on Impact of Air Pollution on Vegetation, Toronto, Ontario.

Dudka, S., R. Ponce-Hernandez, and T.C. Hutchinson. In press. Current level of total element concentrations in the surface layer of Sudbury's soils. Sci. Total Environ.

Freedman, B. 1989. Environmental Ecology. The Impacts of Pollution and Other Stresses on Ecosystem Structure and Function. Academic Press. San Diego.

Freedman, B., and T.C. Hutchinson. 1980. Pollutant inputs from the atmosphere and accumulation in soils and vegetation near a nickel-copper smelter at Sudbury, Ontario, Canada. Can. J. Bot. 58(1): 108–132.

French, H.F. 1990. Cleaning the air, pp. 98–118. *In* L.R. Bown et al. (eds.). State of the World 1990. W.W. Norton, New York.

Galloway, J.N. 1989. Atmospheric acidification: projections for the future. Ambio 18:161–166.

Gunderman, D.G., and T.C. Hutchinson. 1993. Changes in soil chemistry 20 years after the closure of a nickel-copper smelter near Sudbury, Ontario, pp. 559–562. R.L. Allan and J.O. Nriagu (eds.). Proceedings of the International Conference on Heavy Metals in the Environment, vol. 2. Toronto. CEP Consultants, Edinburgh.

Haywood, J.K. 1907. Injury to vegetation and animal life by smelter fumes. J. Am. Chem. Soc. 29:998–1009.

Hazlett, P.W., G.K. Rutherford, and G.W. Van Loon. 1983. Metal contaminants in surface soils and vegetation as a result of nickel/copper smelting at Coniston, Ontario. Reclamation Revegetation Res. 2(2): 123–137.

Hedin, L.O., G.E. Likens, and F.H. Borman. 1987. Decrease in precipitation acidity resulting from decreased SO_4concentration. Nature 325: 244–246.

Hutchinson, T.C., and L.M. Whitby. 1974. Heavy metal pollution in the Sudbury mining and smelting region of Canada. I. Soil and vegetation contamination by nickel, copper and other metals. Environ. Conservation 1:123–132.

Jeffries, D.S. 1984. Atmospheric deposition of pollutants in the Sudbury area, pp. 117–154. *In* J. Nriagu (ed.). Environmental Impacts of Smelters. John Wiley & Sons, New York.

Kato, N., and H. Akimoto. 1992. Anthropogenic emissions of SO_2 and NOx in Asia: emission inventories. Atmos. Environ. 26A(16):2997–3017.

Katz, M. 1954. Atmospheric sulphur dioxide conditions in the Sudbury region, May to October, 1953. Report of Special Sulphur Dioxide Investigating Committee. Ottawa. Unpublished report.

Katz, M., (ed.). 1939. Effect of Sulphur Dioxide on Vegetation. National Research Council, Bill 815. Ottawa, Canada.

Linzon, S.N. 1971. Economic effects of SO_2 on forest growth. J. Air Pollut. Control Assoc. 21:81–86.

Linzon, S.N. 1978. Effects of airborne sulphur pollutants on plants, pp. 109–162. *In* J.O. Nriagu (ed.). Sulphur in the Environment: Part 2, Ecological Impacts. John Wiley & Sons, New York.

Lusis, M.A., A.J.S. Tang, W.H. Chan, D. Yap, J. Kurtz, P.K. Misra, and G. Ellenton. 1986. Sudbury impact on atmospheric deposition of acidic substances in Ontario. Water Air Soil Pollut. 30:897–908.

Malhotra, S.S., and R.A. Blauel. 1980. Diagnosis of Air Pollutant and Natural Stress Symptoms on Forest Vegetation in Western Canada. Publ. no. NOR-X-228. Northern Forest Research Centre, Canadian Forestry Service, Environment, Edmonton, Alberta, Canada.

McCallum, A.W. 1944–1964. Report on Field Examination of the Sudbury Area. Yearly reports 1944–1964. Ontario Department of Mines, Sudbury, Ontario.

McGovern, P.C., and D. Balsillie. 1975. Effects of Sulphur Dioxide and Heavy Metals on Vegetation in the Sudbury Area 1974. Ontario Ministry of the Environment, Northeastern Region, Sudbury, Ontario.

McIlveen, W.D., and D. Balsillie. 1978. Air Quality Assessment Studies in the Sudbury Area. Vol. 2. Effects of Sulphur Dioxide and Heavy Metals on Vegetation and Soils, 1970–1977. Ontario Ministry of the Environment, Northeastern Region, Sudbury, Ontario.

Negusanti, J.J., and W.D. McIlveen. 1990. Studies in the Terrestrial Environment in the Sudbury Area. Ontario Ministry of the Environment, Northeastern Region, Sudbury, Ontario.

OECD. 1991. The State of the Environment: Air Quality. Paris.

Ontario/Canada Task Force. 1982. Report of the Ontario/Canada Task Force for the Development and Evaluation of Air Pollution Abatement Options for Inco Limited and Falconbridge Nickel Mines Limited in the Regional Municipality of Sudbury, Ontario. Intergovernment Task Force Report. Toronto, Ontario.

Ontario Ministry of the Environment. 1978. Report on Particulate Deposition in Sudbury, August 30, 1978 and September 8, 1978. Ontario Ministry of the Environment, Northeastern Region, Sudbury, Ontario.

Ontario Ministry of the Environment. 1982. Sudbury Environmental Study Synopsis, 1973–1980. Ontario Ministry of the Environment Acidic Precipitation in Ontario Study Coordination Office, Toronto.

Ontario Ministry of the Environment. 1985. Ontario's Acid Gas Control Program for 1986–1994, Countdown Acid Rain. Ontario Ministry of Environment and Energy, Toronto.

Ozvacic, V. 1982. Emissions of Sulphur Oxides, Particulate and Trace Elements in the Sudbury Basin. Air Resources Branch, Report ARB-ERTD-09–82. Ontario Ministry of Environment and Energy, Toronto.

Peterson, D.J. 1993. Troubled Lands: The Legacy of the Soviet Environmental Destruction. Westview Press, Boulder, CO.

Potvin, R.R., and D. Balsillie. 1976. Air Quality Monitoring Report for the Sudbury Area 1975. Ontario Ministry of the Environment, Northeastern Region, Sudbury, Ontario.

Potvin, R.R., and D. Balsillie. 1978. Air Quality Assessment Studies in the Sudbury Area—Vol. 1. Air Quality Monitoring for the Sudbury Area (1976–1977). Ontario Ministry of the Environment, Northeastern Region, Sudbury, Ontario.

Saunders, A. 1990. Poisoning the arctic skies. Arctic Circle 1(2):22–31.

Skelly, J.M., D.D. Davis, W. Merrill, E.A. Cameron, H.D. Brown, D.B. Drummond, and L.S. Dochinger (eds.). 1987. Diagnosing Injury to Eastern Forest Trees. Forest Response Program. U.S.D.A. Forest Service, Pennsylvania State University, University Park, PA.

Summers, P.W., and D.M. Whelpdale. 1976. Acid precipitation in Canada. Water Air Soil Pollut. 6:447–455.

Tebbins, B.D., and D.H. Hutchinson. 1961. Application of air quality standards to a community problem. J. Air Pollut. Control Assoc. 11:53–56.

United States Environmental Protection Agency. 1990. National Air Pollutant Emission Estimates 1940–1988. Publication EPA-450/4–90–001. Office of Air Quality Planning and Standards, Research Triangle Park, NC.

Wright, R.F., and M. Hauhs. 1991. Reversibility of acidification: soils and surface waters. Proc. R. Soc. Edinburgh 97B:169–191.

Appendix 4.1. SO_2 Abatement at Falconbridge Sudbury Operations (Mike Kozlowski, Falconbridge Limited)

Falconbridge's efforts to address the sulfur dioxide (SO_2) problem commenced in 1972 with the closure of a plant that had been used to burn sulfur and recover nickel from a pyrrhotite (Fe_7S_8) concentrate made at that time.

The biggest single reduction in SO_2 emissions was the result of building a new $300 million (in 1994 dollars) smelter in 1978. The dramatic reduction was due largely to the construction of a new acid plant that converts the SO_2 to sulfuric acid—a salable commodity.

The processing of ore at the present smelter at Falconbridge Ontario begins with slurry received from the Strathcona Mill (Fig. A4.1). The moisture content is adjusted and the concentrate, together with sand, is fed to two fluid bed roasters. Currently about 60% of the sulfur is eliminated (oxidized) in the roasters, and the resultant calcine/flux fed to two electric furnaces. An intermediate or "furnace" matte containing about 35% nickel is produced during electric smelting. This matte is processed in the converter aisle where more sulfur and iron are oxidized in conventional Peirce-Smith converters to yield a cast matte containing about 52% nickel for shipment to the company's refinery in Norway. The key change that has reduced SO_2 emissions is that off-gases from the roasters are tightly contained and therefore rich in SO_2—about 10%. This gas is suitable for treatment in the single-pass acid plant, whereas the gases from the old smelting process were too dilute to do this.

With the new smelter, most of the rest of the SO_2 is produced in the converter vessels, although the SO_2 from the electric furnaces and the acid plant tail gas make up a significant proportion. All three sources are vented to the atmosphere through the same 93-m stack. In addition to the annual SO_2 emission limits imposed by the control orders shown in Figure A4.2, there is a control order regarding the permissable concentration measured at ground level. Currently, this is 0.5 ppm, calculated on an hourly basis. Periods of high ground-level concentration are usually due to adverse weather conditions. The smelter uses a monitoring system and will shut down the converter aisle if an exceedance is anticipated. Complaints from nearby citizens are investigated immediately and appropriate action, including shutdown if necessary, is taken.

A continuing decline in emissions of SO_2 has occurred since the new smelter was built. The decline from 1988 to 1993 occurred even during a period of a 22% increase in nickel production. This recent continued decrease in SO_2 for each tonne of nickel produced was largely the result of recent changes at the Strathcona Mill and the mines. These changes have been directed at rejecting more of the pyrrhotite in ore and making greater quantities of copper concentrate for sale (thereby reducing the sulfur load to the smelter).

A new magnetic separator allowed more pyrrhotite to be passed to the pyrrhotite rejection circuit. Improvements in the rejection circuit further increased the amount of pyrrhotite removed. Unfortunately, these changes also led to an increase in the nickel lost in this circuit. In 1993, a recycle loop was added to the rejection circuit, which reduces these losses to traditional levels.

At the same time that these improvements in pyrrhotite rejection were being made, less pyrrhotite was being supplied in the ore from the mines. The combined effects were to decrease the pyrrhotite and increase the nickel content of concentrate going to the smelter. The high-grade concentrate allowed an increase in the electric furnace matte grade, which in turn led to a significant reduction in the amount of SO_2 produced per tonne of cast matte.

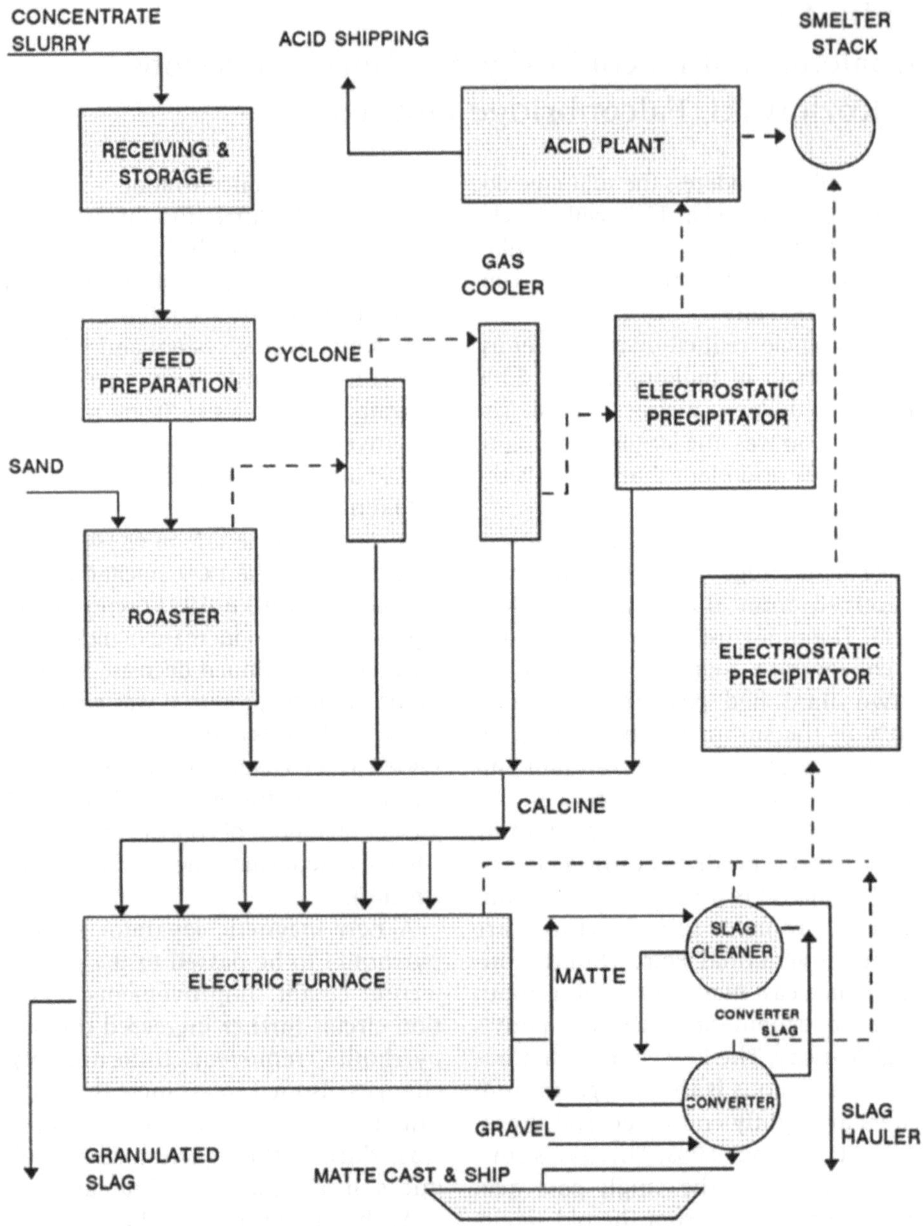

FIGURE A4.1.

Production of copper concentrate leads to nickel losses. To counteract this, several million dollars were spent on flotation columns and press filters in the Strathcona Mill. This had led to a valuable reduction in those losses.

Regulations and Future Developments

The Falconbridge Sudbury Operations has consistently stayed below government control orders. In addition to complying with the new

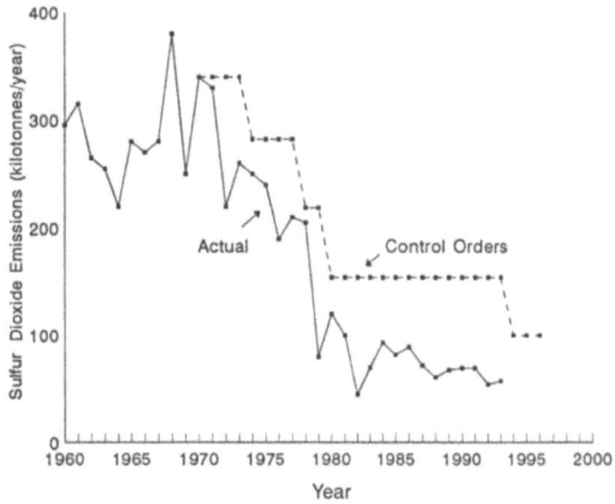

FIGURE A4.2. Changes in emissions of sulfur dioxide from the Falconbridge smelter. Note that the limit set for emissions in 1994 was required under Regulation 661/85.

100-ktonne order for 1994, Falconbridge has voluntarily committed to keep emissions below 75 ktonnes by 1998.

Future efforts in the mill will focus on increasing both pyrrhotite rejection and copper concentrate production, thus increasing the concentrate grade to the smelter. Some of the issues that the research and development work will need to address include:

1. increased complexity of the mill circuits and the need for reliable advanced process control
2. impoundment of pyrrhotite tailings to prevent acid run-off
3. increased sensitivity to the mine feed grade

In the smelter, more SO_2 will be captured for acid production by increasing the amount of sulfur burned off in the roaster. Engineering studies are needed to evaluate capacity shortcomings in the roasters and acid plant. Increasing the degree of roast has a negative impact on metal recoveries, which has to be counteracted by adding more reductant in the electric furnaces. This, in turn, has serious implications for furnace operations (bottom build-up, higher temperatures, increased off-gas volume), which require extensive research and development programs and additional capital expenditures.

Falconbridge has spent more than $40 million in the past 5 years on SO_2 abatement and continues to work on a long-term basis with all levels of government to reduce the environmental impact of its operations.

5

Lake Water Quality Improvements and Recovering Aquatic Communities

W. (Bill) Keller and John M. Gunn

In the remote areas to the northeast and southwest of Sudbury, the environmental damage is less visible than in the denuded landscapes near the smelters but is no less severe. These hilly forested areas, underlain by granitic bedrock (Fig. 5.1), contain most of the more than 7000 lakes estimated to have been damaged by smelter emissions (see Fig. 3.2).

Perhaps the most striking example of biological damage to these lakes is the loss of sportfish species, such as lake trout (*Salvelinus namaycush*), brook trout (*S. fontinalis*), walleye (*Stizostedion vitreum*), and smallmouth bass (*Micropterus dolomieui*) (Beamish and Harvey 1972; Matuszek et al. 1992). Most of Ontario's estimated sportfish losses from acidification occurred in this area (Matuszek et al. 1992). In fact, excluding the losses of Atlantic salmon (*Salmo salar*) from Nova Scotia rivers (Watt et al. 1983), almost all Canada's well-documented cases of fisheries losses from acidification are in the Sudbury area (Kelso et al. 1990). However, a century of industrial emissions has resulted in far more extensive biological damages than the loss of sportfish. Losses of acid or metal-sensitive species, leading to reduced community richness, occurred for organisms at various aquatic trophic levels, including zooplankton (Sprules 1975; Keller and Pitblado 1984), phytoplankton (Kwiatkowski and Roff 1976; Nicholls et al. 1992), and benthic invertebrates (Roff and Kwiatkowski 1977). Effects of acidification on waterfowl (see Chapter 16) and amphibians (Glooschenko et al. 1992) have also been reported for the area.

A simplified illustration of some of the biological changes that occur in a typical Sudbury lake between pH 5.0 and 6.0 is provided in Figure 5.2.

A New Question—Reversibility

More than 35 years ago, Gorham and Gordon (1960) began scientific studies of lakes and ponds near Sudbury. Since then, a vast amount of information has been collected that clearly established the damaging effects of smelter emissions on the chemistry and biology of water bodies. This information has been widely used nationally and internationally in the debate for cleaner air. In recent years, research findings from studies of Sudbury lakes have been used to answer another controversial question: Are acidification damages reversible?

In the 1970s, there was uncertainty, which continues to some degree, whether acidified lakes would recover without the addition of lime or other artificial sources of alkalinity. It was assumed that the acid-neutralizing capacity of watershed soils, once exhausted, could not recover or that the renewal of buffering capacity through the geochemical weathering

Figure 5.1. Acidified lake within the area affected by the Sudbury smelter emissions. (Photo by W. Keller.)

of watershed minerals would take a very long time (Barth 1987; Steinburg and Wright 1994).

Recent studies, however, have shown that some natural systems are highly resilient and can recover rapidly. It appears that the buffering capacity of many acidified aquatic systems was overloaded, not exhausted.

Lake Chemistry

The condition of lakes around Sudbury has been improving in response to declining emissions (Fig. 5.3) at the smelters. Results from detailed studies of individual lakes (LaZerte and Dillon 1984; Hutchinson and Havas 1986; Gunn and Keller 1990) as well as extensive survey data (Keller and Pitblado 1986; Keller et al. 1992b) have demonstrated this trend. These observed chemical changes have provided probably the best direct evidence in the world that abatement programs not only protect sensitive systems but allow for the recovery of damaged ecosystems (Wright and Hauhs 1991). The box inserts in this chapter (Boxes 5.1 and 5.2) provide some experimental evidence from lake and watershed manipulation studies supporting the findings of lake monitoring programs in the Sudbury area.

Figure 5.4 illustrates some of the general chemical trends observed in remote Sudbury lakes since the mid-1970s. As emissions of sulfur dioxide declined (see Fig. 5.3), lakes have shown marked increases in pH and alkalinity and declines in sulfate. Reduced acidic deposition has also reduced the mineral leaching of watershed soils, resulting in reductions in calcium and magnesium concentrations in lake waters. Very similar results have been reported from a regional survey of lakes in southwestern Scotland (Battarbee et al. 1988) and from Nova Scotia and Newfoundland rivers (Thompson 1987) after reductions in acid deposition.

Reductions in the degree of metal contamination of lake waters (Keller and Pitblado

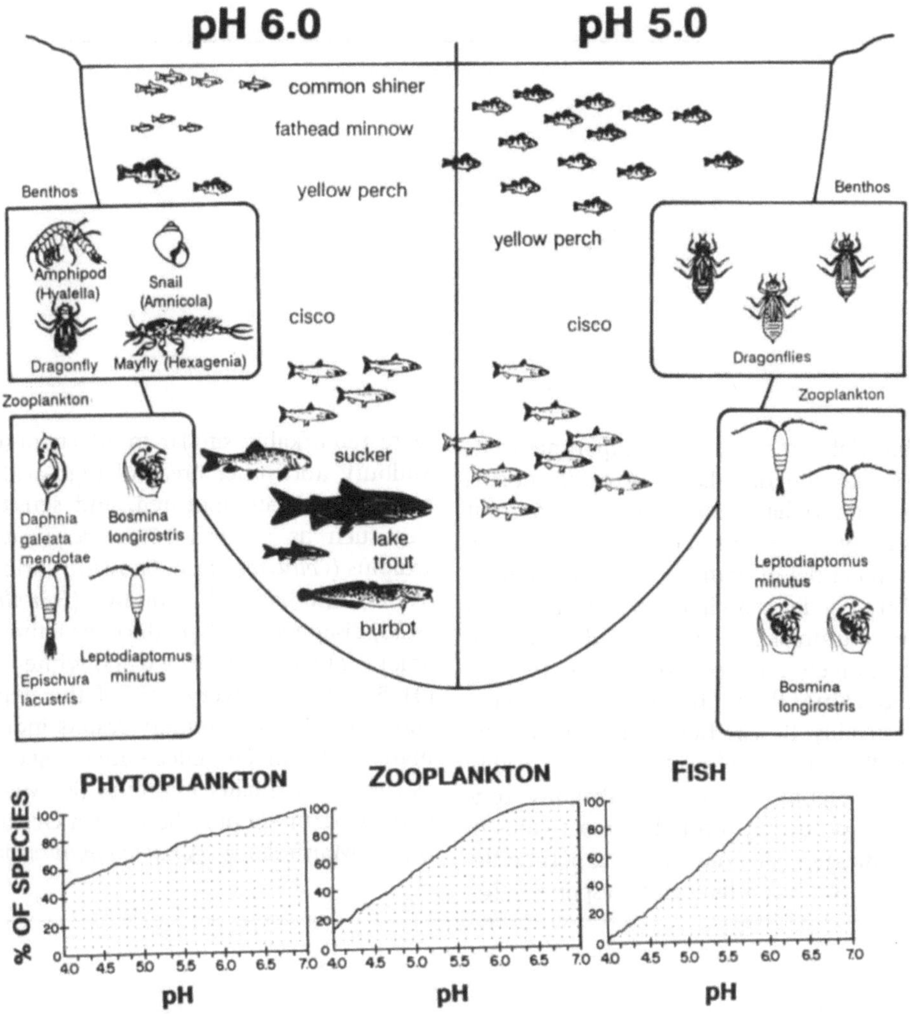

FIGURE 5.2. Simplified illustration of some of the biological changes that accompany lake acidification. Patterns shown are based on findings from Sudbury area studies and the general literature. The dominant trend is the loss of acid-sensitive species at various aquatic trophic levels, leading to an impoverished community.

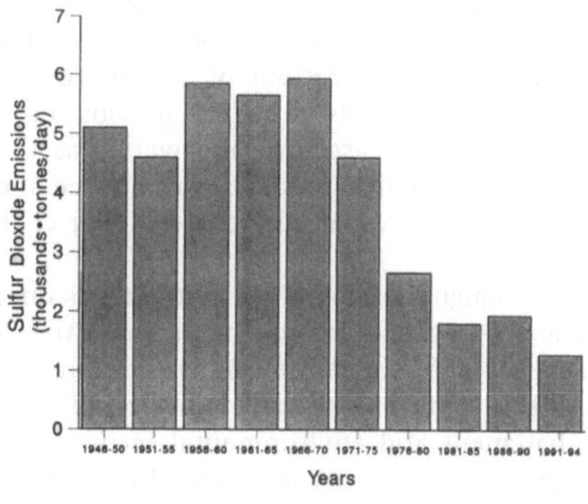

FIGURE 5.3. Average daily sulfur dioxide emissions for metal smelters in the Sudbury area for 5-year intervals beginning in the mid-1940s. The final interval is an average for 1991–1994 using projected emissions of 1000 tonnes/day for 1994.

Box 5.1.

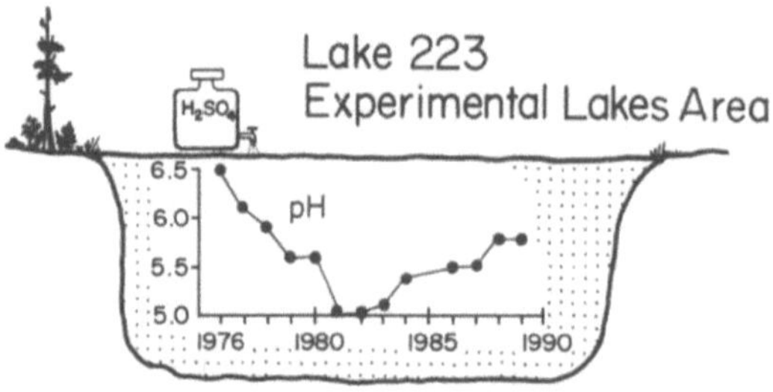

The whole-lake manipulation experiments at Canada's Experimental Lakes Area in northwestern Ontario have provided an important demonstration of acidification effects and the recovery potential of damaged lakes (Schindler et al. 1991). By direct addition of sulfuric acid to the water column of Lake 223, the pH of the lake was progressively reduced from 6.5 in 1976 to 5.0 in 1981. Since then, pH has been allowed to increase steadily. Researchers were initially surprised by the amount of acid needed to acidify the system. Significant internal alkalinity generation was occurring through the activities of sulfate-reducing bacteria. The experimental acidification produced biological responses that were remarkably similar to effects observed in Sudbury and other areas of high-acid deposition. Algal mats appeared, and sensitive species such as crayfish (*Orconectes virilis*), slimy sculpins (*Cottus cognatus*), opossum shrimp (*Mysis relicta*), and fathead minnows (*Pimephales promelas*) disappeared. Reproductive failure occurred among lake trout at pH 5.6 and white suckers at pH 5.0. The recovery part of the experiment continues, but preliminary results indicate that many of the adverse effects are being eliminated as the water quality improves (K. Mills, Canadian Department of Fisheries and Oceans, Winnipeg, Manitoba, *personal communication*).

1986; Fig. 5.4) and surficial sediments (Nriagu and Rao 1987) have also been reported. The changes in concentrations of potentially toxic metals deposited by the atmosphere have been most dramatic near the smelters, where concentrations of nickel and copper are highly elevated. Concentrations of aluminum and manganese, metals abundant in soils, have generally decreased in lake waters as the acid leaching of watersheds was reduced and the pH of acidic lakes increased (Keller et al. 1992b).

Although the general chemistry trends since the 1970s reflect improvement, these trends leveled off by the mid-1980s as emissions stabilized, confirming the link between pollution abatement and environmental recovery. In the late 1980s, many lakes actually showed a reversal of the pattern of recovery; a decline in water quality, probably caused by climatic factors, occurred after 1987 (Fig. 5.4; Keller et al. 1992b). It appears that extensive storage and oxidation of sulfur occurred in lake watersheds during the very dry years of 1986 and 1987. After abundant precipitation in 1988, lake pH values decreased somewhat as sulfate concentrations increased, apparently in response to high sulfur and acid exports from lake watersheds. Further details of the mechanism of sulfur storage and release are provided in Chapter 24.

Nevertheless, despite weather-related fluctuations and the stabilization of emissions, the overall chemistry pattern through the late 1970s and the 1980s demonstrates dramatic improvements in water quality. This greatly

Box 5.2.

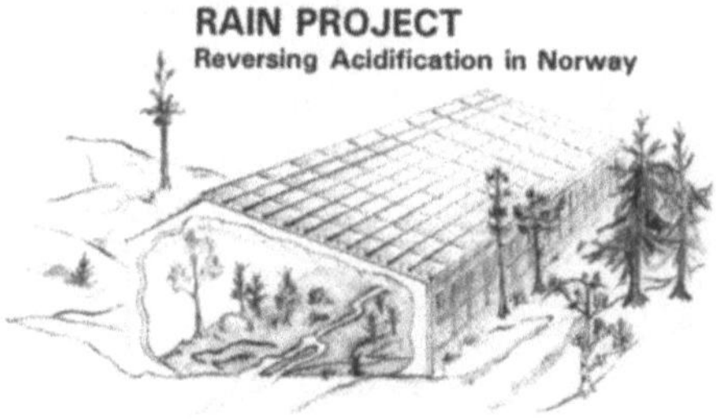

Domed cities may be a far-fetched solution to survival in a contaminated atmosphere, but greenhouses or other such shelters are often used to conduct controlled experiments on the effects of acidic precipitation. In Norway, researchers conducted a unique experiment in which the greenhouses were moved to the site with the soil and plants rather than vice versa (Wright et al. 1988). In a long-term experiment funded by Norwegian, Swedish, and Canadian environmental agencies, scientists from the RAIN (Reversing Acidification in Norway) project altered the chemical composition of precipitation falling on small experimental watersheds. Using covered and uncovered watersheds, sprinklers, irrigation systems, and snow-making machines, a variety of experiments were conducted: (1) sulfuric acid addition to a pristine site; (2) 1:1 mixture of sulfuric and nitric acid addition to a pristine site; (3) acid removal from a contaminated site; and (4) control watersheds, with and without greenhouses. These experiments showed that acid precipitation led to the rapid leaching of base cations (e.g., calcium, magnesium) and toxic forms of aluminum from watershed soils. Reduction in acid inputs reduced these effects. The RAIN project continues to generate important findings on recovery rates and processes. A similar approach is now also being used to assess the effects of other atmospheric contaminants such as carbon dioxide.

expanded the potential resource base of acid-sensitive species such as lake trout (Fig. 5.5). However, the restoration of aquatic systems around Sudbury is still at an early stage. Many lakes remain acidic and metal-contaminated. Some lakes more than 100 km from Sudbury are still highly acidified, with pH less than 5.0. Highly elevated concentrations of copper and nickel are currently restricted to nearby lakes. However, within about 20 km of Sudbury, concentrations of these metals in most lakes exceed suggested safe values for the protection of fish and other aquatic life (see Chapter 20).

Biological Recovery

Observations of biological improvements in Sudbury lakes have been steadily mounting (Keller et al. 1992a). These findings are very important, because the protection/restoration of aquatic communities is a major objective of emission abatement programs.

Algae

Fossil records of algal remains in sediments provide clear evidence of the rapid recovery of planktonic diatoms and chrysophytes with increased pH and reduced metal levels in Sudbury lakes (see Chapter 3). Examination of the changes in phytoplankton community composition in lakes sampled in the mid-1970s or early 1980s, and again in the mid-1980s, also indicates that positive responses of phytoplankton communities have accompanied water quality improvements (Nicholls et al. 1992). Examples of the changes in phytoplankton community richness observed in several of these lakes are shown in Figure 5.6.

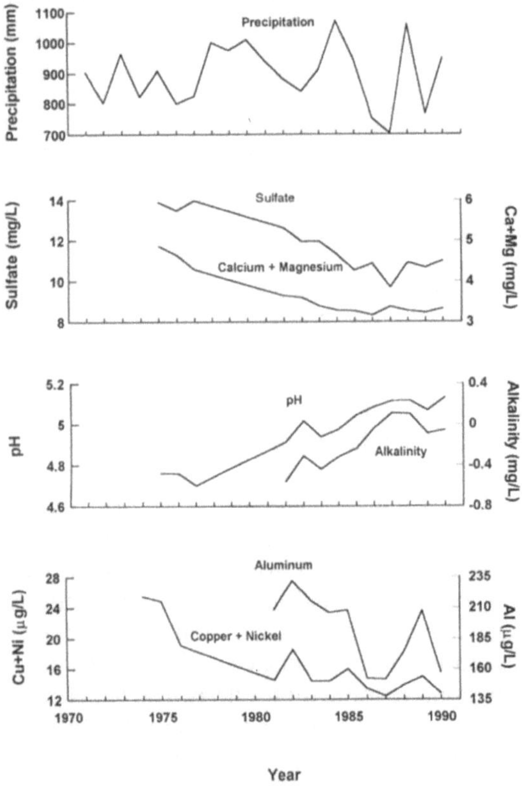

FIGURE 5.4. Time trends in lake chemistry as Sudbury area smelter emissions declined. Values for chemical characteristics are annual averages for a group of seven acidic lakes distant (48–105 km) from Sudbury but within the zone of smelter influence. Total annual precipitation is also shown.

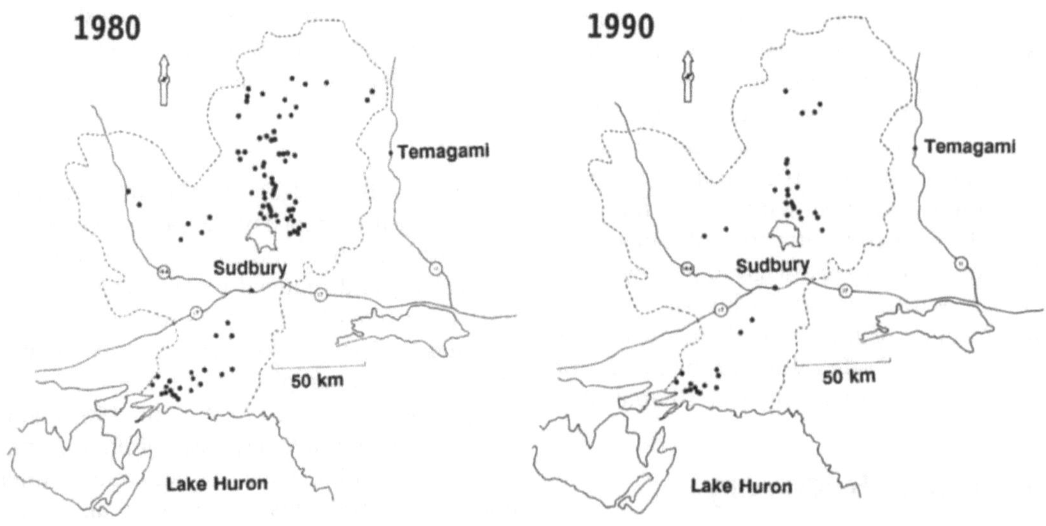

FIGURE 5.5. Number of lake trout lakes too acidic to support viable populations (*solid dots*) has declined from 86 in 1980 to 38 in 1990. Most of these changes occurred early in the 1980s.

FIGURE 5.6. Changes in annual average pH (*solid line*) and the richness (*shaded bars*) of phytoplankton (Nicholls et al. 1992) and zooplankton (Keller and Yan 1991) communities in three Sudbury lakes.

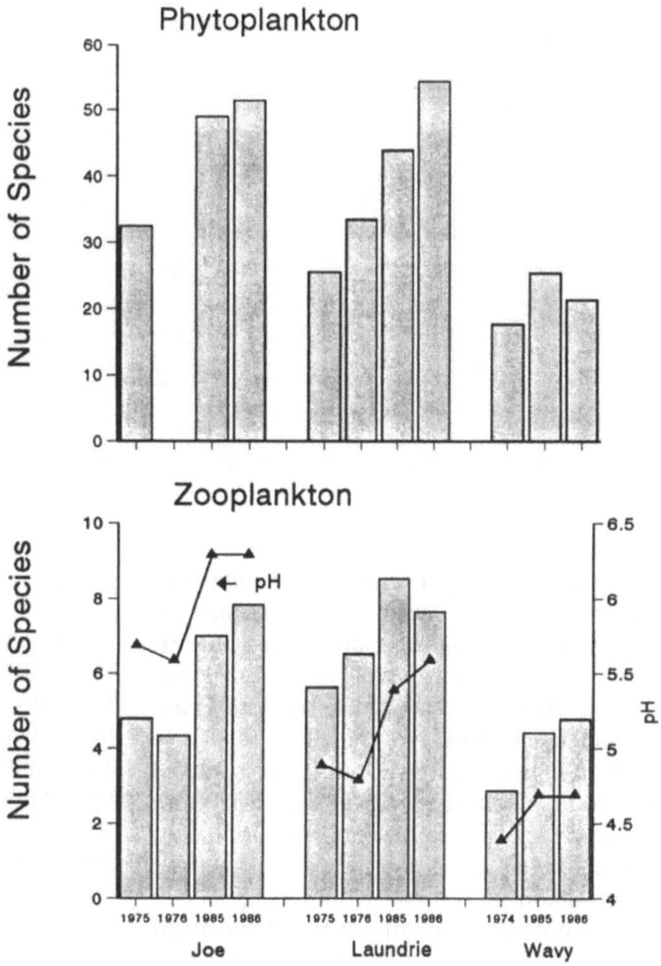

Many acidified Sudbury lakes have extensive benthic growths of filamentous algae, a response to acidified conditions (Fig. 5.7). During the natural recovery of Swan Lake from pH 4.8–5.4 (1982–1987), the community composition of benthic algae changed, and the extent of shoreline coverage declined greatly (Vandermeulen et al. 1993). As pH increased, the dominance of *Zygogonium* gave way to dominance by a mixture of *Spirogyra, Oedogonium,* and *Desmidium.*

Zooplankton

Some natural recovery of planktonic rotifer communities was documented in Swan Lake, as pH rose from 4.0 in 1977 to 4.8–5.1 in 1982–1984 (MacIsaac et al. 1986). Dominance shifted from *Keratella taurocephala* to *Polyarthra* and other species as water quality improved. This change is consistent with survey results indicating dominance by a variety of species, particularly *Keratella cochlearis* and *Polyarthra,* in non-acidic lakes, and strong dominance by *K. taurocephala* and *Gastropus* in acidic lakes (MacIsaac et al. 1987).

Increased average species richness of crustacean zooplankton has been observed in many lakes (Locke et al. 1994), including Laundrie, Wavy, and Joe (Keller and Yan 1991; Fig. 5.6). These patterns followed the negative relationship generally observed between species richness and lake acidity (see Fig. 5.2). New species that became important in these lakes as chem-

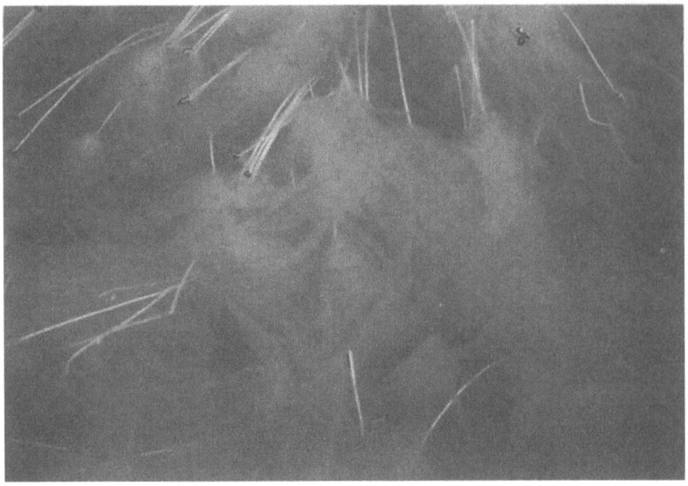

Figure 5.7. Filamentous green algae accumulation in the near-shore zone of an acidic lake. Such growths are widespread in acidified lakes and are a major concern for recreational users. (Photo by W. Keller.)

istry improved included *Holopedium gibberum, Eubosmina, Daphnia pulex, Skistodiaptomus oregonensis, Epischura lacustris,* and *Tropocyclops extensus.* With the exception of *H. gibberum,* which is metal-sensitive, these recolonists are considered acid-sensitive, and all are relatively common in northeastern Ontario lakes. Based on a comparison of species richness in these lakes and nonacidic southcentral Ontario lakes, natural recovery was not complete. However, recolonization by common widespread species made zooplankton community composition much more typical of natural Precambrian Shield lakes.

Benthic Invertebrates

Recovery of the invertebrate communities that occupy the lake bottom has been reported for Whitepine Lake and Laundrie Lake (Gunn and Keller 1990; Griffiths and Keller 1992). In these lakes, as pH increased, the abundance and the number of species collected generally increased, suggesting higher survival under less acidic conditions. Acid-sensitive organisms first collected in the most recent survey included several mayfly species (*Hexagenia, Ephemerella, Caenis*) in Whitepine Lake and the amphipod *Hyalella azteca* in Laundrie Lake.

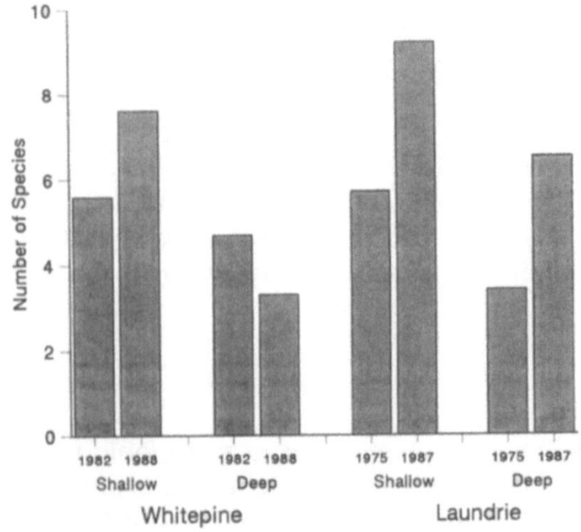

Figure 5.8. Changes in the average number of benthic invertebrate species per sample in littoral (*shallow*) and profundal (*deep*) soft-sediment habitats of Whitepine and Laundrie lakes (Griffiths and Keller 1992). Over the period, pH increased from about 5.4 to 5.9 in Whitepine Lake and 4.9 to 5.6 in Laundrie Lake (northern basin). Note that species richness did not increase in the deep areas of Whitepine Lake, where predation by an expanding lake trout population was intense.

Figure 5.9. Spawning brook trout in Joe Lake, 28 km north of Sudbury. The historical trout population was eliminated from Joe Lake by acidification and metal contamination, but water quality improvements have allowed successful restocking and reproduction. (Photo by V. Liimatainen.)

Other acid-sensitive recolonists included the worms *Arcteonais lomondi*, *Stylaria lacustris*, and *Dero nivea*, the midge *Cladotanytarsus*, and the clams *Musculium securis* and *Pisidium*. Changes in the species richness of these lakes are shown in Figure 5.8.

Fish

Improved water quality has allowed successful restocking of extinct fish species in several Sudbury lakes (Fig. 5.9). But there are relatively few cases of natural recovery among fish communities. The best-documented case is the resumption of successful reproduction by remnant lake trout after pH increased in Whitepine Lake (Gunn and Keller 1990). In this lake, the very abundant population of acid-tolerant yellow perch (*Perca flavescens*) decreased rapidly as lake trout populations were re-established. Not all residual populations, though, responded to improved water quality. The very sparse remnant populations of white suckers (*Catostomus commersoni*) and burbot (*Lota lota*) became extinct despite the pH increase.

Expectations

It is expected that further chemical and biological improvements will result from the large-scale additional abatement measures being implemented by the Sudbury smelting industry. The ultimate environmental effect of these emission control programs cannot be predicted using present knowledge. Close monitoring of environmental responses will be required to assess the adequacy of control measures and determine the need for additional steps. Based on previous findings, the interpretation of monitoring data must carefully consider climatic fluctuations, which can greatly affect time trends in lake chemistry.

In time, under a regime of reduced deposition of atmospheric contaminants, many currently acidic Sudbury lakes should improve such that water quality conditions will no longer constrain aquatic biological communities. Other lakes may show less significant chemical changes, but even in these lakes, some improvement in biological communities is expected. Given suitable water quality conditions, biological improvements will be substantial, based on evidence from experimental neutralization studies (see Chapter 15) as well as observations of natural recovery. Benthic filamentous algae respond almost immediately to decreased lake acidity, through dramatically reduced abundance and shifts in community composition. Phytoplankton communities also respond rapidly to chemical improvements, with increased community richness and shifts to types considered more typical of near-neutral Precambrian Shield lakes. Many common zooplankton species and mobile species of benthic invertebrates show relatively rapid (within a decade, and often much less) recolonization rates.

Complications

It has been suggested that the benefits of reduced acid emissions may be temporarily offset by corresponding decreases in calcium concentrations in lake waters, resulting from reduced leaching rates of watershed minerals (Henriksen et al. 1989; Skeffington and Brown 1992). Declining calcium is a concern because calcium reduces the toxic effects of acid and various metals in soft waters. Calcium limitation, although perhaps of concern for some Ontario lakes or in other regions with extremely dilute lakes such as southern Norway (calcium <1 mg/L), does not appear to be a major problem for Sudbury lakes or Precambrian Shield lakes in general, where calcium concentrations are usually higher. Of 250 lakes sampled within about 250 km of Sudbury, only 9 had calcium concentrations of less than 2 mg/L and the average for these lakes was 1.7 mg/L (Pitblado and Keller 1984). These values are well above the requirements for normal physiological functions in fish.

Communities in recovering Sudbury lakes are becoming more typical of those in nonacidic Precambrian Shield lakes, although it is impossible to demonstrate a return to the exact community that prevailed historically in individual lakes. The establishment of typical communities is expected, because species important in the natural restructuring of communities are residual acid-sensitive species that have persisted in reduced abundance and recolonizing species that are common in the area. Predatory and competitive interactions, as well as acid-sensitivity and recolonization sources, may, however, influence the recovery process (Fig. 5.10).

In the transition phase, between communities typical of acidic lakes and those typical of nonacidic lakes, community structure may be variable (Locke et al. 1994) and somewhat atypical. For example, based on both observations of natural recovery (Keller and Yan 1991) and experimental lake neutralization studies (Yan et al., under review), among crustacean zooplankton, some species normally considered characteristic of nearshore (littoral) habitats, including *Sida crystallina*, *Chydorus sphaericus*, *Orthocyclops modestus*, and *Simocephalus serrulatus*, may invade open water (limnetic) habitats early in the recovery process. These opportunistic invaders are ultimately eliminated when more usual open-water species re-invade the plankton (Fig. 5.10). The occurrence of such transition communities among other groups of organisms has not yet been documented for Sudbury lakes, but they may occur.

In some cases, the biological communities existing in acidic lakes may resist recovery to more typical communities, even under suitable water quality conditions (Fig. 5.10). In fishless lakes, high abundances of predatory invertebrates such as larvae of the phantom midge *Chaoborus*, which are normally controlled by fish predation, may prevent or retard the re-establishment of normal invertebrate communities (Nyberg 1984; Stenson et al. 1993). In particular, zooplankton in small, nutrient-rich, fishless lakes with high *Chaoborus* abundance may be affected by invertebrate predation (Yan et al. 1991). In such lakes, the recovery of natural invertebrate communities may depend on the re-establishment of fish communities that will control invertebrate predators. In the case of fish communities themselves, it appears that in lakes with high abundances of acid-tolerant fish species, interspecific competition may inhibit the successful re-establishment of some species. Examples of this type of negative competitive interaction between acid-sensitive and acid-tolerant species include brook trout and yellow perch and lake trout and cisco (*Coregonus artedii*). Expanding fish populations may limit the recovery of some invertebrate prey species (see Fig. 5.8).

Many species, if given suitable water quality and sufficient time, are undoubtedly capable of re-establishing themselves in aquatic ecosystems. Other species are much less mobile and may not be able to recolonize former habitats. For many fish species, natural recolonization may depend on the existence of connections with unaffected water bodies or other refuge areas (Bergquist 1991). The re-establishment of recreational fisheries will, in many cases, require restocking and the development of protective management strategies. For some nonmobile invertebrate or small fish species, re-introductions may also be necessary to achieve a desired aquatic community structure within a reasonable period of time.

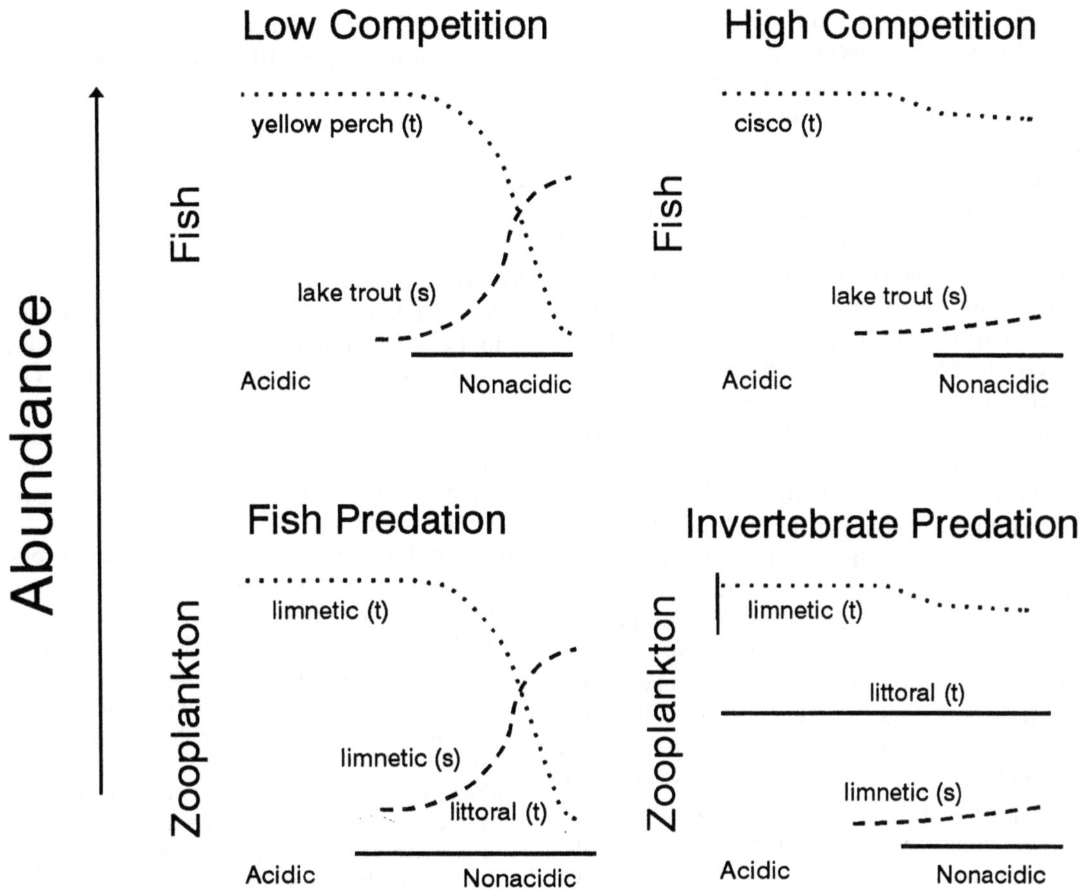

FIGURE 5.10. Some of the expected changes and interactions in fish and zooplankton communities that may occur during recovery in lakes with different biological characteristics. Conditions are low competition between acid-tolerant (*t*) forage fish and acid-sensitive (*s*) sportfish (*upper left*); high competition between acid-tolerant forage fish and acid-sensitive sportfish (*upper right*); changing fish predation from small acid-tolerant planktivores to large acid-sensitive piscivores (*lower left*); and, fishless conditions with intense invertebrate predation (*lower right*).

Summary and Conclusions

Studies in the Sudbury, Ontario, area have shown that water quality changes have occurred in many lakes in response to substantial reductions in sulfur and associated metal emissions in the 1970s. These changes, including decreased acidity, and decreased concentrations of sulfate, calcium, magnesium, aluminum, and manganese, continued into the mid-1980s. By the late 1980s, these trends leveled off or reversed somewhat. During the 1980s, no clear trends in concentrations of copper or nickel, metals directly associated with the smelter emissions, were observed, although decreases in concentrations of these metals did occur between the mid-1970s and early 1980s.

Biological changes have accompanied these chemical improvements. At least partial recovery has been observed for benthic filamentous algae, phytoplankton, zooplankton, benthic invertebrates, and some fish species. The patterns of recovery show movement toward the re-establishment of communities typical of Precambrian Shield lakes in this area. For some nonmobile species, however, natural recolonization is unlikely (Bergquist 1991) and restocking efforts will be necessary.

Based on our current understanding of aquatic systems, a pH of at least 6.0 is needed for the protection of healthy aquatic communities (see Fig. 5.2). To date, only a small number of Sudbury lakes have recovered above this chemical threshold. In those lakes nearest Sudbury, the biological effects of elevated concentrations of metals related to smelter emissions, particularly copper and nickel, must also be considered. Large reductions in metal concentrations as well as decreases in acidity will be required to permit ecosystem restoration in some lakes. Such concurrent elevations in surface water acidity and concentrations of some airborne metals, however, occur in other regions experiencing high deposition of atmospheric contaminants, including areas of Scandinavia and Russia near smelters (Nost et al. 1991). Thus, observations in Sudbury provide very useful comparative data for areas influenced by regional sources of acid and metal deposition as well as areas subjected to the long-range transport and deposition of contaminants.

The widespread chemical and biological improvements seen in lakes of the Sudbury area demonstrate the resiliency of aquatic systems and provide strong support for the use of emission controls to combat aquatic acidification. However, many area lakes are still acidic and metal-contaminated. Further recovery is expected to result from the additional, recently implemented controls (Ontario Ministry of the Environment 1987).

Acknowledgments. Much of the work described here was supported under the Acidic Precipitation in Ontario Study (APIOS) of the Ontario Ministry of Environment and Energy and the Ontario Ministry of Natural Resources. Many researchers were involved in the work summarized. In particular, we acknowledge the following individuals for their contributions to an understanding of lakes in the Sudbury area: Jim Carbone, Nels Conroy, Peter Dillon, Sushil Dixit, Peggy Gale, Ron Griffiths, Harold Harvey, Hugh MacIsaac, Ken Nicholls, Roger Pitblado, John Smol, and Norm Yan.

References

Barth, H. 1987. Reversibility of Acidification. Elsevier, London.

Battarbee, R.W., R.J. Flower, A.C. Stevenson, V.J. Jones, R. Harriman, and P.G. Appleby. 1988. Diatom and chemical evidence for the reversibility of acidification of Scottish Lochs. Nature 332: 530–532.

Beamish, R.J., and H.H. Harvey. 1972. Acidification of the La Cloche Mountain lakes, Ontario, and resulting fish mortalities. J. Fish. Res. Board Can. 29:1131–1143.

Bergquist, B.C. 1991. Extinction and natural colonization of fish in acidified and limed lakes. Nordic J. Freshwater Res. 66:50–62.

Glooschenko, V., W. Weller, P.G.R. Smith, R. Alvo, and J.H.G. Archbold. 1992. Amphibian distribution with respect to pond water chemistry near Sudbury, Ontario. Can. J. Fish. Aquat. Sci. 49 (Suppl. 1):114–121.

Gorham, E., and A.G. Gordon. 1960. The influence of smelter fumes upon the chemical composition of lake waters near Sudbury, Ontario and upon the surrounding vegetation. Can. J. Bot. 30:477–487.

Griffiths, R.W., and W. Keller. 1992. Benthic macroinvertebrate changes in lakes near Sudbury, Ontario following a reduction in acid emissions. Can. J. Fish Aquat. Sci. 49(Suppl. 1):63–75.

Gunn, J.M., and W. Keller. 1990. Biological recovery of an acid lake after reductions in industrial emissions of sulphur. Nature (Lond.) 345:431–433.

Henrikson, A., L. Lien, B.O. Rosseland, T.S. Traaen, and I.S. Sevaldrud. 1989. Lake acidification in Norway: present and predicted fish status. Ambio 18:314–321.

Hutchinson, T.C., and M. Havas. 1986. Recovery of previously acidified lakes near Coniston, Canada following reductions in atmospheric sulphur and metal emissions. Water Air Soil Pollut. 29:319–333.

Keller, W., J.M. Gunn, and N.D. Yan. 1992a. Evidence of biological recovery in acid-stressed lakes near Sudbury, Ontario, Canada. Environ. Pollut. 78: 79–85.

Keller, W., and J.R. Pitblado. 1984. Crustacean plankton in northeastern Ontario lakes subjected to acidic deposition. Water Air Soil Pollut. 23: 271–291.

Keller, W., and J.R. Pitblado. 1986. Water quality changes in Sudbury area lakes: a comparison of synoptic surveys in 1974–76 and 1981–83. Water Air Soil Pollut. 29:285–296.

Keller, W., J.R. Pitblado, and J. Carbone. 1992b. Chemical responses of acidic lakes in the Sudbury, Ontario, area to reduced smelter emissions, 1981–89. Can. J. Fish. Aquat. Sci. 49(Suppl.1): 25–32.

Keller, W., and N.D. Yan. 1991. Recovery of crustacean zooplankton species richness in Sudbury area lakes following water quality improvements. Can. J. Fish. Aquat. Sci. 48: 1635–1644.

Kelso, J.R.M., M.A. Shaw, C.K. Minns, and K.H. Mills. 1990. An evaluation of the effects of acid deposition on fish and the fisheries resources of Canada. Can. J. Fish. Aquat. Sci. 47:644–655.

Kwiatkowski, R.E., and J.C. Roff. 1976. Effects of acidity on the phytoplankton and primary productivity of selected northern Ontario lakes. Can. J. Bot. 54:2546–2561.

LaZerte, B.D., and P.J. Dillon. 1984. Relative importance of anthropogenic versus natural sources of acidity in lakes and streams of central Ontario. Can. J. Fish. Aquat. Sci. 41:1664–1677.

Locke, A., W.G. Sprules, W. Keller, and J.R. Pitblado. 1994. Zooplankton communities and water chemistry of Sudbury area lakes: changes related to pH recovery. Can. J. Fish. Aquat. Sci. 51:151–160.

MacIsaac, H.J., T.C. Hutchinson, and W. Keller. 1987. Analysis of planktonic rotifer assemblages from Sudbury, Ontario area lakes of varying chemical composition. Can. J. Fish. Aquat. Sci. 44:1692–1701.

MacIsaac, H.J., W. Keller, T.C. Hutchinson, and N.D. Yan. 1986. Natural changes in the planktonic Rotifera of a small acid lake near Sudbury, Ontario following water quality improvements. Water Air Soil Pollut. 31:791–797.

Matuszek, J.E., D.L. Wales, and J.M. Gunn. 1992. Estimated impacts of SO_2 emissions from Sudbury smelters on Ontario's sportfish populations. Can. J. Fish. Aquat. Sci. 49(Suppl. 1):87–94.

Nicholls, K.H., L. Nakamoto, and W. Keller. 1992. Phytoplankton of Sudbury area lakes (Ontario) and relationships with acidification status. Can. J. Fish. Aquat. Sci. 49(Suppl. 1):40–51.

Nriagu, J.O., and S.S. Rao. 1987. Response of lake sediments to changes in trace metal emissions from the smelters at Sudbury. Environ. Pollut. 44:211–218.

Nost, T., V. Yakovlev, H.M. Berger, N. Kashulin, A. Langeland, A. Lukin, and H. Muladal. 1991. Impacts of Pollution on Freshwater Communities in the Border Area between Russia and Norway. 1. Preliminary Study in 1990. Report 26. Norsk Institutt for Natureforskning. NIVA, Trondheim, Norway.

Nyberg, P. 1984. Impact of *Chaoborus* predation on planktonic crustacean communities in some acidified and limed forest lakes in Sweden. Rep. Inst. Freshwat. Res. Drottningholm. 61:154–166.

Ontario Ministry of the Environment. 1987. Countdown Acid Rain, Summary and Analysis of the Second Progress Reports by Ontario's Four Major Sources of Sulphur Dioxide. Ontario Ministry of the Environment Report, Toronto, Ontario.

Pitblado, J.R., and W. Keller. 1984. Monitoring of Northeastern Ontario Lakes, 1981–1983. Technical report. Ontario Ministry of the Environment, Sudbury, Ontario.

Roff, J.C., and R.E. Kwiatkowski. 1977. Zooplankton and zoobenthos communities of selected northern Ontario lakes of different acidities. Can. J. Zool. 55:899–911.

Schindler, D.W., T.M. Frost, K.H. Mills, P.S.S. Chang, I.J. Davies, L. Findlay, D.F. Malley, J.A. Shearer, M.A. Turner, P.G. Garrison, C.J. Watras, K. Webster, J.M. Gunn, P.L. Brezonik, and W.A. Swenson. 1991. Comparisons between experimentally and atmospherically acidified lakes during stress and recovery. Proc. R. Soc. Edinburgh 97b:193–227.

Skeffington, R.A., and D.J.A. Brown. 1992. Timescales of recovery from acidification: implications of current knowledge for aquatic organisms. Environ. Pollut. 77:227–234.

Sprules, G.W. 1975. Midsummer crustacean zooplankton communities in acid stressed lakes. J. Fish. Res. Board Can. 32:389–395.

Steinburg, C.E.W., and R.F. Wright. 1994. Acidification of freshwater ecosystems, implications for the future. Proceedings of the 69th Dahlem Workshop. Wiley, New York.

Stenson, J.A.E., J.-E. Svensson, and G. Cronberg. 1993. Changes and interactions in the pelagic community in acidified lakes in Sweden. Ambio 22:277–282.

Thompson, M.E. 1987. Comparison of excess sulphate yields and median pH values of rivers in Nova Scotia and Newfoundland. Water Air Soil Pollut. 35:19–26.

Vandermeulen, H., M.B. Jackson, A. Rodrigues, and W. Keller. 1993. Filamentous algal communities in Sudbury area lakes: effects of variable lake acidity. Cryptogamic Bot. 3: 123–132.

Watt, W.D., C.D. Scott, and W.J. White. 1983. Evidence of acidification of some Nova Scotian rivers and its impact on Atlantic salmon, *Salmo salar*. Can. J. Fish. Aquat. Sci. 40:462–473.

Wright, R.F., and M. Hauhs. 1991. Reversibility of acidification: soils and surface waters. Proc. R. Soc. Edinburgh 97b:169–191.

Wright, R.F., E. Lotse, and A. Semb. 1988. Reversibility of acidification shown by whole-catchment experiments. Nature (Lond.) 334: 670–675.

Yan, N.D., W. Keller, H.J. MacIsaac, and L.J. McEachern. 1991. Regulation of zooplankton community structure of an acidified lake by *Chaoborus*. Ecol. Appl. 1:52–65.

Yan, N.D., W. Keller, K.M. Somers, T.W. Pawson, and R. Girard. (submitted). The recovery of zooplankton from acidification: Comparing manipulated and reference lakes.

6

Lichens: Sensitive Indicators of Improving Air Quality

Peter J. Beckett

Biological monitoring is the application of assessment techniques using plants or other biological material to gain information about the quality and condition of the environment (Cairns 1980). Plants are greatly affected by the physical and chemical environment in which they live. If conditions become altered, the exposed plant community can accurately reflect these changes and can thus be important indicators of the state of the environment (Nash 1988). Plants also collect contaminants from the air and soil and can be sampled from various geographic locations to assess the amount of contaminant present.

This chapter discusses the value of lichens as biomonitors and documents their increased abundance in Sudbury after reductions in sulfur dioxide emissions. This reinvasion by lichens represents some of the best evidence of the natural recovery of the terrestrial ecosystem and thus the importance of the industrial pollution abatement programs that began in the early-1970s.

Lichens

Lichens form conspicuous gray, green, orange, or red patches on trees or rocks (Fig. 6.1). They are composite symbiotic organisms with both an algal and a fungal partner; approximately 90% of the mass is made up of the slow-growing fungal partner. The fungus supplies structural support, and the algal cells support nutrition through photosynthesis.

There is a long history of lichens being used as a sensitive indicator of air quality, particularly with regard to sulfur dioxide in urban areas or near point source emissions (Ferry et al. 1973; Burton 1986; Richardson 1992). Naturalists first observed the disappearance of lichens from polluted areas soon after the start of the Industrial Revolution (Turner and Borrer 1839). In 1859, Grindon observed that "the quality [of lichens near Manchester] has been much lessened . . . through the influx of factory smoke which appears to be singularly prejudicial to these lovers of pure air." As air quality has improved in many industrial centers in western Europe, there has been a marked improvement in lichen abundance (Seaward 1989).

Although lichens are generally very sensitive to air pollutants, not all lichen species are equally affected. At a given sulfur dioxide concentration, certain species disappear while others remain. In Europe, Hawksworth and Rose (1976) developed a scale that relates the occurrence of particular lichens to winter sulfur dioxide levels.

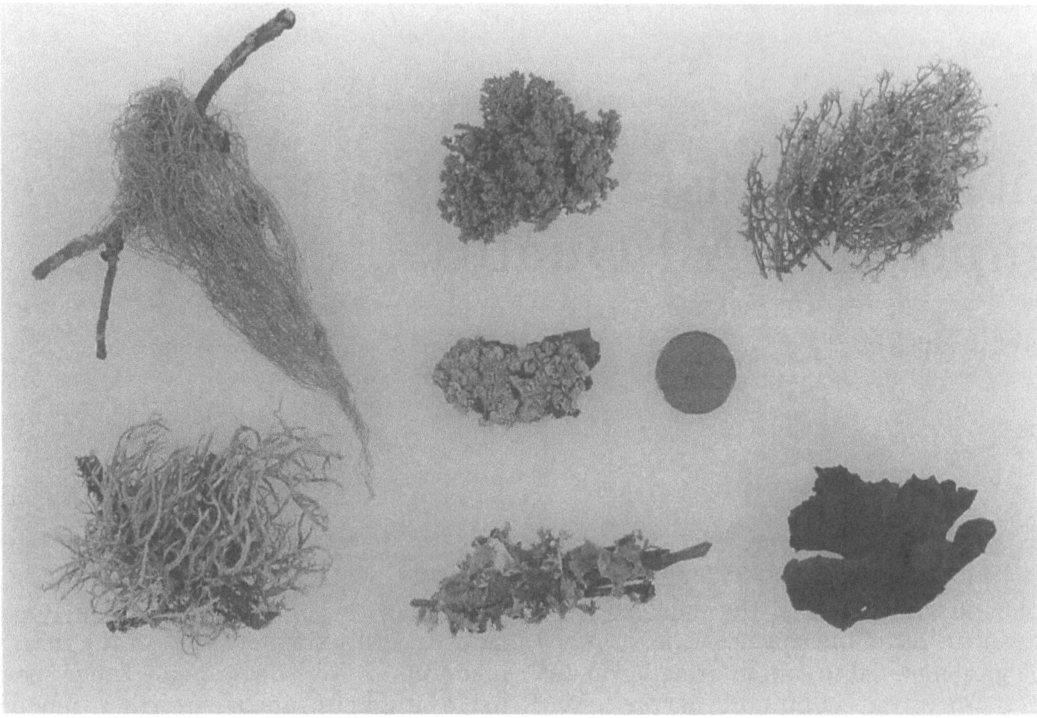

FIGURE 6.1. Examples of typical lichens associated with pollution studies in the Sudbury area. *Top row* (*L–R*): *Usnea hirta* (epiphyte), *Stereocaulon* sp. (rock), *Cladina rangiferina* (soil). *Center*: *Parmelia sulcata* (epiphyte). *Bottom row* (*L–R*): *Evernia mesomorpha* (epiphyte), *Cetraria ciliaris* (epiphyte), *Umbilicaria mulhenbergia* (rock). The coin provided for scale is a $1 Canadian "Loonie."

Characteristic of Lichens as Sensitive Biomonitors

The suitability of using lichens in biomonitoring studies is summarized in Table 6.1 and briefly discussed in the following section.

Lichen species may be conveniently grouped into three morphological forms that are related to their sensitivity to air pollution. Crustose forms cover and tightly adhere to the substrate; foliose lichens are leaflike with many lobes and loosely adhere to the substrate; fruticose lichens grow vertically upward or hang down from twigs (Fig. 6.1). Because of their elongated growth form, fruticose lichens (see Plate 8 following page 182) are generally more exposed to contaminants than crustose and foliose lichens.

TABLE 6.1. Advantages of using lichens as biomonitors

1.	Accumulate substances in measurable quantities
2.	Available in sufficient quantities over a wide geographic area to ensure unbiased sampling
3.	Present throughout the year with relative ease of collection. May be sampled repeatedly
4.	Have various sensitivities to a contaminant to allow a series of communities to form in response to the concentration of the contaminant

A particularly important morphological characteristic of lichens, with regard to their use as biomonitors, is the absence of an outer protective waxy cuticle; this allows contaminants to move freely into the lichens. Because lichens are adapted to obtaining nutrients from rain, they possess the ability to take up elements both passively and directly from the surrounding environment. The cell walls act like the ion-exchange resins found in water softeners and bind with substances found in rainwater. Also, the open spongelike structure of the fungal partner allows gases such as oxygen, carbon dioxide, and sulfur dioxide to readily diffuse into the body of the lichen. The surface structure of the lichen also traps a variety of airborne particulates that range from harmless dust particles to radionuclides (Box 6.1).

The outstanding ability of lichens to accumulate substances either from the air or other parts of the environment has allowed wider application than indication of air quality (Nieboer and Richardson 1981). For instance, these organisms have been exploited for geobotanical prospecting (Box 6.2).

Studies Involving Lichen Biomonitors in the Sudbury Area

Lichen Distribution and Changes over Time

Three surveys have been made of lichens growing on mature balsam poplar (*Populus balsamifera*) tree trunks in the Sudbury area. The first survey was conducted in 1968 (Leblanc et al. 1972) 4 years before the major pollution control initiatives of 1972 (see Chapter 4). The post-emission-reduction surveys were conducted in 1978 (Beckett 1984) and 1989–1990 (Pappin and Beckett in press). Consistent survey techniques were maintained throughout. Ten trees were chosen at each site, and each tree was carefully examined from the base to a height of 2 m. Lichen species present and percentage cover of each species were recorded. Species richness was estimated for each site using the method of Leblanc and De Sloover (1970) to provide an index value for classifying air quality effects on lichen communities. This classification method assigned a score or index of atmospheric purity (IAP) values to lichen communities at each site. The IAP values were then plotted and joined by isometric lines to delineate various zones (1–5) of atmospheric contamination.

The results of the LeBlanc et al. (1972) survey in 1968 (pre-emission reduction) are shown in the upper panel of Figure 6.2. The most heavily affected area, zone 1, covered much of the barren area described in Chapter 2. Zero to four species of lichen were found in this zone, which is commonly referred to as a "lichen desert." In the next zone, zone 2, fewer than 10 species of lichens were observed. This slightly better but still heavily damaged area corresponds generally to the semibarren vegetation

---

Box 6.1. Chernobyl Accident and Lichens

Lichens can effectively monitor radioactive isotopes. This was shown on April 26, 1986, when an explosion occurred in the Chernobyl nuclear-powered electricity generating plant in the Ukraine (Smith and Clark 1986). The explosion and subsequent fire released a plume of radioactive particles for several days before controls were implemented. The main plume moved west over Poland and the Alps and reached Britain 1 week later. It then moved north to deposit significant amounts of radioactive material over Scandinavia (Steinnes and Njastad 1993), with subsequent economic consequences for Lapp reindeer herders. High levels of cesium (^{137}Cs) were found in the meat of many reindeer; up to 10,000 Bq cesium per kilogram was documented. The legal limit for sale is 300 Bq/kg. Most of this radionuclide was ingested by reindeer feeding on highly contaminated lichens. In Poland, there was a startling increase of cesium (165-fold) in *Umbilicaria* after the passage of the Chernobyl cloud across the country. In this case, heavy rain washed out some of the radioactive material, which was subsequently taken up by the lichen. The radioactive cloud arrived in Canada 11 days after the accident. Samples of reindeer/caribou lichen, *Cladina rangiferina*, were collected across the Maritime provinces. Using computer simulation models for deposition, investigators estimated that the cloud crossed the Maritimes at a height of 10,000 m (Smith and Ellis 1990). Once again, lichens were shown to be excellent collectors of aerosol-size particles owing to their high surface/mass ratio and slow growth rates.

Box 6.2. Geobotanical Prospecting

The recognition that some plants grow on soils rich in metals (metallophytes) has led to the development of the science of geobotany, whereby a prospector searches for particular plants while investigating the geology in the anticipation of finding mineral deposit. Another strategy is to determine the metal content of plants as an indication of nearby mineral deposits. The physical appearance of the plant can also be an important clue. For example, crustose lichens growing near mineralized rocks may develop characteristically colored thalli (Easton in press), whereas the dark green thalli of certain *Lecanora* species is associated with copper-rich rock.

Foliose lichens are often used for geobotanical prospecting because they are easy to sample. Lichen samples taken from serpentine rocks are often found to accumulate nickel. Around Contwoyoto Lake, Northwest Territories, high levels of copper were found in *Cetraria* (Tomassini et al. 1976). Later, a mining company independently found an economic copper deposit in the same area. *Cladina* (reindeer or caribou lichens) removed from a rocky outcrop in the Elliot Lake area of Ontario contained high levels of uranium that had been accumulated from a vein containing high levels of the radioactive metal.

zone described in Chapter 2 and by Amiro and Courtin (1981). Species diversity and numbers of particularly sensitive species increased progressively through zones 3 to 5. Zone 5 showed only minor effects caused by sulfur dioxide emissions from Sudbury and was described by Leblanc et al. (1972) as the undisturbed or background condition.

The survey of 1978, which occurred 6 years after the major emission reductions and plant closures, dramatically demonstrated the biological benefits of improving air quality (middle panel Fig. 6.2). At this time, the lichen desert (zone 1) had shrunk dramatically (20% of 1968 area). There were now 10 species growing in zone 2. Farther away from the working smelters (zone 3), 20 species were found. Still farther away in zone 4, 25–30 species were present, and in the outer area of lowest pollution, more than 35 species were present (Beckett 1984).

By 1990 (lower panel Fig. 6.2), the lichen desert had completely disappeared, and zone 2 was reduced to two small areas around each of

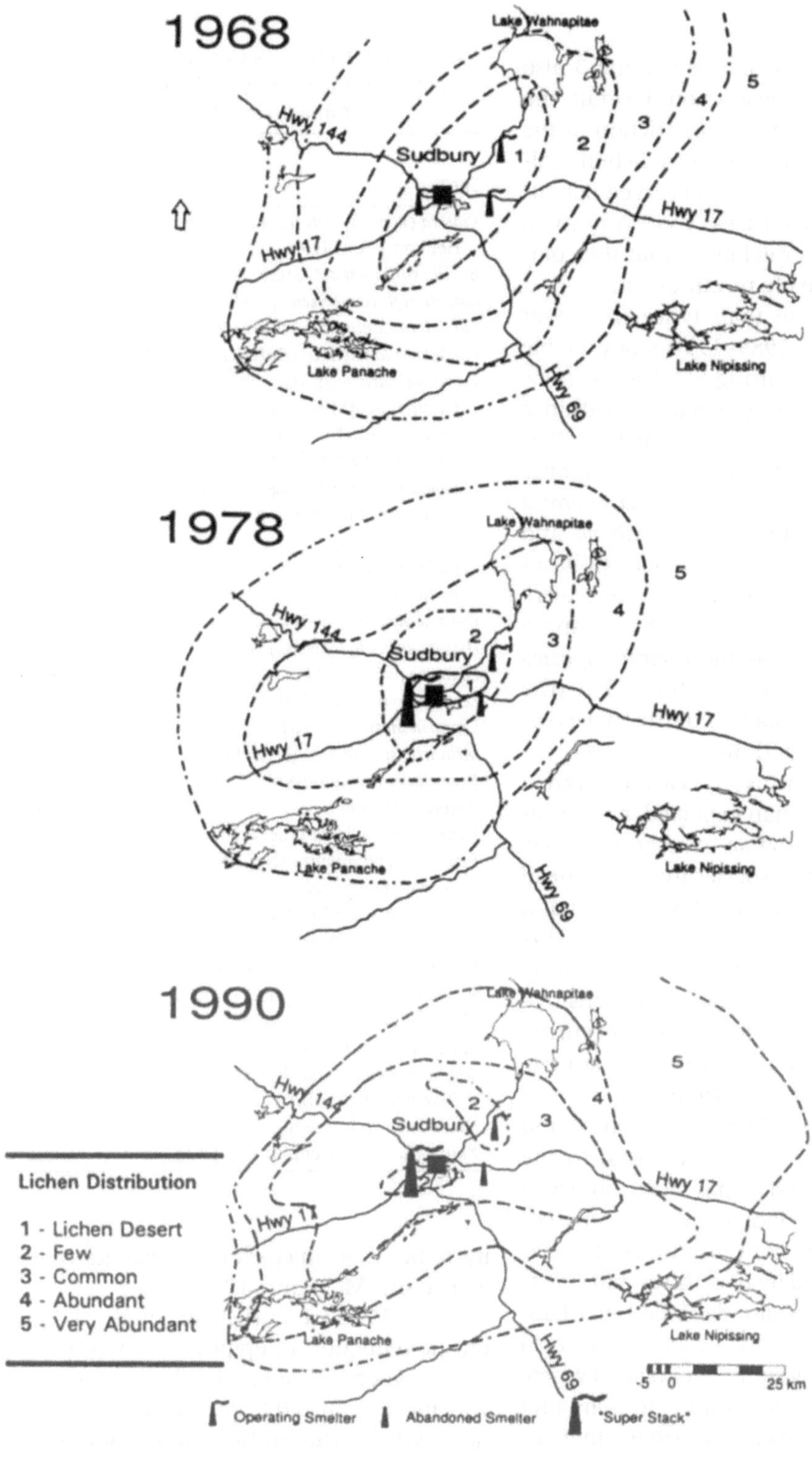

Figure 6.2. Zones of atmospheric purity from the 1968, 1978, and 1989–1990 lichen surveys. Note the large zone 1, lichen desert, around the three operating smelters in 1968 (*upper panel*). In 1978, zone 1 has decreased and has been replaced by zone 2 through lichen re-invasion (*middle panel*). In 1989–1990, zone 1 has been eliminated and zone 2 has split into two small zones around the existing two smelters (*lower panel*).

the currently operating smelters. Zone 3 also showed a significant constriction. Overall, the effect was that of a general constriction of the classified zones as more and more lichens colonized the area. Many of the observed lichens in the newly established areas were very small young lichens; further evidence that the colonization was a relatively recent event.

Near Sudbury more than twice the cover was recorded in the 1989–1990 study (Table 6.2) as was observed during the 1968 survey. In 1968, no lichen species occurred within a radius of 7 km from the three smelters (Leblanc et al. 1972). Between 7 and 15 km, only the crustose lichens *Bacidia chlorococca, Lepraria aeruginosa* (*incana*), and *Lecanora sàligna* and the foliose lichen *Parmelia sulcata* (foliose) were present. By 1990, no sampled trees were found to be devoid of lichen epiphytes, and a majority of the sulfur dioxide-tolerant species (listed above) were found within 2 km (the minimum sampling distance) of the two existing smelters (Pappin and Beckett in press).

Some pollution-sensitive species re-invaded the area much faster than expected, and recolonization did not follow an orderly sequence, with pollution-tolerant species invading ahead of sensitive species (also see Hawksworth and McManus 1989). Results from the 1990 survey indicated that fruticose species, previously reported as rare and sulfur dioxide-sensitive (*Usnea hirta* and *Evernia mesomorpha*), occurred much closer to the smelter (5 km) than expected (Sigal and Johnston 1986).

Overall, an increase in the abundance and diversity of lichens in the area closest to the sources of sulfur dioxide has been observed during the past two decades. During this period, annual average concentrations of sulfur dioxide have dropped by two-thirds and short-term fumigations have also declined (see Chapter 4). Recolonization by lichens was rapid (<6 years); confirming studies in England where lichens re-invade an area within 5–10 years after pollution reductions (Seaward 1989). In London, there was a marked increase in lichens through 1970–1988, and recolonization occurred at a faster rate than expected. This was attributed to rapid improvement in London's air quality rather than microclimatic changes (Hawksworth and McManus 1989).

TABLE 6.2. Lichen species observed on balsam poplar in the vicinity of Sudbury, Ontario, Canada, in 1989–1990

Species	Growth form
Biatora (*Lecidea*) *helvola* (Körber ex Hellbom) H. Olivier	Crustose
Buellia stillingiana J. Steiner	Crustose
Caloplaca flavovubescens (Hudson) Laundon	Foliose
Candelaria concolor (Dickson) B. Stein	Foliose
Candelariella vitellina (Hoffm.) Muell. Arg.	Foliose
Cetraria ciliaris Ach. (var. *halei*)	Foliose
C. pinastri (Scop.) S. Gray	Foliose
C. sepincola (Ehrh.) Ach.	Foliose
Cladina rangiferina (L.) Nyl.	Fruticose
Cladonia botrytes (K. Hagen) Willd.	Fruticose
C. coniocraea (Flk.) Spreng.	Fruticose
C. cristatella Tuck.	Fruticose
C. fimbriata (L.) Fr.	Fruticose
C. rei Schaer.	Fruticose
Evernia mesomorpha Nyl.	Fruticose
Hypogymnia physodes (L.) Nyl.	Foliose
Lecanora pulicaris (Pers.) Ach.	Crustose
Lecanora symmictera Nyl.	Crustose
Lepraria incana (L.) Ach.	Crustose
Melanelia subaurifera (Nyl.) Essl.	Foliose
Parmelia exasperatula Nyl.	Foliose
P. sulcata Taylor	Foliose
Parmelina aurulenta (Tuck.) Hale	Foliose
Pertusaria ophthalmiza (Nyl.) Nyl.	Crustose
Phaeophyscia adiastola (Essl.) Essl.	Foliose
P. pusilloides (Zahlbr.) Essl.	Foliose
P. rubropulchra (Degel.) Essl.	Foliose
Physcia adscendens (Fr. H. Olivier)	Foliose
P. aipolia (Ehrh. ex Humb.) Furnr.	Foliose
Physconia detersa (Nyl.) Poelt	Foliose
Rinodina dakotensis Magn.	Foliose
Scoliciosporum bacidia chlorococcum (Graewe ex Stenh.) Vezda	Crustose
Usnea hirta (L.) Weber ex Wigg.	Fruticose
Xanthoria fallax (Hepp in Arn.) Arn.	Foliose

In the 1968 study (Leblanc et al. 1972), the drastic reduction in epiphytic lichens was observed where the mean level of sulfur dioxide was more than 0.02 ppm, with slight reductions where the sulfur dioxide levels were about 0.01 ppm. The improvement from 1978 to 1990 coincides with sulfur dioxide levels across the whole area, dropping to about 0.01 ppm on an annual basis (see Chapter 5). How-

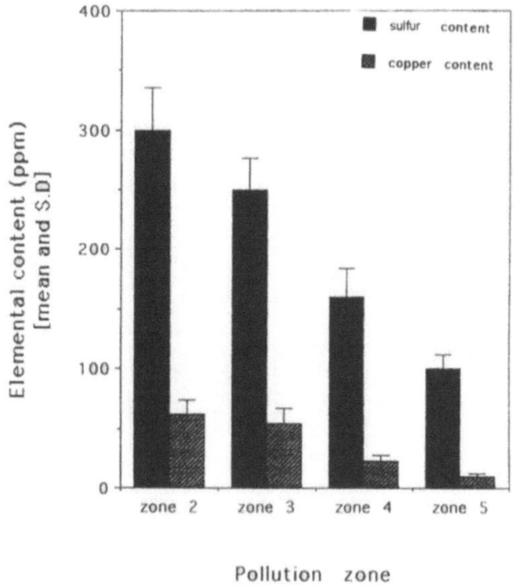

FIGURE 6.3. Sulfur and copper content plus standard deviation of *Parmelia sulcata* collected from 10 trees in each of the 1990 pollution zones.

ever, substrate pH can also influence the richness of the lichen community. Bark with more alkaline pH is better able to buffer the acidity and supply calcium ions to the lichens. The acidity of balsam poplar bark has only decreased slightly from 1978 to 1990 (pH 3–4) and is still considered inhibitory to lichen growth.

Although improvements are apparent in sites close to the smelters, there is evidence that at sites in zone 5, there is a reduction in total lichen diversity. This may be due to changes in the habitat, including age of trees, shading, or effects of long-range transport of contaminants from locations other than Sudbury. No simple causal relationships for the observed changes in lichen communities can be established, but there are strong correlations between the described IAP zones and air quality parameters related to smelter emissions. For example, there is a direct relationship between sulfur or copper content of lichens and the pollution zone from which the lichens were sampled (Fig. 6.3).

There are few epiphytic lichen studies around nickel smelters with which to compare Sudbury. In Russia, around the Severonickel smelter, there is an epiphytic lichen desert (no lichens) within 4 km of the works. Between 8–12 km only 4 species are found. Closer than 12 km is considered as complete destruction of the epiphytic lichens by Gorshkov (1993a) and corresponds to IAP zone 1 in Sudbury. Between 12 and 60 km an increasing diversity of lichens is found, corresponding to IAP zones 2–4. Typical species richness (over 70 species) occurs at distances greater than 60 km (equivalent to IAP zone 5). This pattern corresponds to the Sudbury pattern when emissions were at a maximum in the 1960s.

Comparison to Lichens Growing on Soil or Rock

Although epiphytic lichens suspended in the air have shown a recovery in the past 20 years, lichens on rocky outcrops (saxicolous) and on soil (terricolous) have been much slower to respond (Fig. 6.4). In 1945, Cain reported that only crustose lichens and *Stereocaulon* were found in the most highly polluted zone (approximately 20 km around the smelters). Permanent plots (established in 1977) throughout the inner Sudbury area have demonstrated the slow colonization of rocks by tolerant crustose species of *Lecidea, Lecanora, Porpidia,* and *Rhizocarpon* and by the nitrogen-fixing *Stereocaulon* (fruticose). Other fruticose species

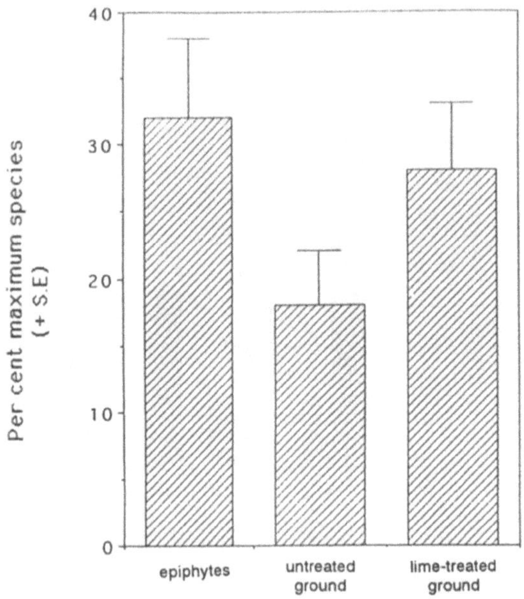

FIGURE 6.4. Comparison of re-invasion rates of lichens for 15 tree plots and nearby ground plots within 25 km of the Copper Cliff smelter between 1978 and 1993. Results are expressed as percentage of the maximum number of expected species for each community type.

such as *Cladonia rei, C. pleurota, C. deformis,* and *C. cristatella* have invaded soil, especially where some moss (*Pohlia nutans*) is present. Reindeer or caribou lichens (*Cladina rangiferina* and *C. mitis*) occur 20–25 km from the smelters (Cox 1993) with little change in distribution since 1977. This sequence is similar to the distribution that was observed by Folkeson (1984) around a brass foundry in Sweden. In the Kola peninsula (Russia), no soil or rock lichens occur within 3 km of the Severonickel smelter complex (Gorshkov 1993a). *Cladonia* are abundant beyond 8 km and *Cladina* is found at 12 km (Gorshkov 1993b), a situation similar to Sudbury.

The iron-rich black encrustation on the rock, a characteristic of sites within 25 km of the Sudbury smelters, may inhibit lichen colonization. Continuing high soil acidity and metal contamination (see Chapter 4) may contribute to much slower ground recolonization rates as compared with recolonization by epiphytes. Despite having similar acidity, metals, in particular aluminum, are less available in bark to inhibit colonization. On soils treated with limestone, where metals are less available, many more lichens are found compared with untreated soil (Fig. 6.4).

Metal Particulates in Lichens

Lichens are able to accumulate metals by trapping particulates (normally sulfates, sulfides, and oxides). Also, dissolved metal ions bind to exchange sites on cell walls, and there is a slow uptake of metals into the algal cells and fungal hyphae (Richardson et al. 1980). Less than 50% of the total metal content is located within the cell or hypha (Brown 1985). From industrial processes, metal-rich emissions are typically in the form of insoluble particulates. The small particulates become entrapped between the hyphae and often build up in the central (medulla) region of the thallus. The linear correlation between metal content of lichens and that found in collected particulates of filters (Saeki et al. 1977) is used as evidence of particulate accumulation in lichens. Further evidence of particulate trapping is the comparison of iron/titanium ratios. In lichens, these ratios are generally of the same magnitude as those found in regional rocks or in industrial areas as detected in industrial emissions (Richardson and Nieboer 1981;

FIGURE 6.5. Changes in nickel concentration in the foliose lichen *Stereocaulon* collected along a transect northwest of the Copper Cliff smelter in 1972, 1984, and 1991 (1972 data from Tomassini et al. 1976).

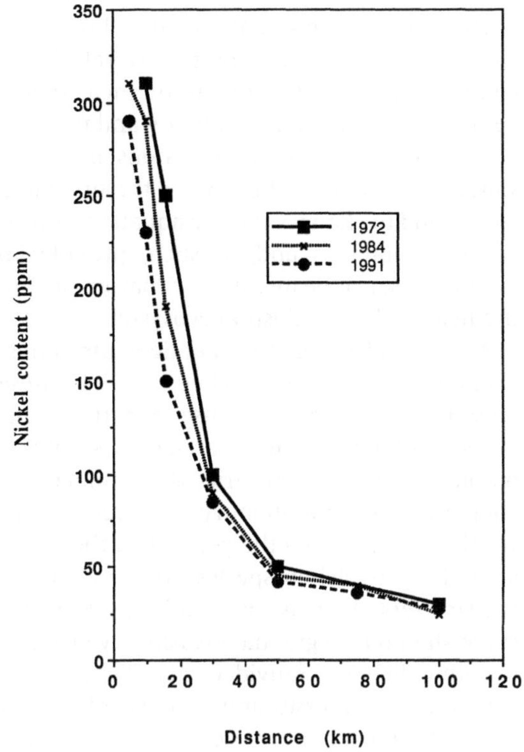

Nieboer et al. 1982). Scanning electron microscopy has facilitated examination of particulates trapped in thalli.

Figure 6.5 shows the typical trend for nickel content of *Stereocaulon*, a fruticose lichen common on rock outcrops, at various distances northwest of the Copper Cliff smelter. Close to Copper Cliff, the nickel content is high (approximately 300 ppm) but then drops rapidly with distance, reaching background levels of a few parts per million (Richardson et al. 1980; McIlveen and Negusanti 1994) beyond 100 km. In addition to distance from sources, prevailing wind direction and topography may also affect the pattern of fallout. Prevailing winds in the area are from the north in the winter and southwest in the summer months (Chapter 1).

Over the past 20 years, there has been a small but steady decline in the metal content of the lichens (Fig. 6.5). The observed change (by 25–30% near the smelter and by 10% at distant localities) is less than might be expected given the reduction in metal emissions from the smelters. Suspended particulates in air and associated nickel content is estimated to have decreased by 50% (see Chapter 5). This discrepancy may be partly explained by previously emitted particles containing metals being redistributed as wind-blown dust rather than just reflecting the actual long-term changes in deposition of metals. Studies by Walthier et al. (1990) demonstrated that residence time for metals in lichens is between 2 and 4 years. Thus, lichens can be expected to show significant changes in elemental content 2–3 years after a reduction in metal emissions.

Summary

The improvement in lichen abundance is attributable to the atmospheric cleanup, particularly the reductions in sulfur dioxide emissions that have occurred since 1972. The changes in lichen abundance clearly demonstrate that terrestrial systems show resiliency and can exhibit positive responses to emission controls.

This study shows that the distribution of sensitive lichen species and the overall lichen community structure were useful measures to delineate areas of poor air quality and to monitor improvement after reductions in atmospheric emissions. Lichens were the first biota to respond to changes (before aquatic communities, Chapter 5) and, as such, provide an important sensitive measure or "barometer" of the health of the industrial ecosystem.

Additional legislation further tightening emission controls became effective in 1994. These new measures should cause a further improvement of the Sudbury area ecosystem as a whole and further constrict the damaged lichen zones around Sudbury. If current trends continue, perhaps by the year 2000, there will be at least 10 lichen species on trees around the smelters. Also, as soil conditions improve, there should be a gradual invasion by ground-dwelling lichens, provided that lichens can compete with mosses and vascular plants for space. Wherever there are significant reductions in sulfur dioxide emissions and, to a lesser extent, metals, there should be a recovery in the lichen flora. This process appears to be rapid provided there are sources of colonizing lichens within a reasonable distance from the damaged area. Not only would this recovery be expected around smelters but also around other similarly affected areas when clean air programs are put into operation.

Acknowledgments. I thank I. Brodo and P. Wong for their assistance in identification of critical specimens. The project could not be completed without the assistance of S. Pappin and many other student summer assistants. Financial contributions were gratefully received from the Ontario Ministry of Northern Development and Mines (Environmental Youth Corps) and Laurentian University Research Fund. The assistance of M. Courtin (Elliot Lake Field Research Station) in preparation of lichen distribution maps and the Ontario Ministry of Environment and Energy with elemental analyses is appreciated. Thanks to the manuscript reviewers, G.M. Courtin and J.M. Gunn, and for suggestions made by D.H.S. Richardson.

References

Amiro, B.D., and G.M. Courtin. 1981. Patterns of vegetation in the vicinity of an industrially disturbed ecosystem, Sudbury, Ontario, Canada. Can. J. Bot. 59(9):1623–1639.

Beckett, P.J. 1984. Using plants to monitor atmospheric pollution. Laurentian Univ. Rev. 16:50–57.

Brown, D.H. (ed.). 1985. Recent Advances in Lichen Physiology. Plenum Press, London.

Burton, A. 1986. Biological Monitors of Environmental Contamination (Plants). MARC report 32. King's College Monitoring Assessment Centre, London.

Cain, R.F. 1945. First Report of the Subcommittee on the Investigation of Sulphur Smoke Conditions and Alleged Forest Damage in the Sudbury Region. Unpublished report.

Cairns, J. 1980. Scenarios on alternative futures for biological monitoring, 1978–1985. pp. 11–21. *In* D.L. Worf (ed.). Biological Monitoring for Environmental Effects. Lexington Books, Lexington.

Cox, J.D. 1993. Survival strategies of lichens and bryophytes in the mining region of Sudbury, Ontario. M.Sc. thesis, Laurentian University.

Easton, R.M. (in press). Lichens and rocks: a review. Geosicence Canada.

Ferry, B.W., M.S. Baddeley, and D.L. Hawksworth (eds.). 1973. Air Pollution and Lichens. Athlone Press, University of London, London.

Folkeson, L. 1984. Deterioration of the moss and lichen vegetation in a forest polluted by heavy metals. Ambio 13(1):37–39.

Gorshkov, V.V. 1993a. Epiphytic lichens in polluted and unpolluted pine forests of the Kola peninsula. pp. 279–289. *In* M.V. Kozlov, E. Haukioja, and V.T. Yarmishko (eds.). Aerial Pollution in Kola Peninsula: Proceedings of the International Workshop, April 14–16, St. Petersburg. Kola Scientific Center, Apatity, Russia.

Gorshkov, V.V. 1993b. The state of moss-lichen cover in polluted and unpolluted pine forests of the Kola peninsula. pp. 290–298. *In* M.V. Kozlov, E. Haukioja, and V.T. Yarmishko (eds.). Aerial Pollution in Kola Peninsula: Proceedings of the International Workshop, April 14–16, St. Petersburg. Kola Scientific Center, Apatity, Russia..

Grindon, L.H. 1859. The Manchester Flora. W. White, London.

Hawksworth, D.L., and P.M. McManus. 1989. Lichen re-colonization in London under conditions of rapidly falling sulphur dioxide levels and the concept of zone skipping. Bot. J. Linn. Soc. 100: 99–109.

Hawksworth, D.L., and F. Rose. 1976. Lichens as Pollution Monitors. Studies in Biology 66. Edward Arnold, London.

Leblanc, F., and J. De Sloover. 1970. Relation between industrialization and distribution and growth of epiphytic lichens and mosses in Montreal. Can. J. Bot. 48:1485–1496.

Leblanc, F., D.N. Rao, and G. Comeau. 1972. The epiphytic vegetation of *Populus balsamifera* and its significance as an air pollution indicator in Sudbury, Ontario. Can. J. Bot. 50:519–528.

McIlveen, W.D., and J.J. Negusanti. 1994. Nickel in the terrestrial environment. Sci. Total Environ. 148:109–138.

Nash, T.H. 1988. Correlating fumigation studies with field effects. *In* T.H. Nash and V. Wirth (eds.). Lichens, Bryophytes and Air Quality. Bibliotheca Lichenologica 30. J. Cramer in der Borntraeger Verlagsbuchhandlung, Berlin-Stuttgart.

Nieboer, E., and D.H.S. Richardson. 1981. Lichens as monitors of atmospheric deposition. *In* S.J. Eisenreich (ed.). Atmospheric Pollutants in Natural Waters. Ann Arbor Science Publishers, Ann Arbor, MI.

Nieboer, E., D.H.S. Richardson, L.J.R. Boileau, P.J. Beckett, P. Lavoie, and D. Padovan. 1982. Lichens and mosses as monitors of industrial activity associated with uranium mining in northern Ontario Canada—Part 3: accumulations of iron and titanium and their mutual dependence. Environ. Pollut. (B)4:181–190.

Pappin, S., and P.J. Beckett. (in press). Changes in epiphytic lichen abundance: tracking air quality improvements in the mining region of Sudbury, Ontario, Canada. Lichenologist.

Richardson, D.H.S. 1992. Pollution Monitoring with Lichens. Naturalists' Handbooks 19. Richmond Publishing, Slough, England.

Richardson, D.H.S., P.J. Beckett, and E. Nieboer. 1980. Nickel in bryophytes, fungi and algae. pp. 367–406. *In* J.O. Nriagu (ed.). Nickel in the Environment. Wiley, New York.

Richardson, D.H.S., and E. Nieboer. 1981. Lichens and pollution monitors. Endeavour (New Series) 5:127–133.

Saeki, M., K. Kunii, T. Seki, K. Sugiyama, T. Suzuki, and S. Shishido. 1977. Metal burdens in urban lichens. Environ. Res. 13:256–266.

Seaward, M.R.D. 1989. Lichens as monitors of recent changes in air pollution. Plants Today 2:64–69.

Sigal, L.L., and J. Johnston. 1986. Effects of simulated acid rain on one species each of *Pseudoparmelia, Usnea* and *Umbilicaria*. Water Air Soil Pollut. 27:315–322.

Smith, F.B., and M.J. Clark. 1986. Radionuclide deposition from the Chernobyl cloud. Nature (Lond.) 332:690–691.

Smith, J.N., and K.M. Ellis. 1990. Time dependent transport of Chernobyl radioactivity between atmospheric and lichen phases in eastern Canada. Environ. Radioactivity 11:151–168.

Steinnes, E., and O. Njastad. 1993. Use of mosses and lichens for regional mapping of ^{137}Cs fallout from the Chernobyl accident. J. Environ. Radioactivity 21:65–73.

Tomassini, F.D.K., K.J. Puckett, E. Nieboer, and D.H.S. Richardson. 1976. Determination of copper, iron, nickel and sulphur by X-ray fluorescence analysis in lichens from the Mackenzie Valley, and the Sudbury District, Ontario. Can. J. Bot. 54:1591–1603.

Turner, D., and W. Borrer. 1839. Specimen of a Lichenographica Britannica. Yarmouth, privately printed.

Walthier, D.A., G.J. Ramelow, J.N. Beck, J.C. Young, J.D. Calahan, and M.F. Marcon. 1990. Temporal changes in metal levels of the lichens *Parmotrema praesorediosum* and *Ramalina stenospora*, southwest Louisana. Water Air Soil Pollut. 53:189–200.

7

Natural Recovery of Vascular Plant Communities on the Industrial Barrens of the Sudbury Area

Keith Winterhalder

In 1972, the Coniston smelter and the Falconbridge iron ore sintering plant were closed, Inco Limited commissioned its 381-m Superstack, and other emission cutbacks came into effect. This led to expectations of rapid reestablishment of vegetation on the barrens, because it was widely believed that sulfur dioxide fumigation was the main factor directly impeding vegetation recovery. It was observed, however, that immediate recovery in the barren zone was at first confined to moist, sheltered, nutrient-enriched sites, such as stream channels (Fig. 7.1). In more exposed barren areas, recolonization did not begin until at least 10 years after the initiation of atmospheric improvement. The foreground of Figure 7.2 shows a site close to the Coniston smelter, photographed at 19-year intervals, in which the only change has been minimal colonization by tickle grass (*Agrostis scabra*), tufted hairgrass (*Deschampsia caespitosa*), and sorrel (*Rumex acetosella*).

In the case of the relict woody plants on the barrens, described in Chapter 2, all species except red maple either maintained or increased their size and vigor in the 20 years after 1970, whereas the red maple continued to undergo "regressive dieback."

The first hint that certain native vascular plant species might be capable of establishing themselves on acid, metal-contaminated soils became evident in 1972, when tickle grass began to colonize plots that had been treated with an N-P-K fertilizer, a treatment that was ineffective in detoxifying the soil for the nonnative experimental grass species (Winterhalder 1974). Tickle grass had formerly been noted in the Sudbury area along roadsides and creeks and in wet depressions. During the next 20 years, this weak perennial became increasingly common in the barren zone, at first appearing mostly on the flood plains of creeks, where it formed a cover so dense that potential new growth in the second year was smothered by the dead remains of the previous year's growth. It was also found under relict birches, poplars, and blueberries, where the soil had been enriched with organic matter, as well as in rock crevices on patches of the metal-tolerant moss *Pohlia nutans* (Beckett 1986). In 1991, Archambault demonstrated that the Sudbury population of tickle grass possessed enhanced metal tolerance.

The next native grass to colonize the barrens was tufted hairgrass (Fig. 7.3). The appearance of this grass in the Sudbury area was first noticed in 1972 and was documented by Cox and Hutchinson in 1980. Even as early as 1974, dense, partly senescent stands of this grass occurred in moist depressions in the barren zone, and it is likely that it began to colonize certain sites several years before its formal documentation. Hutchinson (*personal communication*) has hypothesized that the origin of

FIGURE 7.1. (*Upper photo*) Valley of Coniston Creek, 4 km north of the Coniston smelter in 1972, showing a narrow zone of unidentified sedges bordering the creek, with scattered relict shrubs in the background. (*Lower photo*) The same site 12 years later (August 1984), the edge of the stream colonized by *Carex aquatilis, C. retrorsa,* tufted hairgrass, field horsetail, *Solidago graminifolia,* prairie willow, and meadow willow.

the Sudbury population of *Deschampsia caespitosa* might be 80 km to the southwest, on Goat Island, near Little Current on Manitoulin Island. He suggested that seeds may have been transported to Sudbury with coal shipments on their way from the Lake Huron port of Little Current to the Sudbury smelters. An isozyme variation study by Bush and Barrett (1993) has cast some doubt on the likelihood of such an origin but has not disproved the hypothesis. Another metal-tolerant population of tufted hairgrass, at the mining center of Cobalt 150 km northeast of Sudbury, has been shown by Bush and Barrett (1993) to be genetically distinct from the Sudbury and Goat Island populations.

Two introduced grass species that have colonized barren sites to a more modest degree are redtop (*Agrostis gigantea*) and Canada bluegrass (*Poa compressa*). Although commercially available seed of both of these species is currently used in the revegetation operations described in Chapters 8 and 10, the populations seen colonizing barren land are often distant from revegetation sites, and it is likely that they have arisen from populations that predate revegetation activities. Both species favor heavier soils, and Canada bluegrass is a major colonist of the sides of the deeply gully-eroded "badlands" landscapes that characterize soils dominated by

FIGURE 7.2. (*Upper photo*) Barren site 2 km north of the Coniston smelter in July 1967. (*Lower photo*) The same site in July 1986, showing colonization by tickle grass, tufted hairgrass, and sorrel.

silty clays, whereas redtop is often found in the bottom of erosion gullies.

Enhanced metal tolerance has been demonstrated in both of these grasses. Hogan et al. (1977) found copper-tolerant strains of redtop on the copper- and nickel-rich surface of a roast bed west of Sudbury, some of which were later found (Hogan and Rauser 1979) to be nickel-tolerant, whereas Rauser and Winterhalder (1985) found enhanced zinc tolerance in some Canada bluegrass individuals near the old Coniston roast bed. Rauser and Winterhalder also found the alien grass foxtail barley (*Hordeum jubatum*) invading the Coniston roast bed surface itself, but this species proved not to be metal-tolerant; its presence had presumably been facilitated by the high pH and calcium content of this particular roast bed surface.

A *Carex* from the species group *Ovales* sometimes invades the drier lowland barren soils with the metal-tolerant grasses and, on occasion, can even colonize rocky slopes. It is not always easy to distinguish the species in the field, but in most cases, it appears to be *C. aenea*.

The sedges commonly colonizing the creek flood plains are wool sedge (*Scirpus cyperinus*),

FIGURE 7.3. (*Upper photo*) Barren area 2.5 km northeast of the Copper Cliff smelter in July 1979. (*Lower photo*) The same site in August 1990, showing colonization by tufted hairgrass.

Carex retrorsa, and a *Carex* from the species group *Ovales,* probably *C. scoparia.* These sedges are often accompanied by willows, especially balsam willow (*Salix pyrifolia*), prairie willow (*S. humilis*), meadow willow (*S. gracilis*), and shining willow (*S. lucida*).

On metal-contaminated organic barren sites, the principal invaders since the 1970s have included a rush (*Juncus brevicaudatus*), wool sedge, and tickle grass. Figure 7.4 shows the changes that have taken place on a large barren peatland since 1974.

Although grasses and grasslike plants are the most common colonists of barren sites, a few broadleafed plants also play a role. Sorrel, a highly acid-tolerant plant, occasionally colonizes barren ground along with tickle grass. This species is a common colonizer of industrially disturbed sites in other parts of the world, invading very acidic (pH 3.4) coal measure spoils in the United Kingdom (Rees 1953–1954), and sulfur dioxide-fumigated land near Trail, British Columbia (Archibold 1978). It also occurs on metal-rich soils of the Hartz region of Germany (Schubert 1953–1954) and has been used as a geobotanical tool in northern Greece in exploration for metal sulfide deposits (Kelepertsis and Andrulakis 1983).

Figure 7.4. (*Upper photo*) A large barren peatland 4.5 km northwest of the Coniston smelter, July 1974, with cattail (*Typha latifolia*) in the moat, concentric rings of rattlesnake grass (*Glyceria canadensis*) and wool sedge around the margin, and a predominantly barren center. (*Lower photo*) The same peatland in July 1993, showing almost complete colonization of the center by *Juncus brevicaudatus* and tickle grass. Note the increased size and vigor of the white birches in the foreground. By 1993, the largest of these birches was dead, presumably as the result of bronze birch borer infestation of a drought-weakened tree.

Bladder campion (*Silene cucubalus*), known in Europe to be capable of selection for metal tolerance (Ernst 1974; Lolkema et al. 1986), sometimes becomes established on silty subsoils exposed by erosion. Among the non-flowering vascular plants, both field horsetail (*Equisetum arvense*) and wood horsetail (*E. sylvaticum*) are occasional colonizers of barren soils of a silty texture. It is clear that research is needed on the above species with respect to the possibility of genetic-based metal tolerance.

The shrub species dwarf or bog birch (*Betula pumila*) has shown a spectacular ability to colonize barren land. It began to move onto barren stony slopes from a small fen in the early 1980s (Fig. 7.5). Somewhat later, the same species began to spread onto barren soil and also into an open tufted hairgrass meadow from a single relict individual in a different locality. Roshon (1988) has shown that there has been some genetic selection for metal tolerance in the Sudbury population of dwarf birch, but it is suspected that its success is at least partly due to lack of competition and the enhanced moisture supply provided by run-off from the many rock outcrops.

The colonization phenomenon that is closest to normal boreal zone vegetation succes-

FIGURE 7.5. Dwarf birch colonizing a barren slope 4 km northeast of the Copper Cliff smelter, August 1988.

sion is that which is taking place in the vicinity of the Coniston smelter, which has been closed since 1972. Here, white birch (see Plate 7, following page 182), a typical boreal forest pioneer, began to colonize vigorously in the mid-1980s. Figure 7.6 shows colonization by white birch in a grove of dead maples near Coniston, where scattered relict birches were the seed source. Presumably, the cessation of emissions not only improved the quality of the atmospheric environment to which the aboveground component of vegetation was exposed but also reduced the level of dry deposition of sulfur dioxide and copper, nickel, and iron particulates into the barren and partially barren soils around the smelting operations. Soil pH changes of up to one unit have been observed in the vicinity of the Coniston smelter (K. Winterhalder, *unpublished data*). Although the predominant cause was probably the leaching of free acids, it is possible that weathering of residual glacial till material released bases such as calcium and potassium, which displaced some of the adsorbed hydrogen ions, which were, in turn, lost through leaching. The role of microorganisms in regulating pH in well-drained upland soils is likely to be insignificant when compared with that in lake sediments and wetland soils, which act as sulfur sinks and experience alkalinity generation through the activities of sulfur-reducing microorganisms.

While attempting to explain the greater tendency of tufted hairgrass to spread near the Coniston smelter than near the operating smelters, Cox and Hutchinson (1981) found that the soluble copper and nickel content of Coniston area soils were just as high 5 years after closure as those near active smelters, but that the soluble aluminum content was lower—an observation that would correlate well with the reduced acidity of the Coniston soils observed by the author. Presumably, the differential would be even greater 10–15 years after closure, facilitating white birch colonization. Cox and Hutchinson (1981) also suggested that the atmospheric deposition of copper and nickel particles onto the leaves of plants in an acid environment might have a direct toxic effect through the foliage.

It is interesting to contrast the behavior of white birch and red maple seedlings on barren sites under current improved soil conditions. The birch seedlings develop in a normal fashion, except for the development of the marginal chlorosis syndrome described in Chapter 2, whereas the maple seedlings produce a single pair of foliage leaves that turn red within a month and rarely survive past the first summer dry period. Healthy red oak seedlings are found within oak stands but not on the barrens, probably as a result of their large seed size and specialized mode of dispersal by ro-

FIGURE 7.6. (*Upper photo*) Site 3.5 km northwest of the Coniston smelter in July 1980, showing scattered relict white birches with marginal chlorosis syndrome and many dead red maples. (*Lower photo*) Same site in June 1989, showing colonization by white birch seedlings.

dents such as squirrels. Single healthy pine seedlings are occasionally seen on barren sites, sometimes a great distance from the seed source, despite their large small-winged seeds. The rarity of such occurrences may have as much to do with seed predation and soil toxicity as with lack of dispersal. Bare mineral soil is normally considered to provide an ideal seed bed for pines, but in the case of the Sudbury barrens, a relatively non toxic site would also be necessary. As the result of differential erosion, surface soils on the barrens show great variability in their chemistry and phytotoxicity, which may explain why some sites suitable for pine seed germination exist.

It is clear that one of the factors determining the order in which plant species colonize the soil is the composition of the "rain" of seeds that land on the soil surface. Species that have the lighter wind-dispersed seeds and those that are present as relicts each have an initial advantage. However, after seed deposition, a second limiting factor comes into play—that of the plant's ability to grow on the metal-toxic soil. The differentiating effect of the second factor is illustrated well by white birch and red maple. The former has light wind-dispersed

Box 7.1. Colonization of Barren Sites Created by Volcanic Eruption: Krakatau and Mount St. Helens

The most famous historical account of colonization of a newly created patch of barren land is that which followed the eruption of the Indonesian island of Krakatau in 1883. Most of the original island disappeared, leaving only the smaller, barren, pumice-covered island of Rakata. Nine months after the explosion, the only living organism seen by a French expedition was a small spider, presumably carried to the island as part of the "plankton of the wind." By 1886, there were 15 plant species, by 1897 49, and by 1928 nearly 300. Yet even today, although the island is covered by what seems to be typical tropical rainforest, not a single tree species characterizing the primary forest of neighboring Java and Sumatra is to be found (Wilson 1992).

On May 18, 1980, Mount St. Helens in Washington State, U.S.A., erupted for the first time since 1921. The north side of the mountain collapsed into the north fork of the Toutle River Valley, and 550 km^2 of forest was destroyed by the force of the explosion. Additional areas scorched by the heat of the blast or buried by mud flows (lahars) resulting from the melting of glaciers brought the total extent of the damage to 600 km^2. In August 1982, the U.S. Congress set aside 44,000 ha of the affected area for education and scientific study, to be known as the Mount St. Helens National Volcanic Monument (Franklin et al. 1988).

In the extensive area of forest blow-down, recovery from surviving root systems began almost immediately. In the 96-km^2 area of scorched trees, many of the deciduous trees were able to leaf out fairly normally, whereas the conifers suffered relatively high mortality. Of particular interest to the restoration ecologist, however, is the 20-km^2 barren area to the north of the volcano, known as the Pumice Plains, which was first buried by a debris slide, then by pumice, leading to a complete destruction of plant life. In a small spring-fed oasis on the Pumice Plains, willow (*Salix commutata*), with its light wind-dispersed seeds, was a rapid colonizer. The only species that has established itself abundantly on the drier pumice is the sub-alpine lupine (*Lupinus lepidus*), a rather surprising occurrence because its seeds are relatively large and not wind-dispersed. It is a symbiotic nitrogen-fixer and, as such, plays an important role in nitrogen buildup in the ecosystem that develops on new growth material.

Against the advice of scientists, the Soil Conservation Service (SCS) has dispersed seeds of introduced revegetation species such as birdsfoot trefoil (*Lotus corniculatus*) by helicopter over 32,000 ha, including 2400 ha of the barren land, in an attempt to stabilize the surface and reduce soil erosion. Not only were the introduced plants ineffective in mitigating channel erosion, but they attracted small rodents, which, during the winter, killed established conifers by chewing the bark. Also, the Army Corps of Engineers, the Weyerhaeuser Corporation, and the Washington State Department of Natural Resources dispersed seeds of exotic species to the west of Mount St. Helens, and seeds were carried onto the slopes by wind and elk. The presence of these non-native plants, especially birdsfoot trefoil, proved to have a negative effect on conifer establishment and survival. As a consequence of these results, the SCS has established a native plant seed collection and nursery program.

The inhibitory effect of vigorous nitrogen-fixers in the early stages of succession may not be confined to exotics. Morris and Wood (1989) found that *Lupinus lepidus* had an effect that could either facilitate or inhibit survival and growth of two other herbaceous colonizers—fireweed (*Epilobium angustifolium*) and pearly everlasting (*Anaphalis margaritacea*), and del Moral and Bliss (1993) suggested that the lupine only facilitates colonization after its demise. Overall, primary succession on the pumice was slow, and there was a surprising paucity of algal, lichen, or moss pioneers (Dale 1992).

An interesting observation has been that "late successionals" (i.e., plants that would not normally be expected to attain dominance in the forests of the Pacific Northwest for centu-

Box 7.1. (continued).

ries after disturbance) such as hemlock and true fir are able to colonize the debris slide within the first decade after the eruption. This phenomenon is in keeping with Egler's theory of initial floristic composition, which suggests that plant succession is not always the orderly sequence from pioneer to climax that Clements hypothesized. To quote del Moral and Bliss (1993): "Stochastic elements and low-probability events play a greater role than has been realized in determining the early development of devastated landscapes."

seeds and sufficient tolerance to become established on Coniston area soils, whereas red maple, with its heavier seeds but more frequent relict individuals, is relatively metal-intolerant.

The richer of the wooded sites discussed later, whether dominated by birch or by oak, are currently characterized by the presence of wavy hairgrass in the understory, and this species is just beginning to appear in previously barren sites that have been colonized by white birch. Wavy hairgrass was not mentioned by Gorham and Gordon (1960) as an understory component northeast of the Falconbridge smelter, although they commented on the presence of less conspicuous grasses such as poverty grass (*Danthonia spicata*) and rice grass (*Oryzopsis asperifolia*). It seems likely, therefore, that the wavy hairgrass is a relatively recent arrival in Sudbury's partially barren woodlands. In contrast, Gordon and Gorham (1963) recorded wavy hairgrass as present in the "fume-kill" area near the iron smelter at Wawa, Ontario, but only beyond 6.4 km from the smelter, whereas today it is a widespread dominant at that site. In comparing Sudbury and Wawa seed sources of wavy hairgrass, Archambault (1989) found no difference in copper or nickel tolerance between the two populations. It is likely that it is the species' well-known tolerance of low pH (Larcher 1975) rather than of metals that allowed its spread on both of these sites after the reduction of atmospheric pollutant inputs to the soil.

Conclusions and Prognosis

Natural recovery of plant communities on acid, metal-contaminated soils is likely to continue at a very slow rate compared with the rapid colonization that occurs after soil amelioration (see Chapters 8 and 14). Based on the present state of knowledge, grasses and deciduous trees and shrubs such as birch, poplar, and willows will take the lead. Conifers are likely to be later colonists, especially those that require a mineral soil seed bed, because much of the mineral soil is still heavily contaminated with toxic metals. However, there is already tangible evidence of conifers such as white pine becoming established from seed at distances far removed from a seed source. The vector is not known, but it is unlikely to be birds because most birds digest the seeds that they eat. It is equally unlikely that pine seeds would remain attached to animal fur, but a distinct possibility is transport by wind drift over the surface of hard-packed snow in winter.

It should not be forgotten that colonization (Box 7.1) and possibly ecological succession are not the only dynamic processes occurring during the natural recovery of a damaged ecosystem. Genetic selection in one or more of the colonizing species may also be taking place, the implications of which are thoroughly discussed by McNeilly (1987). The impoverished plant communities that are currently found in the Sudbury area are not only structurally and floristically different from the normal plant communities in the region, but they are likely to have a different genetic make-up.

As the unweathered glacial till components still present in these contaminated soils break down and release their bases and contribute to the soil's sand, silt, and clay fractions, the soils will become less toxic and there will be the opportunity for colonization by less metal-tolerant species and less tolerant ecotypes. It is difficult to estimate the amount of time that a naturally recovering contaminated site might take to reach a stable quasi-natural community structure and floristic composition, especially in view of the possibility of global climate change. It would seem likely, however, that the time frame might be in the order of at least a century.

References

Archambault, D.J.-P. 1989. Metal tolerance studies on populations of *Deschampsia flexuosa* (L.) Trin. (wavy hair grass) from northern Ontario. Honours B.Sc. thesis, Laurentian University.

Archambault, D.J.-P. 1991. Metal tolerance studies on populations of *Agrostis scabra* Willd. (tickle grass) from the Sudbury area. M.Sc. thesis, Laurentian University.

Archibold, O.W. 1978. Vegetation recovery following pollution control at Trail, British Columbia. Can. J. Bot. 56(14):1625–1637.

Beckett, P.J. 1986. *Pohlia* moss tolerance to the acidic, metal-contaminated substrate of the Sudbury, Ontario, Canada, mining and smelting region, pp. 30–32. *In* Environmental Contamination—Second International Conference. CEP Consultants Ltd., Edinburgh.

Bush, E.J., and S.C.H. Barrett. 1993. Genetics of mine invasions by *Deschampsia cæspitosa* (Poaceae). Can. J. Bot. 71:1336–1348.

Cox, R.M., and T.C. Hutchinson. 1980. Multiple metal tolerances in the grass *Deschampsia cespitosa* (L.) Beauv. from the Sudbury smelting area. New Phytol. 84:631–647.

Cox, R.M., and T.C. Hutchinson. 1981. Environmental factors influencing the rate of spread of the grass *Deschampsia cespitosa* invading areas around the Sudbury nickel-copper smelters. Water Air Soil Pollut. 16:83–106.

Dale, V.H. 1992. The recovery of Mount St. Helens. World and I 7(6):262–267.

del Moral, R., and L.C. Bliss. 1993. Mechanisms of primary succession: insights resulting from the eruption of Mount St. Helens. Adv. Ecol. Res. 24:1–66.

Ernst, W.H.O. 1974. Schwermetallvegetation der Erde. Fischer, Stuttgart.

Franklin, J.F., P.M. Frenzen, and F.J. Swanson. 1988. Re-creation of ecosystems at Mount St. Helens: contrasts in artificial and natural approaches. pp. 1–37 *In* J. Cairns, Jr. (ed.). Rehabilitating Damaged Ecosystems. Vol. 2, CRC Press Inc., Boca Raton, FL.

Gordon, A.G., and E. Gorham. 1963. Ecological aspects of air pollution from an iron-sintering plant at Wawa, Ontario. Can. J. Bot. 41:1063–1078.

Gorham, E., and A.G. Gordon. 1960. Some effects of smelter pollution northeast of Falconbridge, Ontario, Canada. Can. J. Bot. 38:307–312.

Hogan, G.D., G.M. Courtin, and W.E. Rauser. 1977. Copper tolerance in clones of *Agrostis gigantea* from a mine waste site. Can. J. Bot. 55:1043–1050.

Hogan, G.D., and W.E. Rauser. 1979. Tolerance and toxicity of cobalt, copper, nickel and zinc in clones of *Agrostis gigantea*. New Phytol. 83:665–670.

Kelepertsis, A.E., and I. Andrulakis. 1983. Geobotany-biogeochemistry for mineral exploration of sulphide deposits in northern Greece—heavy metal accumulation by *Rumex acetosella* L. and *Minuartia verna* (L.) Hiern. J. Geochem. Exploration 18:267–274.

Larcher, W. 1975. Physiological Plant Ecology. Springer-Verlag, New York.

Lolkema, P.C., M. Doornhof, and W.H.O. Ernst. 1986. Interaction between a copper-tolerant and a copper-sensitive population of *Silene cucubalus*. Physiol. Plantarum 67:654–658.

McNeilly, T. 1987. Evolutionary lessons from degraded ecosystems. pp. 271–286. *In* W.R. Jordan, M.E. Gilpin, and J.D. Aber (eds.). Restoration Ecology: A Synthetic Approach to Ecological Research. Cambridge University Press, New York.

Morris, W.F., and D.M. Wood. 1989. The role of lupine in succession on Mount St. Helens: facilitation or inhibition? Ecology 70(3):697–703.

Rauser, W.E., and E.K. Winterhalder. 1985. Evaluation of copper, nickel, and zinc tolerances in four grass species. Can. J. Bot. 63:58–63.

Rees, W.J. 1953 to 1954. Some preliminary observations on the flora of derelict land. Proc. Birmingham Nat. History Philos. Soc. 18(5):119–129.

Roshon, R.D. 1988. Genecological studies on two populations of *Betula pumila* var. *glandulifera*, with special reference to their ecology and metal tolerance. M.Sc. thesis, Laurentian University, Sudbury, Ontario.

Schubert, R. 1953 to 1954. Die Schwermetallpflanzengesellschaften des östlichen Harzvorlandes. Wissenschaftliche Zeitschrift der Martin-Luther-Universität Halle-Wittenberg 3:51–70.

Wilson, W.O. 1992. The Diversity of Life. W.W. Norton and Co., New York.

Winterhalder, K. 1974. Reclamation studies on industrial barrens in the Sudbury area. Proceedings of the Fourth Annual Workshop, Ontario Cover Crop Committee, Guelph.

Section C

Goals of Restoration

Anthony D. Bradshaw

If we were to take a broad view of the world, there is no reason why the degradation of Sudbury could not be left as it is, as a monument to the destructive way in which resources were exploited in the past. Although the degraded area is very large in relation to other degraded areas, it is small in relation to the whole of North America, and although there is some pollution leaving the area, most of the effects of the degradation are self-contained. To many visitors, the area is extraordinary and fascinating. At the same time, it is a sharp lesson for us all about the ease with which we can degrade our environment completely.

But it is unlikely to be acceptable to most people, particularly those who live in the Sudbury area. There is plenty of experience from other areas that because, above all else, environmental degradation leads to loss of attractiveness of a region, there is a vicious spiral of economic and social degradation. Industries leave, new industries find sites elsewhere, unemployment increases, the more active members of the work force move out, incomes decline, and the total economy of the area crumbles. As a result, the area becomes less and less able to improve itself.

The primary goal of restoration is therefore, for most people, an aesthetic one—to restore the visible environmental quality of the area. This could be done either by restoring the quality of the built environment or the quality of the natural environment. However, the former is unlikely to happen without a restored confidence in the area bringing in new capital. It is therefore restoration of the quality of the natural environment that is the key.

But, except in exceptional circumstances and at great expense, the natural environment cannot be bought off the shelf and carried in by truck; it has to be carefully created from what is there. This is not easy. The primary goal is made up of several separate but related goals, all aimed toward the re-creation of a viable functioning ecosystem—the community of soils, water, plants, and animals that live and interact together in one place.

The first of these goals must be a chemical one—to restore the invisible quality of the environmental background of the area. In an intensely degraded area such as Sudbury, the soil has been almost completely lost and with it the store of organic matter and plant nutrients without which plants cannot grow. These have to be painstakingly rebuilt by the use of fertilizers and the growth of the plants themselves. At the same time, there can be continuing pollution—problems of acidity and metal contamination in the soils and tailings and in leachates from them and dust-blow from the tailings and degraded land surfaces. These have to be eliminated by soil treatments and plant growth so that the release of pollutants is prevented at source. This will then improve the quality of the water bodies that receive their water from the polluted land.

The second goal is then a biological one—to set about creating ecosystems that grow and prosper. This means establishing plants that will grow vigorously and clothe the denuded hillsides. In doing this, they will stabilize the eroded soils, add organic matter, tie up the metal pollutants, and so progressively make conditions better for themselves as well as for other more-sensitive species. In the water bodies, the communities have to re-establish similarly.

The third goal, which stems from the first two, is biological and aesthetic—to restore the biological diversity of the area. In its degraded state, the area supports only a very few terrestrial plants and animals, those tolerant of the extreme conditions. The missing plant and animal species must be encouraged to recolonize. In part, this will be brought about by achievement of the first two goals. In particular, most of the animals will come back if the conditions are right, but some of the plants and other less mobile organisms such as the fish may require help. None of the missing species should be considered as unimportant; the fish, for example, have profound effects on the feeding (trophic) structure of the aquatic communities.

In the end, the measure of success is the degree to which ecosystems are created that have satisfactory structure and function. What this entails can be most easily demonstrated in the Fig. C.1, in which there are two dimensions, representing the negative changes in structure and function that have occurred in the degradation and therefore the positive changes that restoration must entail. Onto this diagram, the goals that have been discussed can be readily superimposed, because improvement in ecosystem structure and function are the two processes that underlie all the other goals.

The contributions in this section show very clearly and elegantly what this entails in practice and how possible it is to achieve the ultimate goal of total environmental improvement,

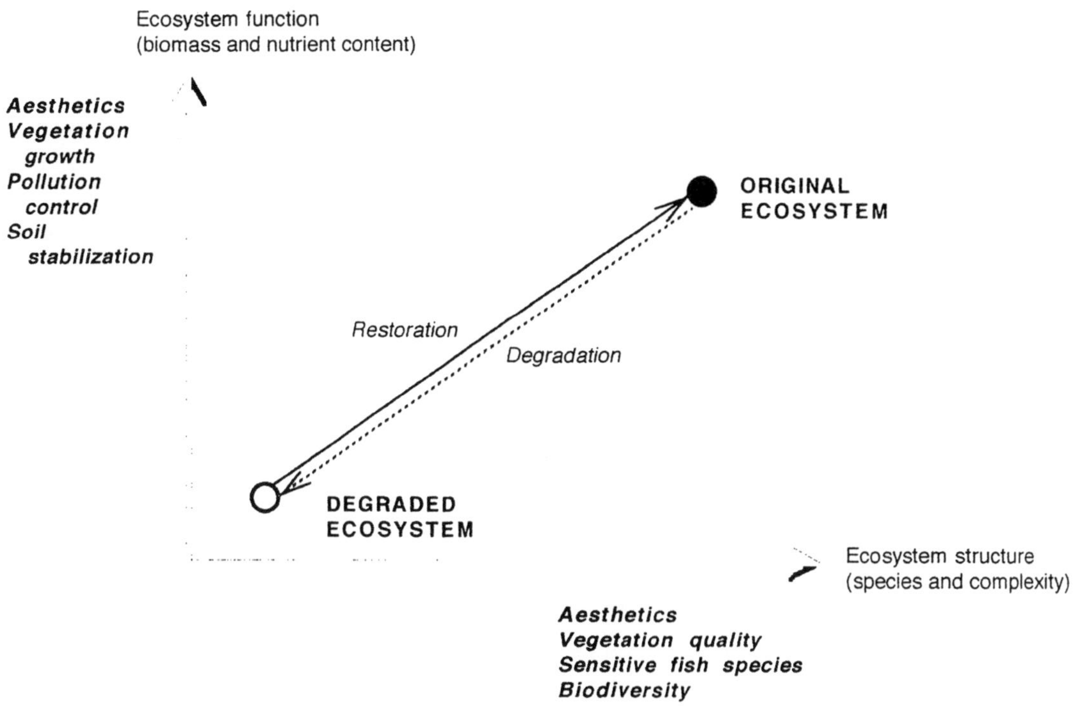

FIGURE C.1. Degradation and restoration of an ecosystem (developed from Bradshaw, 1987, in Restoration Ecology, W.R. Jordan et al. (eds.), Cambridge University Press).

even from such an extreme starting point as that to be found at Sudbury. But the contributions also show that achievement depends on people. Goals can only be reached if the community takes part, and if people, whether in industry, townships, or schools, understand the goals and are prepared to participate.

8

Municipal Land Restoration Program: The Regreening Process

William E. Lautenbach, Jim Miller, Peter J. Beckett, John J. Negusanti, and Keith Winterhalder

Many cities and towns throughout the world suffer from environmental problems that affect the quality of life of their residents. Unfortunately, there are relatively few examples in which the communities have been effective at reducing or eliminating these problems. Sudbury's land reclamation and air pollution reduction program, Los Angeles's (Leuts and Kelly 1993) and Tokyo's (Nishimura 1989) smog reduction programs, the cleaning of the Thames River in London (Andrews 1984), Singapore's multifaceted environmental improvement program (Jee 1988), and land reclamation in the lower Swansea Valley of Wales (Box 8.1) are some examples where measurable success has been achieved.

Community Challenge

The destructive influence of past mining activities not only left Sudbury with a severe environmental problem, but its 160,000 inhabitants also inherited conditions that greatly restricted their socio-economic prospects. In the early 1970s, approximately 17,400 ha of land lacked vegetation, much of the soil had been acidified and metal-contaminated, and severe erosion had occurred on steep slopes and hilltops (Amiro and Courtin 1981; Freedman 1989). An additional 72,100 ha of land was semibarren, consisting mainly of stunted birch (*Betula papyrifera*) and red maple (*Acer rubrum*). Most highway and railway view corridors and many areas around neighborhoods and towns were severely affected. This blackened landscape, described in the media as "moonscape" and "dead zone" (Young 1992), limited the ability of the municipality to attract people and corporations to the area.

Despite the air pollution controls initiated in 1972, there was little positive response in vegetation growth (see Chapter 7). The damages to the vegetation and soil that had occurred over the past century were not going to recovery quickly without direct human actions. A large-scale remedial program was needed.

Program Beginnings

In early 1974, the Regional Municipality of Sudbury, a political entity that brought together the City of Sudbury and six adjoining towns (an area of 2800 km^2) began its restoration effort with the creation of a technical advisory committee composed of representatives of university, industry, government, and the general public. This group was given the challenge of providing the elected members of the regional council with advice on how to restore area vegetation.

One of the first tasks was to identify and map the areas needing treatment. At the same

Box 8.1. Lower Swansea Valley Project

From 1717 to the end of the First World War, Swansea, in southern Wales, was the world center of non-ferrous metal smelting (Weston et al. 1965). At different times, copper, lead, zinc, silver, and arsenic were smelted in 22 works along the tidal area of the River Tawe. Later, 10 steel and tin plants were established, making this one of the highest concentrations of metal works in Britain. Ore came from various parts of Britain as well as Chile, Mexico, Norway, Spain, Portugal, and Australia. A small amount of ore was also shipped to Swansea from Sudbury in 1900 by the Vivian Company that operated the Murray Mine. After the collapse of the metallurgical industries, the city of Swansea was left with the legacy of 325 ha of completely derelict land, consisting of huge piles of slag, ash, and coal shale. The dust, fumes, and physical appearance of the area were a depressing mixture. In 1961, a community-based program was begun to reclaim this derelict land in the lower Swansea Valley. The program succeeded, over the next 30 years, after a great deal of hard work, money, and technical assistance to revegetate the land while revitalizing the community (Bridges 1991).

time, experimental vegetation plots were installed at several locations to test various soil additives and grass mixtures. These small-scale test plots successfully demonstrated that future large-scale program implementation should be possible (Fig. 8.1). Also, relict patches of vegetation could form a nucleus for spread into treated areas.

Unlike restoration efforts on mine site wastes in North America (Daniels and Zipper 1988; Hossner 1988) and Europe (Bradshaw 1983), Sudbury's damaged landscape still possessed

Figure 8.1.(*a* and *b*) Example of vegetation plots used to test treatment procedures for land reclamation in Sudbury. (Photo and studies by Winterhalder [1984].)

many of the characteristics of its original form. The topography consisted of rocky Precambrian Shield hillsides with patchy soils and wetlands. Standard mechanical reclamation techniques (Down and Stocks 1977; Peters 1984) were not suitable for treating such landscape. Other methods were required. Use of two local elementary school groups to reclaim trial plots near their school site successfully demonstrated that an alternate manual reclamation method was possible.

In 1978, the municipality began a large-scale reclamation program. In that year, the local mining companies laid off 3500 employees and curtailed all summer student hirings. In response, the municipality explored several programs to create or provide summer employment for students. Land reclamation was chosen as one of the principal programs to address these socio-economic needs. Funds were obtained from various government agencies and local mining companies (Lautenbach 1987).

Scientific Underpinnings

The challenges for an effective reclamation program were many (Winterhalder 1984; Freedman 1989). On barren sites near the smelters, soil pH ranged from about 3.0 to 4.5, whereas copper, nickel, and aluminum frequently reached 1000 µg/g (see Chapter 4). Also, the blackened sites had extreme surface temperatures (>50°C) and were arid in summer and subject to severe frost heaving in winter (see Chapter 18). Laboratory experiments demonstrated that root growth of seedlings in these toxic soils was inhibited, resulting in dehydration and death (Whitby et al. 1976; Winterhalder 1984).

There was little scientific information to guide the design and implementation of an effective reclamation program for this type of landscape. Restoration ecology was and still is a very new and developing field of study (Bradshaw 1983; Cairns 1988). Therefore, testing and monitoring were essential for achieving the objectives of the reclamation program. The objectives were to

- **create a self-sustaining ecosystem with minimal maintenance**
- **use plant species that are tolerant of acidic soils and low nutrient concentrations**
- **use seed application rates that allow for natural colonization and thus increase species diversity**
- **give preference to the use of native species**
- **restore nutrient cycles and pools by the use of species that fix nitrogen (legumes)**
- **use species that attract and provide cover for wildlife**
- **undertake initiatives that speed up natural successional changes**

Restoration Process

One of the main reclamation goals was to reestablish a forest similar to that which once covered the barren hills. However, initial field trials had demonstrated that liming was needed to detoxify the soils and that a herbaceous grass cover was desirable before shrubs and trees could be successfully established (Winterhalder 1984). Therefore, two major program components evolved: the first to conduct liming and initiate grass-herb cover, and the second to introduce tree and shrub species that did not colonize spontaneously.

Initial Grassing

Soil pH was first measured to determine the amount of crushed limestone required to neutralize soil acidity. An application of approximately 10 tonnes of agricultural-grade calcitic or dolomitic limestone was required per hectare to raise the pH of these soils from 3.0–4.5 to the desired level of 5.5–6.0.

Limestone was bulk-shipped to reclamation staging areas, where it was bagged for subsequent transport to barren hillsides. The most practical and economical means of transportation to the work site was used. Equipment used included pickup and flatbed trucks, rail flatbeds, helicopters, and all-terrain vehicles. After the bagged limestone was moved to the work site, it was carried by employees up the hills for spreading. Lime bags were placed at 1-m intervals to ensure adequate area coverage, then spread (Fig. 8.2). Approximately 80% of the workforce's time in grassing was required to bag, move, and spread lime. Most of this work took place early in the spring and summer months.

Later in the summer, after the crushed limestone had reacted with the soil for several weeks, workers returned to limed sites and spread a high-phosphorus fertilizer (6N-24P-24K) at a rate of 400 kg/ha. After the application of fertilizer, a seed mixture of grasses and nitrogen-fixing legumes (Table 8.1) was sown (Fig. 8.3) at a rate of 45 kg/ha. Seeding was done in mid-August to coincide with fall rains. No attempt was made to create an even coverage of grasses and legumes (Winterhalder 1983). In fact, a patchy cover of approximately 24–40% of grasses and legumes was preferred because it allowed subsequent invasion and colonization by native herbs, shrubs, and trees, thus encouraging species and genetic diversity.

Between 1978 and 1993, 3070 ha of barren land was limed, fertilized, and seeded. This represented most of the barren lands immediately adjacent to major road corridors and nearly all the areas within urban neigh-

Figure 8.2. Spreading of crushed limestone.

Table 8.1. Summary of land reclamation treatment used on barren land affected by industrial emission in Sudbury

1. Limestone (usually dolomitic limestone)
   Applied at a rate of 10 tonnes/ha to raise pH to more than 5
   Applied several weeks before fertilizer and seed
2. Fertilizer—usually 6N-24P-24K
   Applied at 390–400 kg/ha
3. Seed mixture
   Several grasses and two legumes; applied at 30–50 kg/ha
   Mixture (by weight)

Redtop	*Agrostis gigantea*	20%
Creeping red fescue	*Festuca rubra*	10%
Timothy	*Phleum pratense*	20%
Canada bluegrass	*Poa compressa*	15%
Kentucky bluegrass	*Poa pratensis*	15%
Birdsfoot trefoil	*Lotus corniculatus*	10%
Alsike clover	*Trifolium hybridum*	10%

Grass-legume mixture is sown in mid–late August after start of cooler nights and autumn rains

4. Trees and shrubs—plant 1–2 years later
   Commonly used species include jack pine (*Pinus banksiana*), red pine (*P. resinosa*), white pine (*P. strobus*), white spruce (*Picea glauca*), tamarack (*Larix laricina*), red oak (*Quercus borealis*), and black locust (*Robinia pseudocacia*)

borhoods. In total, it represented approximately 20% of the barren land in the region. Also, both mining companies have operated their own restoration projects to cover tailings with vegetation and to rehabilitate land around their operations (see Chapters 9–11).

Tree Planting

Although grass-legume establishment was rapidly followed by spontaneous colonization of birch, poplar, and willows (see Chapter 13), no coniferous species appeared in the first few years

Figure 8.3. Cyclone seeding of grass and legumes after an area has been treated with limestone.

after treatment. Early attempts to plant conifers directly into untreated barren areas were also largely unsuccessful. However, with liming and the establishment of an herbaceous cover with its associated nutrients, soil moisture, and shade, successful tree planting of conifers and other trees became possible. Test plots in 1978–1982 demonstrated good growth and survival of trees planted on reclaimed sites (Lautenbach 1985).

The first main planting, consisting of 228,000 trees, occurred in 1983. Since then, work crews have planted more than 1.5 million trees on previously grassed sites (Fig. 8.4).

The basic goal of the municipal tree planting program is to create a self-sustaining ecosystem by matching tree species to the unique habitat features (soil moisture, exposure, slope, etc.) of the site being reclaimed. In an effort to create informal natural-appearing landscapes, plantation-style plantings have been avoided, and trees are planted in groups at fairly low densities to allow for natural infilling. Often, several different types of conifer

Figure 8.4. Tree planting.

FIGURE 8.5. Survival rates (+1 SD) after 3 years (1984–1987) of a sample of 300 trees for species used in reclamation work in the Sudbury area. The species included *Ce,* white cedar; *Pj,* jack pine; *Pr,* red pine; *Pw,* white pine; *Sb,* black spruce; *Sn,* Norway spruce; *Sw,* white spruce; *Ta,* tamarack; *Le,* European larch; *Lj,* Japanese larch; *Aw,* white ash; *Mh,* sugar maple; *Ms,* silver maple; *Lb,* black locust; *Or,* red oak. (Data from Beckett and Negusanti [1990].)

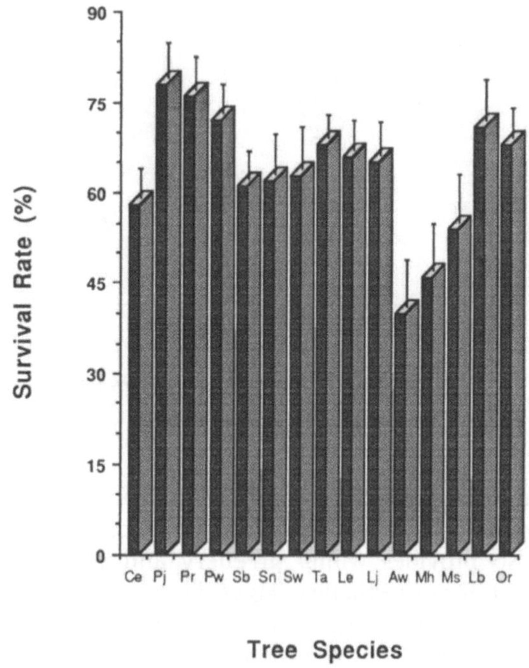

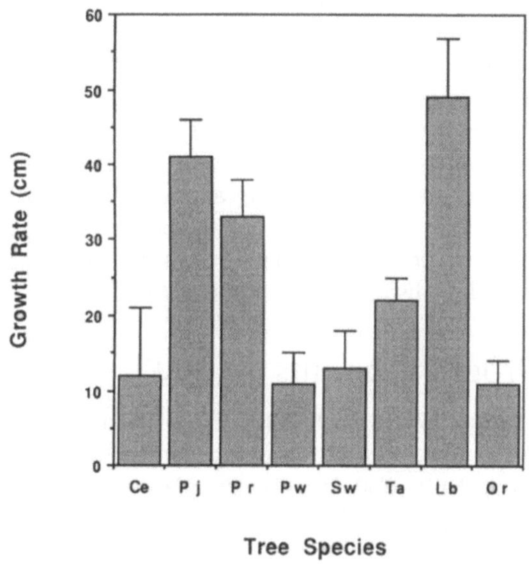

FIGURE 8.6. Mean annual growth in height of major tree species used in reclamation work in the Sudbury area. Measurements made in 1987, 3 years after planting. Means + SE are indicated for a sample of 300 trees for each species. Species included in *Ce,* white cedar; *Pj,* jack pine; *Pr,* red pine; *Pw,* white pine; *Sw,* white spruce; *Ta,* tamarack; *Lb,* black locust; *Or,* red oak. (Data from Beckett and Negusanti [1990].)

and deciduous trees are planted in each location. Species that have shown good survival (Fig. 8.5) and growth (Fig. 8.6) and are readily available from nurseries are selected. Planting stock mainly consists of 2- or 3-year-old bareroot seedlings and 6-month- to 1-year-old containerized or paper pot seedlings. More than three-quarters of the planted material has been conifers, with an emphasis on pines (red pine [*Pinus resinosa*], white pine [*P. strobus*], and jack pine [*P. banksiana*]) that have high survival and growth rates. The general aim is to plant species typical of mature northern Ontario forests to accelerate the slow pro-

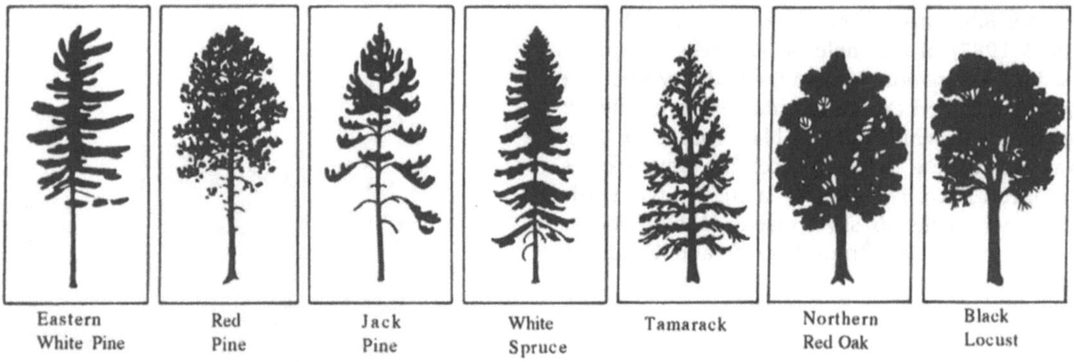

FIGURE 8.7. Principal tree species used in revegetation. The trees are not drawn to scale.

gression through successional birch and poplar woodland stages (Fig. 8.7). Planting of conifers also provides winter greenery and achieves part of the aesthetic goals of the program.

The black locust (*Robinia pseudoacacia*) is used for its nitrogen-fixing ability and tolerance in very dry and exposed sites. Although a non-native species, black locust has been useful in other reclamation efforts (Ashby et al. 1980) because of its rapid growth and good survival characteristics (Figs. 8.5 and 8.6). In addition to improving the nitrogen content of the soil, it also appears to be a useful "nurse" tree, providing shelter for more-sensitive species planted beneath it.

At present, all tree planting activity is limited to the spring season of the year. Trees are shipped from nurseries as soon as the ground thaws and are stored in refrigerated vans for planting. Planting generally occurs during the months of May and June.

In the past few years, many school classes, clubs, and environmental groups have volunteered their time and labor to assist in grassing and tree planting activities. Many of these organizations are now volunteering each year and are being encouraged to adopt areas for longer-term reclamation activity.

Monitoring

Field crews working under the supervision of faculty at Laurentian University monitor results of the reclamation program. Monitoring involves assessment of grass, shrub, and tree planting survival and the measurement of various chemical and physical site characteristics. Key findings include (1) pH values have remained elevated after liming; (2) metal uptake by plants declines after liming; (3) insects, birds, and small mammal populations have increased in reclaimed areas; (4) spontaneous colonization of herbaceous and woody species has occurred on treated sites; and (5) the percentage of grass cover has tended to decrease relative to the percentage cover of legumes and woody species. Greater details of these studies are provided in Chapter 13.

Findings for tree planting activities have been equally encouraging. Survival rates after 3 years have averaged 70% across all species (see Fig. 8.5), and growth rates have been similar to those from less-disturbed areas (Figs. 8.8–8.10).

Program Results

During 1978–1993, 3070 ha of barren land were reclaimed by grassing. During this period, 1,692,000 trees were planted on previously grassed sites (Fig. 8.11).

Total cost of the reclamation program during 1978–1993 has been approximately $15 million Canadian, with about 80% of the funds spent for salaries of the employees. Given the rough topography in which restoration was needed, manual grassing and tree planting techniques appeared to be the most cost-effective means of restoring the landscape. On average, treatment of a hectare of land costs less than $5000.

FIGURE 8.8. Cumulative growth of jack pine planted in 1979 compared with a non-polluted reforestation stand planted in the same year ($n = 100$ at each site). (Unpublished data of P. Beckett and J. Negusanti.)

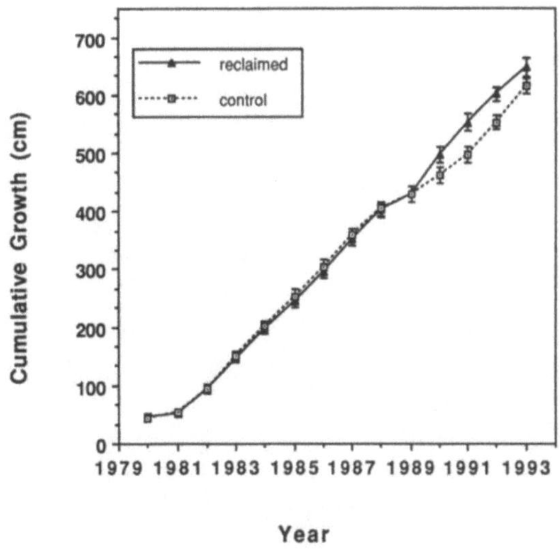

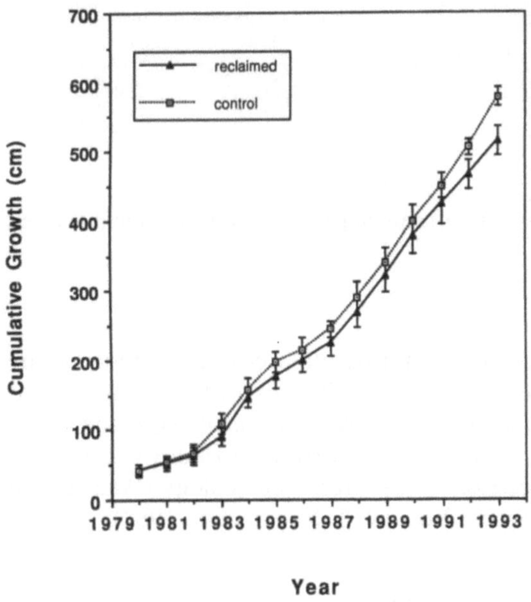

FIGURE 8.9. Cumulative growth of red pine planted in 1979 compared with a non-polluted reforestation stand planted in the same year ($n = 100$ at each site). (Unpublished data of P. Beckett and J. Negusanti.)

In comparison with many mechanized restoration techniques, the program has the added benefit of providing needed employment, being labor- rather than capital-intensive. More than 3340 individuals have been directly employed through the municipality's land reclamation program over the past 15 years. This has included 1600 students and 1740 individuals who were unemployed or on social assistance.

Future Initiatives

Considerable barren land and many other challenges remain for the municipal reclamation program. As the areas near road corridors are completed, additional costs and perhaps new approaches are needed to treat the more-remote areas. The newly forested areas are also potentially vulnerable to insect pests, fire, and expand-

FIGURE 8.10. Picture of reforested area near Hannah Lake (1994) showing natural-type placement of trees. (Photo by P. Beckett.)

ing urban development and need to be protected.

One of the potential benefits of large-scale land reclamation activities that was not recognized at the start of this program is the improvement in the quality of drainage water from the treatment areas (Skraba 1989). To date, watershed liming treatments have contributed to increased pH and alkalinity in two area lakes (see Chapter 15). This discovery has prompted the Technical Advisory Committee to look at reclaiming several whole watersheds in the municipality to achieve the added benefit of improved aquatic environments. Similar watershed liming is being used on an experimental basis in the United States (Gubala and Driscoll 1991), United Kingdom (Howell and Dalziel 1992), and Scandanavian countries (Olem et al. 1991).

The Technical Advisory Committee is also pursuing the idea of conserving or setting aside a major barren area as a reminder of the environmental changes the community has made. Such a preserve will also provide scientists with the opportunity to study natural succession and barren land ecology (Watson and Richardson 1972; Jordan et al. 1987).

Socio-Economic Benefits

The municipality's and mining companies' land reclamation programs have fundamentally changed the physical appearance and psychological "mindscape" of the community. It is now increasingly difficult to find completely barren landscapes within public view corridors. This transformation (see Plates 9–11, following page 182) and the process used to achieve this success have been widely published, and the municipality has been the recipient of several prestigious environmental awards (Table 8.2).

FIGURE 8.11. (*a–d*) A site in Sudbury before and after liming, grassing, and tree planting.

International program recognition and landscape transformation have been a tremendous source of pride for area residents and municipal officials. As the municipality reclaimed barren sites, the community in turn placed a greater emphasis on vegetation enhancement throughout all neighborhoods. This has led many private citizens to improve and green their own properties.

TABLE 8.2. Program awards

1992 United Nations Local Government Honours Award, presented to 12 municipalities at the Earth Summit in Rio de Janeiro in June 1992
1992 Chevron Conservation Award, presented in Washington, D.C.
Government of Canada Environmental Achievement Award for Municipal Leadership 1990, presented in Ottawa by Governor General Ray Hnatyshyn
1990 Lieutenant Governor's Conservation Award from the Conservation Council of Ontario, presented in Toronto by Lieutenant Governor Lincoln Alexander
1990 Arboricultural Award of Merit, presented by the International Society of Arboriculture Ontario Inc.
1986 Community Improvement Award, presented by the Ontario Horticultural Association

Although a direct linkage is difficult to establish, the land reclamation appears to have had a major influence in the economic diversification of the area over the past 15 years. The visual improvement of the area has been significant enough to allow the municipality to now market itself as a summer and winter tourism destination based on the establishment of a northern science center (Fig. 8.12). It has also made it easier to market the community to prospective businesses that are looking to establish facilities in northern Ontario. Land reclamation has removed a negative community stigma that formerly affected some corporate decisions. This program is an important example of how environmental improvement can contribute to economic development.

FIGURE 8.12. Aerial view of Science North, with the Lily Creek wetland in the background and the sailing club in the foreground.

Lessons Learned

Overall, the program has been successful because most program activities drew on the lessons learned from pilot test plots and small demonstration project findings. This has led to a very simple treatment recipe being adopted (see Table 8.1). With continuous monitoring of performance, adjustments have been made as the need arose (e.g., seed mixture altered, and supervisor/field crew ratios adjusted).

The existence of a volunteer Technical Advisory Committee has also demonstrated the value of having a broadly based community support group spearheading problem-solving assignments. This group has also been able to focus on many other vegetation enhancement initiatives in the community.

However, in addition to having the volunteer advisory committee spearheading the problem-solving aspects of the program, it was essential to have a host organization (in this case, the municipality) to carry it out. The municipality, in effect, became the general contractor to ensure that required reclamation tasks occurred. In this role, the municipality provided logistic support for the program and was responsible for hiring and supervising all the program's employees, for purchasing or acquiring all support materials, and for all payroll and accounting functions. It also appointed a land reclamation co-ordinator to oversee each year's program. For the program to progress as far as it has, this organizational and catalytic role was essential.

Finally, this program demonstrates the benefits of being committed to achieving stated goals and objectives and rigorously pursuing them. These goals have included the rehabilitation of barren land, improvement of a negative regional image, and a commitment to a labor-intensive approach. In the process, time was not wasted on attributing blame but on solving a community problem and in providing the necessary community leadership to achieve this. This program also demonstrates the benefits of an ecosystem approach underpinned by scientific principles in recreating a self-sustaining local environment.

Acknowledgments. This Technical Advisory Committee is comprised of individuals from Inco Limited, Falconbridge Limited, Laurentian University, Cambrian College, Ontario Ministry of Natural Resources, Ontario Ministry of the Environment and Energy, Ontario Ministry of Northern Development and Mines, Ontario Hydro, Nickel District Conservation Authority, Sudbury Horticultural Society, City of Sudbury, Town of Walden, Regional Municipality of Sudbury, Sudbury Master Gardeners, and interested local citizens. We thank the members of this committee, many of whom have given more than a decade of volunteer time to this project. A. Bradshaw, J. Gunn, and T. Peters provided a technical review of this manuscript.

References

Amiro, B.D., and G.M. Courtin. 1981. Patterns of vegetation in the vicinity of an industrial disturbed ecosystem, Sudbury, Ontario. Can. J. Bot. 59(9):1623–1639.

Andrews, M.J. 1984. Thames Estuary: pollution and recovery. pp. 195–227. *In* P.J. Sheeham et al. (eds.). Effects of Pollutants at the Ecosystem Level. John Wiley & Sons, Chichester.

Ashby, W.C., C. Kolar, and N.F. Rodgers. 1980. Results of 30 year old plantations on surface mines in the Central States, pp. 99–107. *In* Proceedings of Trees for Reclamation in the Eastern United States Symposium, Lexington, Kentucky. October 27–29. General Technical Report NE-61. USDA Forest Service, Broomal, PA.

Beckett, P.J., and J. Negusanti. 1990. Using land reclamation practices to improve tree condition in the Sudbury smelting area, Ontario, Canada, pp. 307–320. *In* J. Skousen et al. (eds.). Proceedings of the 1990 Mining and Reclamation Conference and Exhibition. West Virginia University, Morgantown, WV.

Bradshaw, A.D. 1983. The restoration of ecosystems. J. Appl. Ecol. 20:1–17.

Bridges, E.M. 1991. Reclaiming contaminated land in the city of Swansea. Agric. Eng. 46:115–117.

Cairns, J., Jr. (ed.). 1988. Rehabilitating Damaged Ecosystems. Vol. 1. CRC Press, Boca Raton, FL.

Daniels, W.L., and C.E. Zipper. 1988. Improving coal surface mine reclamation in the central Appalachian region, pp. 139–162. *In* J. Cairns, Jr. (ed.). Rehabilitating Damaged Ecosystems. Vol. 1. CRC Press, Boca Raton, FL.

Down, C. R., and J. Stocks. 1977. Environmental Impact of Mining. John Wiley and Sons, New York.

Freedman, B. 1989. Environmental Ecology: The Impacts of Pollution and Other Stresses on Ecosystem Structure and Function. Academic Press, San Diego.

Gubala C.P., and C.T. Driscoll. 1991. Watershed liming as a strategy to mitigate acidic deposition in the Adirondack region of New York, pp. 145–159. *In* H. Olem, R.H. Schreiber, R.W. Brocksen, and D.B. Porcella (eds.). International Lake and Watershed Liming Practices. Terrene Institute Inc., Washington, DC.

Hossner, L.R. 1988. Reclamation of Surface Mined-Lands. CRC Press Inc., Boca Raton, FL.

Howell G., and T.R.K. Dalziel. 1992. Restoring Acid Water: Loch Fleet 1984–1990. Elsevier Applied Science, Essex, England.

Jee, K.K. 1988. Environmental improvement in Singapore. Ambio 17:233–237.

Jordan, W.R., M.E. Gilpin, and J.A. Aber (eds.). 1987. Restoration Ecology: A Synthetic Approach to Ecological Research. Cambridge University Press, Cambridge.

Lautenbach, W.E. 1985. Land Reclamation Program 1978–1984. Regional Municipality of Sudbury, Sudbury, Ontario.

Lautenbach, W.E. 1987. The greening of Sudbury. J. Soil Water Conserv. 42(4):228–231.

Leuts, J., and W.J. Kelly. 1993. Cleaning the air of Los Angeles. Sci. Am. 269:32–39.

Nishimura, H. (ed.). 1989. How to Conquer Air Pollution—A Japanese Experience. Elsevier, Amsterdam.

Olem, H., R.H. Schreiber, R.W. Brocksen, and D.B. Porcella. 1991. International Lake and Watershed Liming Practices. Terrene Institute Inc., Washington, DC.

Peters, T.H. 1984. Rehabilitation of mine tailings: a case of complete reconstruction and revegetation of industrially stressed lands in the Sudbury area, Ontario, Canada, pp. 403–421. *In* P.J. Sheehan et al. (eds.). Effects of Pollutants at the Ecosystem Level. Wiley, New York.

Skraba, D. 1989. Effects of surface liming of soils on stream flow chemistry in denuded acid, metals contaminated watershed near Sudbury, Ontario. M.Sc. thesis, Laurentian University, Sudbury, Ontario.

Watson, W.Y., and D.H.S. Richardson. 1972. Appreciating the potential of a devastated land. Forestry Chron. Dec.:313–315.

Weston, R.L., P.D. Gadil, B.R. Salter, and G.T. Goodman. 1965. Problems of revegetation in the lower Swansea Valley, an area of extensive industrial dereliction, pp. 297–325. *In* G.T. Goodman, R.W. Edwards, and J.M. Lambert (eds.). *In* Ecology and the Industrial Society. Blackwell, Oxford.

Whitby, L.M., P.M. Stokes, T.C. Hutchinson, and G. Myslik. 1976. Ecological consequences of acidic and heavy-metal discharges from the Sudbury smelters. Can. Mineral. 14:47–57.

Winterhalder, K. 1983. The use of manual surface seeding, liming, and fertilization in the reclamation of acid metal contaminated land in the Sudbury, Ontario mining and smelting region of Canada. Environ. Technol. Lett. 4: 209–216.

Winterhalder, K. 1984. Environmental degradation and rehabilitation in the Sudbury area. Laurentian Univ. Rev. 16(2):15–47.

Young, J.E. 1992. Mining and the earth, pp. 99–118. *In* L.R. Brown et al. (eds.). State of the World 1992. W.W. Norton, New York.

9

Revegetation of the Copper Cliff Tailings Area

Tom H. Peters

Mining and processing of metal-bearing ores produces vast quantities of hazardous waste material such as tailings, waste rock, slag, and flue dust (Freedman 1989). Of these, tailings (see Plate 16, following page 182), the materials that are discarded after the ore is separated by milling and flotation are probably the largest primary source of contamination associated with metal extraction (Salomons and Forstner 1988). Tailings often contain high concentrations of acid-generating sulfides and quantities of residual metals.

Tailings have produced serious disposal problems throughout the world. In some cases, tailings have been simply dumped as loose piles or even into rivers, such as at the Bougainville site in Papua, New Guinea (540 million tonnes in the Kawerong River) (Moore and Luoma 1990), or into the sea, such as at the copper smelter in Ilo, Peru (40 million m^3/year) (Young 1992). This irresponsible dumping has destroyed aquatic life over large areas. Fortunately, such practices are now relatively rare. However, even the more conventional land-based disposal of tailings in ponds creates substantial environmental effects through the production and escape of acid drainage water (containing high levels of dissolved metals). One of the largest hazardous waste sites in the United States is the Clark Forks complex, a copper mining area in Montana where tailings are stored in ponds that cover at least 35 km^2 and contain 200 million m^3 of waste. Seepage from the tailings ponds and airborne release of flue and surface dusts significantly increase metal concentrations in the Clark Fork River as far as 500 km from the site (Moore and Luoma 1990).

In Sudbury, the mining industries produce about 8–10 million tonnes of tailings per year. Depending on the operation, from 25–75% of the tailings are used to fill mined-out areas underground. However, because of the expansion of the ore due to blasting and grinding, not all the resulting tailings can be put back underground. A large amount, about 450 million tonnes, is therefore stored in ponds on the surface. At Copper Cliff alone, the tailings storage area covers 2225 ha (Fig. 9.1). In addition to the acid mine drainage problem (see Chapters 10 and 21), surface storage of tailings can produce serious dust problems once the ponds are filled and the material begins to dry out.

The need to control dust emissions from tailings areas has been appreciated by Sudbury operators for some time. In the past, temporary crops of fall rye have been seeded on the dry tailings, and various chemical sprays have been applied to bond the surface particles. However, in the early spring and late fall, weather conditions often produced a "freeze-dry" situation. This usually occurred when the surface of the tailings froze at night and then thawed rapidly during sunny days. The moisture released by thawing evaporated and was

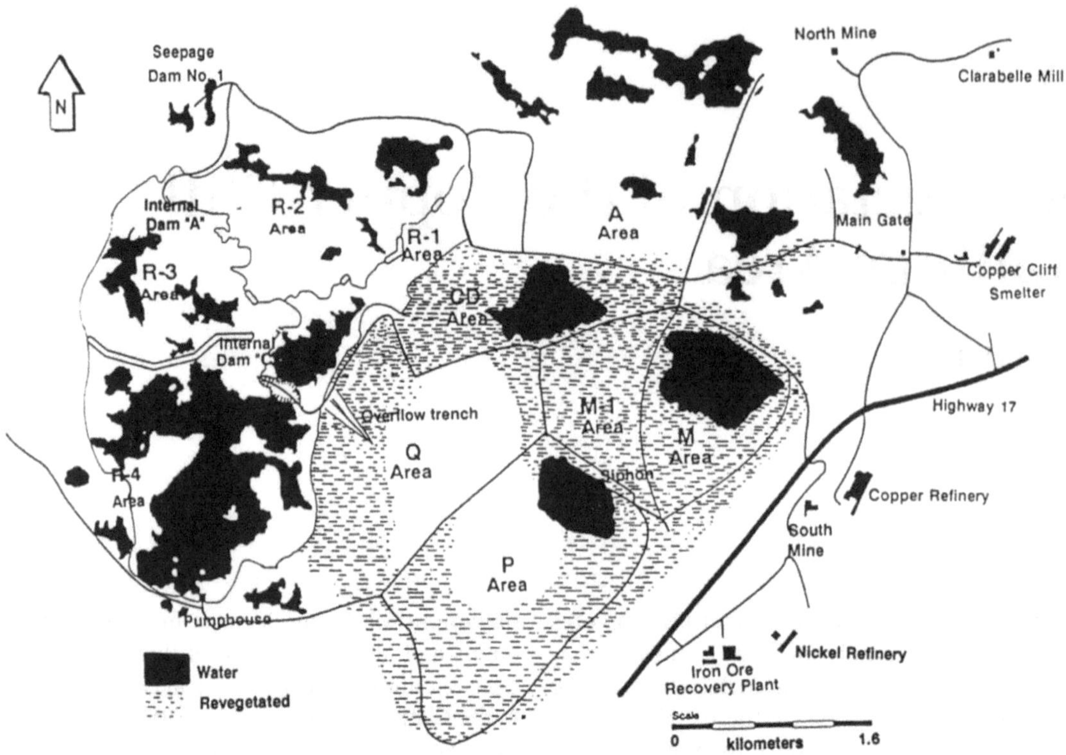

FIGURE 9.1. Diagram of Inco Limited's Copper Cliff tailings area.

not replaced by capillary action, leaving the particles dry and without cohesive binding. With winds of sufficient velocity, these particles readily became airborne.

A review of progress in preventing this dust problem by establishing a vegetation cover at the Copper Cliff tailings area is the topic of this chapter.

History of the Problem

By the late 1940s, the level of tailings in Inco's existing disposal areas at Copper Cliff were equal to or higher than the rocky hills of the local topography. Without shelter, the fine particles of the tailings frequently became airborne (Fig. 9.2). This blowing dust became a nuisance to local residents and adversely affected local industrial operations. For example, the dust contaminated lubricants in equipment and was an impurity that affected the product quality during electrolytic refining of copper.

Development of a Technique for Establishing a Vegetation Cover

It is not surprising that many of the early efforts (in the 1940s and 1950s) at establishing a vegetation cover met with failure. Fresh tailings had no nutritive value, limited water holding capacity, no soil structure, a low pH, and no organic matter (Crowder et al. 1982). However, eventually, from what often seemed like a trial-and-error process, some basic principles and procedures developed that made revegetation possible at the Copper Cliff site. These developments grew mainly from an experimental program begun in 1957 to test seed bed amelioration procedures in large plots. From these studies and earlier experience, it was recognized that the deficiencies of tailings as a growth medium could be overcome if the following principles and procedures were adopted:

FIGURE 9.2. Before stabilization, the surface of Inco's Copper Cliff tailings area was susceptible to wind erosion. (Photo by Tom Peters.)

1. Seeding should begin in the area closest to the source of the prevailing winds to minimize the covering or damaging of young plants by drifting tailings.
2. With sulfide ore tailings, sufficient agricultural limestone should be added to raise the pH of the seed bed strata to approximately 4.5–5.5. Half the limestone should be applied at least 6 weeks before seeding and the balance at the time of seeding.
3. The addition of adequate amounts of nitrogen, phosphate, and potassium fertilizers, to ensure plant establishment, should be made at the time of seeding. The amount of fertilizer should be sufficient to compensate for the loss of nutrient availability due to complexing with other elements (aluminum, iron, calcium, etc.) present in the tailings.
4. In a climate similar to Sudbury, grasses should be planted in the late summer to take advantage of cooler temperatures and available moisture.
5. Legumes should be seeded in the spring, with a power-till seeder, 1 or 2 years after the initial grasses have become established. A full season of growth, along with soil stabilization by grass roots, will provide protection for the legumes against the heaving effects of repeated surface freezing and thawing.
6. A companion crop (fall rye, *Secale cereale*) should be used to provide a quick protective canopy. This crop will reduce surface wind velocity and provide shade for the slower-growing grass seedlings. The companion crop also acts as a seed trap for wind-dispersed seeds from adjacent naturally vegetated areas.
7. Due to the lack of organic matter in the tailings and the subsequent rapid loss of nitrogen from applied commercial fertilizers, frequent small applications of a nitrogenous fertilizer should be made during seedling establishment. The use of a slow-release nitrogen fertilizer is another option, but its use is limited by the season and weather conditions.
8. Mulching (wood fiber, straw, shredded paper mixtures, etc.) should be used, especially when seeding wind-exposed areas or slopes with a south to southwesterly exposure. In areas exposed to the wind, a tackifier (asphalt emulsions, emulsion products from paint, adhesive, and forestry industries, etc.) should be added to the mixture being hydroseeded to stabilize the mulch cover (Brooks et al. 1989).
9. If a layer of natural soil is used to cover the tailings and provide a seed bed, the change in surface porosity, particularly on

FIGURE 9.3. Spreading agricultural limestone on the tailings. (Photo by Tom Peters.)

slopes, must be considered. If the change in surface porosity is great, adequate drainage should be installed to eliminate water erosion.

While revegetation experience was being gained in Sudbury, other specialists throughout the world were also attempting to solve many of the unique challenges posed in the establishment of a vegetation cover on tailings (Marshall 1983; Dean et al. 1986; Australian Mining Industry Council 1987; Powell 1992).

Current Program

The current tailings reclamation and vegetation program at Inco Limited in Copper Cliff differs greatly from the manual land reclamation procedures described in the previous chapter. It is highly mechanized, using a variety of heavy earth-moving equipment for leveling the site and standard agricultural equipment for surface preparation, discing, liming, fertilizing, and seeding.

Once the proper surface contour is established, to ensure drainage and permit the safe operation of equipment, surface treatments begin. Agricultural limestone is first spread and disced into the tailings surface (Fig. 9.3). On average, 25 tonnes of limestone per hectare are used. The seeding then begins in late July and continues until mid-September. At this time of year, moisture is readily available, and temperatures are more suitable for seed germination and seedling establishment.

Inco is now combining a straw mulch and a chemical binder on these freshly seeded areas as a surface stabilizer (Fig. 9.4). This combination of straw and chemical binders, over freshly seeded areas, has several benefits. The straw acts as a trap for seeds of local indigenous species growing on adjacent land. As it decomposes, the straw also acts as a much-needed source of organic matter in the tailings. The straw, obtained from local farmers, carries natural soil particles, which have lodged in it. This soil contains various soil microorganisms, which will, in turn, inoculate the tailings and accelerate the soil-building process.

At the time of seeding, additional agricultural limestone is spread and disced into the surface. A broadcast application of 8-24-24 (nitrogen-phosphorus-potassium) fertilizer at the rate of 740 kg/ha is spread and harrowed into the surface. A conventional farm seed drill follows. Fall rye is seeded at a rate of 60 kg/ha, as a companion crop, along with a grass seed mixture at the rate of 68 kg/ha. At the same time, this seed drill places additional fertilizer at the rate of 350 kg/ha in bands along the seed row. This is followed by a double-corrugated roller (cultipacker) grass seed-

FIGURE 9.4. Spreading straw mulch. Microorganisms in the mulch inoculate the tailings and hasten its development to a natural soil. (Photo by Tom Peters.)

er, which plants additional grass seed (22 kg/ha) and compacts the soil to provide a firm seed bed. After the initial germination, a slow-release nitrogen fertilizer is broadcast over the seedling area at a rate of 90 kg/ha. As part of the long-term maintenance program, additional limestone and fertilizer are applied, as required, based on soil tests. On areas that are not readily accessible, such as steeper outside slopes, or where conventional agricultural equipment cannot be operated safely, a hydroseeder is used for seeding and fertilizing.

The current grass seed mixture, depending on species availability, is 25% each of Canada bluegrass (*Poa compressa*) and redtop (*Agrostis gigantea*), 15% each of Kentucky bluegrass (*Poa pratensis*) and timothy (*Phleum pratense*), and 10% each of tall fescue (*Festuca arundinacea*) and creeping red fescue (*Festuca rubra*). The legume birdsfoot trefoil (*Lotus corniculatus*) is seeded the following spring to allow a full growing season for the young plants to become established (Heale 1991).

Soil Development

In a study conducted by Labine (1971), it was shown that approximately 10 years after seeding, a 2- to 3-cm organic horizon (A-zero) existed and the beginning of a podzolic profile was occurring in the treated tailings area. However, drainage at different slope levels affected soil profile development and resulted in the formation of iron pans at different depths (Peters 1988). This iron pan layer limited root penetration to cracks in the formation.

The Copper Cliff tailings, due to their deficiency of clay-sized particles, behave more like a sandy loam and are prone to moisture deficiency (Dimma 1981). Lacking colloidal moisture absorption, capillary action is the only physical means of retaining water in the tailings (Pity 1979). Maintaining sufficient levels of phosphorus is also difficult because of fixation of phosphates by the high levels of iron oxides in the tailings (Dimma 1981). However, this problem can be readily overcome by adjusting the application rate. This is one of the main reasons for recommending the high fertilizer application rate of 740 kg/ha at the Copper Cliff site.

In contrast to the success at providing sufficient phosphorus, it has been difficult to maintain sufficient nitrogen levels for good plant growth. There was no organic matter in tailings and no microorganisms involved in organic matter decay. Therefore, there was no residual processes to tie up nitrogen from the fertilizer and slowly release it. In the early stages of establishing plants on the tailings, nitrogen had to be repeatedly added as need-

FIGURE 9.5. View of a revegetated area of the Copper Cliff tailings showing the managed gradual transition from a grass and legume cover to a re-established forest cover. (Photo by Tom Peters.)

ed. In addition to the fertilizer, nitrogen-fixing legume species, such as birdsfoot trefoil (*Lotus corniculatus*), were planted in an attempt to provide a continuous source of nitrogen.

Tree Planting

In the early 1960s, white birch (*Betula papyrifera*), trembling aspen (*Populus tremuloides*), and willow (*Salix* spp.) started to invade the grassed areas of the tailings from neighboring land. Mowing of the grassed areas was stopped to encourage the development of a natural tree and shrub cover (Fig. 9.5). In the early 1970s, test plots of trees and shrubs were also established. Based on their performance, a program to plant tree species was initiated (Box 9.1). Jack pine (*Pinus banksiana*) and red pine (*Pinus resinosa*) were two coniferous species selected for the tree planting program (Figs. 9.6 and 9.7). A deciduous species, black locust (*Robinia pseudoacacia*), has also been planted. Locust is a nitrogen-fixing legume, and it has adapted well to local environmental conditions.

Ecosystem Development

At the same time as the plant community was becoming established in the tailings area, various species of insects, birds, and small mammals began to colonize the area. In 1974, after consulting with local wildlife clubs, the decision was made to develop the reclaimed tailings area as a Wildlife Management Area. It was thought that this was the most suitable and practical end-use for the tailings area.

The abundance of wildlife at the tailings area has increased considerably in recent years. The rehabilitated area now supports a diverse community of bird species (Fig. 9.8). The species changed as the vegetative cover evolved from a prairielike grassland, to a scattered tree savanna, to a forest of indigenous species. Ninety-plus avian species have been identified, including 24 shorebirds, 3 gulls, 17 waterfowl, and various meadow- and wood-habiting species (Peters 1984). Eight species were ranked as provincially significant in that they were rare or uncommon (E. Heale, *personal communication*). Many birds nest and raise their young in the area, and the Copper Cliff site serves as a stopover point during spring and fall migrations.

Environmental Concerns

One of the major recommendations of the Wildlife Management Plan was to study possible contamination of the food chain. The

Box 9.1.

Inco Limited has carried out an annual tree planting program since the early 1960s. From 1978 to 1993, more than 900,000 tree seedlings were planted. To ensure that the number of seedlings of the desired species would be available for this program, the company began its own forestry seedling crop production in 1985. The growing of the tree seedlings underground developed from a joint Laurentian University–Inco Limited research project on the possibility of using an underground area of mines as a temperature-controlled, disease- and insect-free site for food production. Since 1985, 87% of the tree seedlings that have been planted out on local stressed land have been grown by the company. Seedlings are grown underground for 16 weeks, brought to the surface, and hardened off for 2 weeks before planting in the spring. The balance of the seedlings are grown in Inco's Copper Cliff greenhouse. Current annual crop production is 250,000 seedlings.

Red and jack pine seedlings are grown underground on the 4600-ft level of Inco Limited's Creighton Mine. At that level, the air is a relatively constant 24°C due to geothermal energy. (Photo by E. Heale.)

metal content of the two principal grass species, redtop and Canada bluegrass, growing on the tailings fell within normal ranges for grasses (Rutherford and Van Loon 1980).

A 6-year study with mallard ducks (*Anas platyrhynchos*) fenced on the revegetated tailings found that metal accumulations were not a cause for concern. Meadow voles (*Microtus pennsylvanicus*) inhabiting the tailings were also found to have no toxic levels of heavy metals in their vital organs (Cloutier et al. 1986). Voles, along with deer mice (*Peromiscus maniculatus*), are important links in the food chain to red fox (*Vulpes vulpes*) and coyotes (*Canis latrans*), which inhabit the tailings area.

Other mammals that have been observed in this rapidly developing ecosystem are snowshoe hares (*Lepus americanus*), eastern chipmunks (*Tamias striatus*), red squirrels (*Tamiasciurus hudsonius*), beaver (*Castor canadensis*), muskrat (*Ondatra zibethicus*), and black bear (*Ursus americanus*) (E. Heale, *personal communication*).

Figure 9.6. Planting 14-week-old seedling conifer trees, which were grown underground in Inco Limited's Creighton Mine, on the Copper Cliff tailings area. (Photo by Tom Peters.)

Figure 9.7. View of a portion of the "CD" area of Inco Limited's Copper Cliff tailings, showing 30 years of ecosystem development. The area was revegetated with a grass and legume mixture in 1960, and the jack and red pine were planted in the early 1970s. (Photo by Tom Peters.)

Figure 9.8. A Canada goose at home on the "CD" area pond. Cattails and bullrushes were introduced into this pond in the early 1970s. (Photo by Mike Peters.)

Other Values of a Vegetated Tailings Area

The original objective of establishing vegetation on the tailings area was to stabilize the surface to control dust emissions. In recent years, the important and pressing problem of the quality of the drainage and seepage water from the tailings has become a major concern. There is an important research need for information on the hydrology of vegetated tailings areas, but at present there appear to be three ways that revegetated tailings can have an effect on the water quality problem:

1. **by intercepting precipitation**
2. **by evapotranspiration**
3. **by forming an oxygen-consuming barrier of decomposing plant residues**

The percentage of precipitation intercepted will vary with the density and the amount of the vegetative cover. Generally, dews are intercepted 100%, whereas only minimal amounts of water are intercepted during heavy rainstorms or during spring melt. In summer, the precipitation on vegetation evaporates and does not enter the ground to become seepage water. On sunny winter days, the stems of the vegetation also act as air passages in accumulated snow and thus increases the area exposed to the evaporation process.

Vegetation can remove a large amount of water through evapotranspiration and thus reduce the amount of water that infiltrates and leads to acid drainage. For example, it has been experimentally shown that to produce 454 g of dry matter, an alfalfa (*Medicago sativa*) plant transpires 364 kg water (Northen 1958). The water removed by evaporation from the surface or by evapotranspiration by the plants will not penetrate into the tailings to create acid mine drainage. However, less water may simply increase the concentration of dissolved constituents, a hydrogeochemical consideration that is part of the needed research in this area (M. Wiseman, *personal communication*).

The third means of preventing acid drainage is by blocking oxygen from reaching the sulfide material. The development of a thick root mass, along with a surface mat of decaying vegetation, may achieve this by acting as an oxygen-consuming barrier that reduces the access of atmospheric oxygen into the sulfide tailings (Peters 1988). In addition to the potential benefits of a vegetation cover, several other covers have been tested as a means of blocking oxygen and thus the production of acid drainage water (Spires 1975; Michelutti 1978). Falconbridge Limited has experimented with several different types of dry and wet covers placed on the tailings before attempting to establish vegetation. The dry covers (rock, gravel, etc.) have the benefit of preventing the upward migration of metallic salts, which can be toxic to vegetation roots. The wet covers, such as a covering of wetland vegetation, offer considerable promise as a more permanent solution to acid mine drainage by blocking oxygen and water penetration into the deeper layers of the tailings. The potential use of these wetland covers is the topic of the next chapter.

Summary

When faced with a challenging restoration job such as establishing a vegetation cover on a large sulfide tailings area, there rarely are textbook solutions that can be used (Bradshaw et al. 1978). Each site has its own unique problems and potential solutions. However, the main goal in tailings restoration is usually the same—to establish a maintenance-free cover. From our experience in Sudbury, the steps in achieving this goal were

1. **Establishment of initial plant communities using available species that are tolerant of factors characteristic of the tailings. These factors include drought, low pH, poor soil texture, and lack of organic materials and nutrients**
2. **Modification of the local microclimate to benefit plant establishment**
3. **Establishment of soil invertebrate and microbial communities. These assist in the soil building process and decompose naturally accumulating organic matter**
4. **Establishment of nutrient cycles**
5. **Establishment of a rich diverse plant community that can support wildlife and other natural assets**

While working along this path, the local practitioners must continue to monitor responses and be flexible enough to "adapt not adopt" the practices developed by others (Peters 1984, 1988).

Acknowledgments. I acknowledge the assistance in the preparation of this chapter from staff at both Inco Limited and Falconbridge Limited. Their willingness to provide information, figures, and pictures is greatly appreciated. Tony Bradshaw, John Gunn, and Marty Puro kindly provided reviews and comments.

References

Australian Mining Industry Council. 1987. Mining and the Return of the Living Environment. Canberra, Australia.

Bradshaw, A.D., R.N. Humphries, M.S. Johnson, and R.D. Roberts. 1978. The restoration of vegetation on derelict land produced by industrial activity, pp. 249–278. *In* M.W. Holgate and M.I. Woodman (eds.). The Breakdown and Restoration of Ecosystems. Plenum, New York.

Brooks, B.W., T.H. Peters, and J.E. Winch. 1989. Manual of Methods Used in the Revegetation of Reactive Sulphide Tailings Basins. CANMET, Ottawa, Canada.

Cloutier, N.R., F.V. Clulow, T.P. Lim, and N.K. Davé. 1986. Metal (Cu, Ni, Fe, Co, Zn, Pb) and Ra-226 levels in the tissues of meadow voles (*Microtus pennsylvanicus*) living on nickel and uranium mine tailings in Ontario, Canada. Environ. Pollut. 41: 295–314.

Crowder, A.A., B.E. McLaughlin, G.K. Rutherford, and G.W. Van Loon. 1982. Site factors affecting semi-natural herbaceous vegetation at Copper Cliff, Ontario. Reclamation Revegetation Res. 1: 177–193.

Dean, K.C.L., J. Froisland, and M.B. Shirts. 1986. Utilization and stabilization of mineral wastes. Bulletin 688. U.S. Department of the Interior, Salt Lake City, UT.

Dimma, D.E. 1981. The pedological nature of mine tailings near Sudbury, Ontario. M.Sc. thesis, Queen's University, Kingston, Ontario.

Freedman, B. 1989. Environmental Ecology: The Impacts of Pollution and Other Stresses on Ecosystem Structure and Function. Academic Press, San Diego.

Heale, E.L. 1991. Reclamation of tailings and stressed lands at the Sudbury, Ontario operations of Inco Limited, pp. 529–541. *In* Proceedings of the Second International Conference on the Abatement of Acidic Drainage. Vol. 1. September 16–18, 1991, Montreal, Canada. MEND, Ottawa.

Labine, C.L. 1971. The influence of certain seeded grasses on the evolution of mine tailings. B.Sc.

thesis, Biology Department, Laurentian University, Sudbury, Ontario.

Marshall, I.B. 1983. Mining, Land Use, and the Environment. II. A Review of Mine Reclamation Activities in Canada. Environment Canada, Ottawa, Canada.

Michelutti, R.E. 1978. The establishment of vegetation on high iron-sulphur tailings by means of overburden, pp. 25–30. *In* Proceedings of the Third Annual Meeting of the Canadian Land Reclamation Association, Sudbury, Ontario. CLRA, Guelph, Ontario.

Moore, J.N., and S.N. Luoma. 1990. Hazardous wastes from large-scale metal extraction. Environ. Sci. Technol. 24:1278–1285.

Northen, H.T. 1958. Introductory Plant Science. Ronald Press Co., New York.

Peters, T.H. 1984. Rehabilitation of mine tailings: a case of complete reconstruction and revegetation of industrially stressed lands in the Sudbury area, Ontario, Canada, pp. 403–421. *In* P.J. Sheehan et al. (eds.). Effects of Pollutants at the Ecosystem Level. Wiley, New York.

Peters, T.H. 1988. Mine tailings reclamation. Inco Limited's experience with the reclaiming of sulphide tailings in the Sudbury area, Ontario, Canada, pp. 152–165. *In* W. Salomons and U. Forstner (eds.). Environmental Management of Solid Wastes. Springer-Verlag, Berlin.

Pity, A.F. 1979. Geography and Soil Properties. Methuen, London.

Powell, L.J. 1992. Revegetation options, pp. 49–91. *In* L.R. Hosner (ed.). Reclamation of Surface Mined Lands. Vol. II. CRC Press, Boca Raton, FL.

Rutherford, G.K., and G.W. Van Loon. 1980. The ecological qualities of some tailings in the Sudbury rim area and the composition of vegetation associated with them, pp. 43–55. *In* Proceedings of the Fifth Annual Meeting of the Canadian Land Reclamation Association, Timmins, Ontario. CLRA, Guelph, Ontario.

Salomons, W., and U. Forstner (eds.). 1988. Chemistry and Biology of Solid Wastes. Springer-Verlag, New York.

Spires, A.C. 1975. Studies on the use of overburden soils in facilitating vegetative growth on high sulphide tailings. M.Sc. thesis, Laurentian University, Sudbury, Ontario.

Young, J.E. 1992. Mining the earth, pp. 100–118. *In* Lester Brown (ed.). State of the Environment 1992, Worldwatch Report. W.W. Norton, New York.

10

Engineered Wetlands as a Tailings Rehabilitation Strategy

Bob Michelutti and Mark Wiseman

Acid mine drainage from sulfur-bearing waste rock and tailings is one of the most serious environmental challenges facing the mining industry today (Campbell and Marshall 1991). It is caused when metal sulfides in the waste material react with water and oxygen to form sulfuric acid, which in turn dissolves residual metals such as nickel, copper, iron, lead, or zinc. This process is dramatically accelerated by bacteria. The composition of acid mine drainage water is highly variable from site to site. The pH can vary from 3 to 6, and the concentration of metals can vary by several orders of magnitude.

While existing surface run-off is presently being treated, acid mine drainage is a particularly serious long-term problem in the Sudbury area. The approximately 3000 ha of acid-generating tailing ponds in Sudbury contain nearly 0.5 billion tonnes of tailings. This is more material than was excavated in the construction of the Panama Canal. It is roughly estimated to represent 25% of all the sulfide tailings in Canada. Conventional treatment costs and financial assurances to mitigate acid mine drainage in Canada are in the billions of dollars. New treatment technologies are therefore being sought by mining companies to address this problem.

Wetlands

Wetlands are one of the new technologies being explored because of their ability to establish conditions to prevent or treat acid mine drainage. Treatment occurs when wetlands are established downstream of tailing ponds to filter and remove contaminants by biological and chemical processes. Prevention of acid mine drainage may occur when wetlands are created on the surface of tailing ponds. Wetlands form the transition zone between land and water and are often highly productive biological ecosystems.

One of the initial studies on the benefits of natural wetlands for controlling acid mine drainage was conducted at the Tub Run Bog in West Virginia (Wildeman 1991). Researchers observed that the wetland affected acid mine drainage water by increasing pH from 3.5 to 5.0, decreasing sulfate from 250 mg/L to 10 mg/L and decreasing iron from 50 mg/L to less than 2 mg/L. Many other researchers have since shown that various chemical characteristics of acid mine drainage water can be improved by passing it through natural or constructed wetlands (Wheeland 1987; Hammer 1989; Wildeman 1991; Davé 1993; Kalin 1993).

Although considerable research is required before all the chemical processes that lead to these improvements are understood, several processes by which wetlands ameliorate acid mine drainage have been identified. Perhaps the most common method of dissolved metal removal is by an ion exchange mechanism

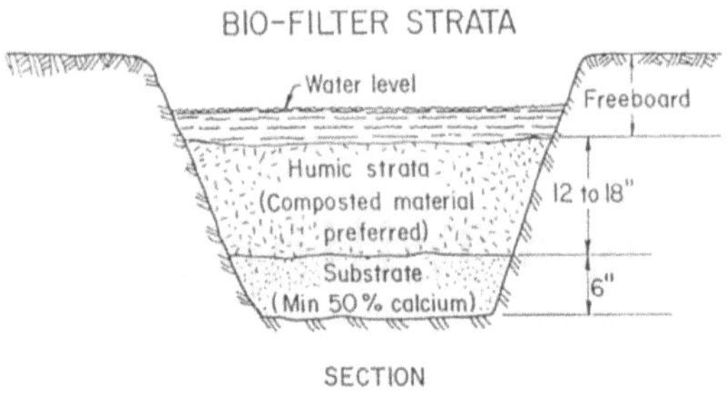

Figure 10.1. Cross section of a constructed wetland designed to promote subsurface flows. In this design the effluent is piped into the substrate below the layers of organic matter (Wildeman 1991).

with the organic material. Metals are complexed and bound to organic molecules or particles as the contaminated effluent passes through the wetland. Falconbridge Limited has studied the binding and release of these metals and has found that they remain complexed only if the conditions are anaerobic (Fyfe 1990).

Another predominant removal mechanism is by bacterially catalyzed sulfate reduction, which results in the precipitation of insoluble metal sulfide. Sulfate-reducing bacteria such as *Desulfovibrio* require low pH values and reducing conditions, which is exactly what is found in swamps or wetlands. These bacteria also require carbon as a nutrient, and this is provided in natural wetlands, by the rotting vegetation. Therefore, in an artificially created or engineered wetland, one must establish plants or incorporate organic material substrates such as compost or peat moss as the carbon source for bacteria.

These processes are simplified in the following equation:

$$\underset{\text{(sulfuric acid)}}{2H_2SO_4} + \underset{\text{(dissolved metals)}}{2M} \underset{\text{Aerobic bacteria}}{\overset{\text{sulfate-reducing bacteria}}{\rightleftharpoons}}$$

$$\underset{\text{(metal sulfides)}}{2MS} + \underset{\text{(water)}}{2H_2O} + \underset{\text{(oxygen)}}{3O_2}$$

Other less significant metal removal mechanisms that may operate in the engineered wetland include the following:

1. **precipitation of ferric and manganese ions as hydroxides**
2. **adsorption of metals by ferric hydroxides**
3. **uptake of sulfur and metals by plants such as cattails or algae**
4. **neutralization and precipitation through the generation of NH_3 and HCO_3 by bacteria**
5. **filtration of suspended solids by plants**

Natural versus Constructed Wetlands

From an environmental and societal standpoint rather than a scientific perspective, it will become increasingly difficult to use natural wetlands for pollution abatement. Wetlands are becoming an endangered ecosystem, and pressure for their protection is increasing. It has been estimated that 440,000 acres of wetlands in North America were annually altered or lost during the 1950s to the 1970s (Wheeland 1987). The answer is to construct artificial wetlands. To do this, natural systems must therefore be studied so that we can apply and perhaps even improve their acid-neutralizing processes in the constructed wetlands.

Water in a natural wetland flows mainly across the surface due to the low permeability of the organic substrata. To improve the contact time between the contaminated effluent being treated and the organic material in the wetland, water in an engineered wetland can be introduced as subsurface flow so that it flows up through the organic layers. An effective distribution and dispersal design can also be engineered to prevent short circuiting of the flow path of the water (Figs. 10.1 and 10.2).

Another limitation to natural wetlands for treatment of acid drainage is that natural wetlands tend to be weakly acidic because they generate large quantities of humic acids. The predominantly surface flow in natural wetlands results in aerobic reactions that tend to produce hydrogen ions or more acidity. Because acidity is one of the concerns that mining companies with sulfide ore bodies are trying to resolve, an improvement in the acid/base conditions of a natural wetland is therefore required. This can be accomplished in some cases by using a bed of limestone in the substratum to neutralize any acid produced (Fig. 10.1). By placing the limestone in the anaerobic substrate and forcing the water to flow up through it, the iron remains in the ferrous form. This prevents the armoring or coating of the limestone by ferric iron precipitates. Successful preliminary field testing of this approach has recently been completed in the United States (Kepler and McCleary 1994).

Constructed wetlands can be designed to operate under anaerobic (reducing) conditions that are required by sulfate-reducing bacteria. These bacteria promote the formation of hydrogen sulfide, which in turn complexes with dissolved metals to form highly insoluble metal sulfides. Some of these microbes release hydrogen gas and lead to the production of hydrogen sulfide gas (Wildeman 1991). These mechanisms increase the pH by complexing and precipitating hydrogen or releasing hydrogen ions into the atmosphere. Bacterially driven denitrification and methanogenesis also lead to hydrogen ion consumption and pH increases (Fig. 10.3).

Other reactions that facilitate the removal of iron and manganese require aerobic conditions to produce iron and manganese hydroxides. Certain bacteria, such as *Thiobacillus ferro-oxidans*, which catalyze the oxidation of iron sulfide, also require oxygen. To facilitate these reactions, a two-step process may be needed in the engineered wetland, with an aerobic system established downstream of an anaerobic system.

In conclusion, constructed wetlands can be designed to reduce the contaminants in the acid mine drainage to environmentally acceptable concentrations. In some cases, depending on the specific contaminants present in the acid mine drainage, a multistage wetland may be required that incorporates an anaerobic wetland design, followed by a system providing aerobic conditions to remove different contaminants under different conditions.

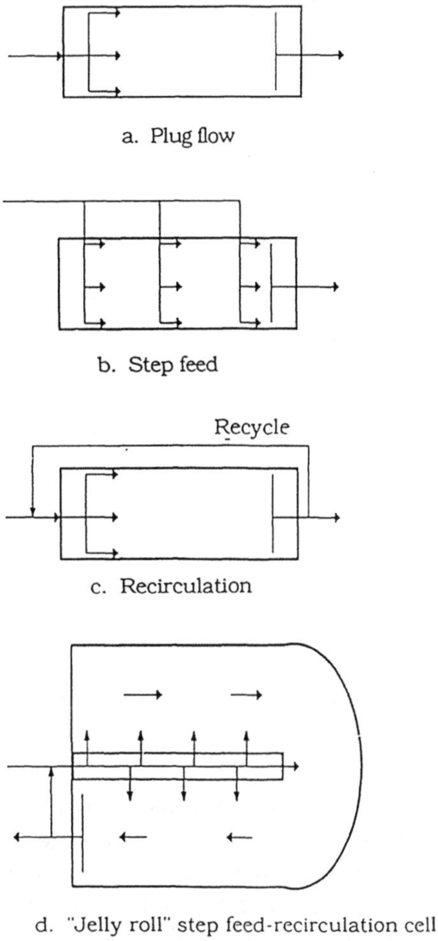

FIGURE 10.2. Various flow patterns for constructed wetlands (Wildeman 1991).

Prevention versus Remediation

Most of the work to date has involved the use of constructed wetlands as a downstream remediation system for contaminated effluent. Many examples of this approach can be found in the literature. For example, the Tennessee Valley Authority is currently operating seven

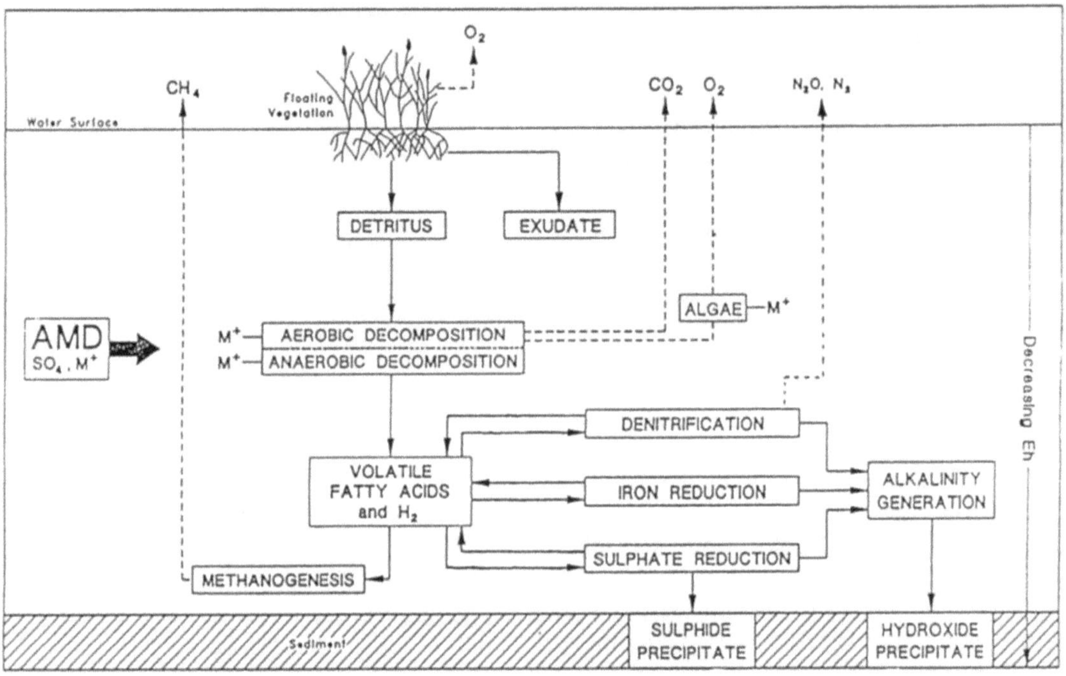

FIGURE 10.3. Schematic of floating vegetation wetland process (Kalin 1993).

constructed wetlands and has plans for eight more to treat acidic coal mine drainage. Typically, required treatment areas vary from 300–15,000 m^2 of wetland per liter of effluent per second, depending on the degree of contamination. The average constructed wetland has a treatment area of about 1000 $m^2/L/s$ (Hammer 1989).

In Idaho Springs, the Big Five Wetland was constructed in 1987 to treat acid mine drainage from metal mining operations. The wetlands were able to remove almost 100% of the copper and zinc from the contaminated seepage, 63% of the iron but virtually none of the manganese. Interestingly, the pH rose from 3.0 to 6.2, indicating that wetlands can be very effective in reducing acidity (Wildeman 1991).

An extensive field evaluation was conducted using test cells to treat Inco's tailings seepage at Makela, Sudbury. This project, sponsored by the Mine Environment Neutral Drainage (MEND) committee, also demonstrated that metals and pH could be ameliorated by the wetlands. It was discovered that the most active metal removal processes were taking place in the sediments (Kalin 1993). This led to a floating cattail design, so that the cattails suspended from rafts would not deplete the nutrients in the sediments required by the bacteria but would still provide them with a source of carbon. The Makela test cells removed 80–87% of the nickel, 77–98% of the copper, 10–20% of the sulfur, and 47–73% of the acidity from the loading water. A schematic diagram of the process is shown in Figure 10.3.

Falconbridge Limited has developed a 150-ha wetland downstream of their smelter waste water treatment system (Fig. 10.4). Nickel concentrations were observed to decrease 0.5 mg/L to 0.2 mg/L during the summer months. Preliminary investigations by Laurentian University have indicated that some of the metals are being removed by algae. Future studies will investigate the role and relative contributions of bacteria in the removal of heavy metals. The effluent from this system consistently meets the new provincial limits set under the

Figure 10.4. Wetland area below the Falconbridge smelter.

Municipal Industrial Strategy for Abatement limits and is non-toxic to fish and zooplankton (*Daphnia magna*). Various fish species such as minnows, white suckers, stickleback, and perch now inhabit the wetland.

During the 1980s, Falconbridge Limited started looking for alternative methods to grassing for tailings as an abandonment strategy. The downstream wetland approach was known, but this was considered a remediation technique, not a permanent solution. It was theorized that a more permanent solution could be achieved by covering the tailing with water. By creating an oxygen barrier by this means, the oxidation of iron sulfide could be reduced, thereby preventing acid mine drainage and metal leaching (literature reviewed by Itzkovich 1993). Falconbridge Limited started testing this approach in 1986 using a shallow water cover and aquatic plants over highly oxidized tailings. Although the results showed some improvement, especially with respect to nickel concentrations, they were not adequate for a "walk-away strategy" (Dave and Michelutti 1991). It was hypothesized by the researchers and supported by the literature that oxidation was continuing without the presence of atmospheric oxygen. Because the test wetlands were established on highly oxidized pyrrhotite, ferric ion was most likely continuing the oxidation reaction (Wheeland 1987; Wildeman 1991).

$$FeS_2 + 14Fe^{3+} + 8H_2O \rightleftarrows 15Fe^{2+} + 2SO_4^{2-} + 16H^+$$

Using isotope data, it was found that 35–80% of the dissolved sulfate is produced by a reaction not involving free (dissolved) oxygen (Wheeland 1987). Therefore, these initial experiments by Falconbridge Limited suggested that it was important to prevent the tailings from oxidizing before covering with a wetland. If the tailings are already in an oxidized state, then considerable time may be necessary to flush out the ferric ions and oxidation products before seeing any improvements. Falconbridge Limited and the University of Waterloo, with MEND support, have developed a model (FALCTAIL) to estimate the number of years required to flush latent oxidation products from a tailings area.

Another wetland that has been extensively studied is the Rio Algom Panel site in Elliot Lake. A uranium tailings spill into a small basin was subsequently recolonized naturally by cattails such that a 13-acre wetland resulted. At the west end of this wetland area, partially exposed tailings resulted in dissolved solids concentrations of 600–2000 mg/L, iron from 1 to 80 mg/L, and sulfates from 50 to 100 mg/L (Dave 1993). In the central to eastern section of the wetland, where the tailings were completely submerged, the pH was 6.2–9.8, dissolved solids were 100–300 mg/L, iron was 0.002–0.4 mg/L, and sulfates were 50–100 mg/L. Of particular interest was that there were no strong seasonal differences in water quality in the submerged areas. It was estimated that due to iron complexing and the extremely slow rate of pyrite oxidation and release, it would take 32,000 years for all the pyrite to oxidize and 926,000 years for all the mobilized iron to leave the system. At these slow rates, impacts to downstream ecosystems would be nil.

Perhaps the most ambitious wetland cover field study is that being conducted by Rio Algom at their Quirke mine tailings. In preparation for flooding, half the 64-ha site was leveled and had limestone incorporated into the top 20 cm. The leveling was performed to reduce the requirement for higher dams to flood the area, and the limestone was added to counter any acidity that may occur in the event of severe droughts that would expose the tailings. After some initial start-up problems, the area was flooded in October 1992 (Vivyurka 1993). As the flooding was taking place, lime slurry at the rate of 100 mg/L of $CaCO_3$ was added to the inflow water to counter the existing acidity. Flooding took place by diverting a portion of the discharge from a small upstream lake. By mid-December 1992, the total area was flooded to a minimum depth of 0.3 m. Subsurface water sampling (using a piezometer nest) is ongoing, and although it is too early to draw any conclusions, preliminary data indicate good water quality. To date, pH values have been all above neutral, but that would have been expected with the limestone and lime that had been initially added. This testwork is also unique in that it will determine the impact of water cover on the fate of radionuclides. A similar large-scale flooding scenario is also being evaluated at Stekenjokk, Sweden (Broman and Goransson 1994). Evaluations of historically flooded tailing are ongoing within the MEND program and also indicate the long-term stability of unoxidized tailing deposited underwater (Fraser and Robertson 1994).

Conclusions/ Recommendations

Although promising results are being obtained in the wetland research, there are still many unknowns.

1. **Are there seasonal variations in treatment capabilities of the wetland?**
2. **Will a wetland in time become totally saturated with metals and cease to be effective?**
3. **Will the process continue to work as the wetland naturally evolves into a dryland ecosystem?**
4. **What is the potential for re-release of organically bound metals?**
5. **What are the fates of metal ions in this ecosystem (i.e., is there bioaccumulation occurring with long-term impacts)?**
6. **Can a self-sustaining biological community be maintained in the constructed wetland?**

Initial studies and field evaluations indicate that wetlands could become an acceptable walk-away strategy for acid-generating sulfide wastes. Two basic approaches to wetlands are downstream remediation and pollution prevention by a water cover. Both concepts could be incorporated into one abandonment strategy. This is perhaps one of the most promising technologies to emerge as a walk-away abandonment strategy for acid-generating sulfide wastes.

Acknowledgments. We thank A. Bradshaw, D. Pearson, and J. Gunn for review comments on the manuscript.

References

Broman, P.G., and T. Göransson. 1994. Decommissioning of Tailings and Waste Rock Areas at Stekenjokk, Sweden. *In:* MEND/Canada, International Land Reclamation and Mine Drainage Conference and Third International Conference on the Abatement of Acidic Mine Drainage. Volume 1. Department of the Interior, Bureau of Mines Special Publication SP06B-94.

Campbell, D., and I. Marshall. 1991. Mining: breaking new ground. *In* The State of Canada's Environment. Government of Canada, Ottawa.

Davé, N.K. 1993. A Case History of Partially Submerged Pyritic Uranium Tailings under Water. Division Report MSL 93–32 (CR). Elliott Lake, Ontario.

Davé, N.K., and R.E. Michelutti. 1991. Field evaluation of wet cover alternatives for high sulphide tailings, pp. 61–82/91. *In* Proceedings of the Second International Conference on Abatement of Acidic Drainage, Montreal, Canada. MEND. Biblioteque Nationale du Quebec, Canada.

Fraser, W.W., and J.D. Robertson. 1994. Subaqueous disposal of reactive mine waste: an overview and update of case studies, pp. 250–269. *In* Bureau of Mines Special Publication SP 06B-94. Vol. 1. MEND/Canada, International Land Reclamation and Mine Drainage Conference and Third International Conference on the Abatement of Acidic Mine Drainage. U.S. Department of the Interior.

Fyfe, J.D. 1990. The Fate of Metal Hydroxides at the Falconbridge Treatment System. Laurentian internal company reports.

Hammer, D.A. 1989. Constructed Wetlands for Wastewater Treatment, Municipal, Industrial and Agricultural. Lewis Publishers, Chelsea, MI.

Itzkovitch, I.J. 1993. MEND 1992 Annual Report. Canmet, Ottawa, Ontario.

Kalin, M. 1993. Treatment of Acidic Seepage Using Wetland Ecology and Microbiology. DSS file 015SQ.23440–2–9217.

Kepler, D.A., and E.C. McCleary. 1994. Successive alkalinity-producing systems (SAPS) for the treatment of acidic mine drainage, pp. 185–194. *In* Bureau of Mines Special Publication SP 06B-94. Vol. 1. International Land Reclamation and Mine Drainage Conference and Third International Conference on the Abatement of Acidic Mine Drainage. U.S. Department of the Interior.

Vivyurka, A. 1993. Flooding of Existing Tailings Site—Quirke Mine. MEND Project 2.13.1.

Wheeland, K.G. 1987. Notes Regarding the Role of Wetlands in Mine Closure, Rats #4.1. Centre de Recherche Noranda, Pointe Claire, Quebec.

Wildeman, T. 1991. Handbook for Constructed Wetlands Receiving Acid Mine Drainage. Risk Reduction Engineering Laboratory, U.S. Environmental Protection Agency, Cincinnati, OH.

11

Preservation of Biodiversity: Aurora Trout

Ed J. Snucins, John M. Gunn, and W. (Bill) Keller

> . . . the worst thing that will probably happen . . . is not energy depletion, economic collapse, conventional war, or even the expansion of totalitarian governments. As terrible as these catastrophes would be for us, they can be repaired within a few generations. The one process now going on that will take millions of years to correct is the loss of genetic and species diversity by the destruction of natural habitats. This is the folly our descendants are least likely to forgive us.
>
> *(Wilson 1984)*

The habitat alteration and destruction caused by Sudbury's metal extraction and smelting industries have contributed to the global depletion of biological resources (Box 11.1). Damage to local terrestrial vegetation and soils, described in Chapter 2, was striking. Less apparent but more widespread was the damage to aquatic ecosystems. Acidification of lakes from atmospheric deposition of smelter emissions occurred over an area of 17,000 km^2 and affected lakes as far as 120 km from the city (Neary et al. 1990). An estimated 134 gamefish populations, as well as many populations of less well-studied fish species were extirpated (Matuszek et al. 1992). The loss of these populations did not endanger entire species, but it did contribute to the loss of unique genetic strains. The losses are part of an alarming global trend to decreasing fish diversity. By region, the percentages of fish species classified as endangered, threatened, or in need of special protection are as follows: South Africa, 63%; Europe, 42%; Sri Lanka, 28%; North America, 31%; Australia, 26%; Iran, 22%; Latin America, 9% (Moyle and Leidy 1992). Within-species genetic diversity is also declining as fish are extirpated from individual lakes and rivers that comprise portions of their native range (Nehlson et al. 1991; Kaufman 1992).

Of the many populations threatened by acidification of lakes in the Sudbury area, only the aurora trout (see Plates 12 and 15, following page 182), a rare strain of brook trout (*Salvelinus fontinalis*), was the subject of extraordinary preservation efforts. It was extirpated from its native habitat and in 1987 was placed on Canada's endangered species list (Table 11.1). This chapter presents the story of the aurora trout restoration program, a combination of personal and agency commitment and perseverance, that saved the fish from extinction and ultimately restored it to its native habitat.

Box 11.1. Global Loss of Biological Diversity

Biological extinction is not a new phenomenon. In fact, it is estimated that more than 90% of all the species that ever existed on earth are now extinct (Simpson 1952). The fossil record indicates that there have been five main periods of mass extinction during the past 600 million years. Although there is some debate over the causes of these extinctions, most authorities seem to agree that each was triggered by a natural catastrophic event in the environment, such as sudden climatic change, drop in sea level, or meteorite impact (Raup 1986).

Averaged over the entire span of life on earth, the rate of extinction amounts to about one species per year. But our current trend far exceeds this rate. Some scientists believe that, on average, several species are disappearing each day, and they estimate that if present trends continue, more than one-quarter of the earth's biodiversity, estimated to be between 3 and 30 million species (May 1990), will be lost in the next 20–30 years (McNeely et al. 1990). The current period of mass extinction is particularly worrisome in that it is caused largely by human activity. Habitat alteration and destruction, chemical pollution, overharvesting, and the introduction of exotic species that displace or eliminate native biota are among the factors contributing to the modern depletion of biodiversity.

The current accelerated loss of biological diversity should concern us. From an ethical standpoint, some people argue that every species has an inherent right to exist independent of its material benefit to humans. There are also many human-centered utilitarian reasons for preserving biodiversity, not the least of which is that our survival depends both directly and indirectly on diversity at all levels of biological organization. Ecosystems with their variety of habitats and communities provide

Box 11.1. (continued).

essential ecological services such as the maintenance of air and water quality, soil formation and protection, climate control, and nutrient cycling. The harvesting of natural resources supplies us with food, clothing, and shelter. Wild plants supply the genetic material for selective breeding of domestic crops to increase yields and enhance pest and disease resistance. Many pharmaceuticals, too, are derived from plants. Within-species genetic diversity provides the many varieties of a species that are each suited to different environmental conditions.

Perhaps the greatest long-term benefit of biodiversity is the supply of the raw materials that enable humans and nature to respond to changing environments and stresses. Our welfare will largely be determined by how we respond to the current period of accelerated biodiversity loss in which this storehouse of potential solutions to current and future problems is quickly becoming depleted.

Description of the Aurora Trout

The aurora trout and its mother species, the brook trout, are both multihued and spectacularly beautiful, although different in the details of their coloration. Brook trout have a dorsal background color of olive green to dark brown, which is mottled by yellow spots and vermiculations (Fig. 11.1). Along the sides, this coloration pales to a snow-white abdomen that is often tinged with pink. Many red spots surrounded by pale blue halos speckle the sides. Pectoral, pelvic, and anal fins have a leading white edge backed by a black bar and orange or red posterior.

In contrast, the aurora trout's dorsal coloration fades along the sides to iridescent steel blue and silver, colors mimicking the shimmering brilliance of the fish's namesake, the aurora borealis, or northern lights. Adult aurora trout do not possess the yellow spots and vermiculations of the brook trout, and there are few, if any, red spots (Figs. 11.2 and 11.3). The coloration of the males intensifies during spawning. The sides and upper abdomen take on a vivid red color, often accented with a band of midnight black along the abdomen.

History of the Aurora Trout

The native range of the aurora trout consists of two small waterbodies: Whirligig Lake (11-ha surface area) and Whitepine Lake (77-ha surface area), located 110 km north of Sudbury (Fig. 11.4). Each is part of a chain of lakes situated on a ridge in an isolated part of Lady Evelyn Smoothwater Wilderness Park. The surrounding terrain is hilly and rough, topography typical of the Precambrian Shield, and access is gained by canoe or aircraft only. Historically, Whitepine Lake also contained a population of brook trout, and both lakes supported white sucker (*Catastomus commersoni*) populations.

The aurora trout likely evolved from a population of brook trout isolated some time after continental glaciers receded, about 10,000 years ago. The brook trout were probably trapped as water levels dropped and the land slowly rebounded upward after being freed of the weight of the glaciers. This strain of fish subsequently evolved in isolation and diverged sufficiently to become distinct from other brook trout.

In 1923, a party of anglers from the United States visiting Whitepine Lake caught some of these fish and took one back to the Carnegie Museum in Pittsburgh. The following year, more specimens were collected, and in 1925, a description of the fish was published in the scientific literature. Subsequently, the lakes were often visited by anglers willing to undertake the 4-day journey by canoe and trail to catch this trophy sportfish renowned for its spectacular coloration and superior fighting ability.

TABLE 11.1. 1993 Canadian Species at Risk

Mammals

Extinct
- Dawson Caribou
- Sea Mink

Extirpated
- Atlantic Walrus (*N.W. Atlantic Pop.*)
- Black-footed Ferret
- Gray Whale (*Atlantic Pop.*)
- Grizzly Bear (*Plains Pop.*)
- Swift Fox

Endangered
- Beluga Whale (*S.E. Baffin Is. Pop., St. Lawrence R. Pop., and Ungava Bay Pop.*)
- Bowhead Whale
- Eastern Cougar
- Peary Caribou (*High Arctic Pop. and Banks Is. Pop.*)
- Right Whale
- Sea Otter
- Vancouver Is. Marmot
- Wolverine (*Eastern Pop. Quebec/Labrador*)

Threatened
- Beluga Whale (*Eastern Hudson Bay Pop.*)
- Harbour Porpoise (*Western Atlantic Pop.*)
- Humpback Whale (*North Pacific Pop.*)
- Peary Caribou (*Low Arctic Pops.*)
- Pine Marten (*Nfld. Pop.*)
- Wood Bison
- Woodland Caribou (*Maritime Pop.*)

Birds

Extinct
- Great Auk
- Labrador Duck
- Passenger Pigeon

Extirpated
- Greater Prairie-Chicken

Endangered
- Eskimo Curlew
- Harlequin Duck (*Eastern Pop.*)
- Henslow's Sparrow
- Kirtland's Warbler
- Loggerhead Shrike (*Eastern Pop.*)
- Mountain Plover
- Peregrine Falcon (subspecies anatum)
- Piping Plover
- Sage Thrasher
- Spotted Owl
- Whooping Crane

Threatened
- Baird's Sparrow
- Burrowing Owl
- Ferruginous Hawk
- Loggerhead Shrike (*Western Pop.*)
- Marbled Murrelet
- Roseate Tern
- White-headed Woodpecker

Fish

Extinct
- Blue Walleye
- Deepwater Cisco
- Longjaw Cisco
- Longnose Dace (*Banff Pop.*)

Extirpated
- Gravel Chub
- Paddlefish

Endangered
- Acadian Whitefish
- Aurora Trout
- Salish Sucker

Threatened
- Black Redhorse
- Blackfin Cisco
- Channel Darter
- Copper Redhorse
- Deepwater Sculpin (*Great Lakes Pop.*)
- Enos Lake Stickleback
- Lake Whitefish (*Lk. Simcoe Pop.*)
- Margined Madtom
- Shorthead Sculpin
- Shortjaw Cisco
- Shortnose Cisco

Reptiles & Amphibians

Extirpated
- Pygmy Short-horned Lizard

Endangered
- Blanchard's Cricket Frog
- Blue Racer
- Lake Erie Watersnake
- Leatherback Turtle

Threatened
- Blanding's Turtle (*Nova Scotia Pop.*)
- Eastern Massasauga
- Spiny Softshell Turtle (*Eastern Pop.*)

Plants

Extirpated
- Blue-eyed Mary
- Illinois Tick Trefoil

Endangered
- Cucumber Tree
- Engelmann's Quillwort
- Furbish's Lousewort
- Gattinger's Agalinis
- Heart-leaved Plantain
- Hoary Mountain Mint
- Large Whorled Pogonia
- Mountain Avens (*Eastern Pop.*)
- Pink Coreopsis
- Pink Milkwort
- Prickly Pear Cactus (*Eastern Pop.*)
- Skinner's Agalinis
- Slender Bush Clover
- Slender Mouse-ear Cress
- Small White Lady's Slipper
- Small Whorled Pogonia
- Southern Maidenhair Fern
- Spotted Wintergreen
- Thread-leaved Sundew
- Water-pennywort
- Western Prairie Fringed Orchid
- White Prairie Gentian
- Wood Poppy

Threatened
- American Chestnut
- American Ginseng
- American Water-willow
- Anticosti Aster
- Athabasca Thrift
- Bird's Foot Violet
- Blue Ash
- Bluehearts
- Colicroot
- Giant Helleborine
- Golden Crest
- Golden Seal
- Kentucky Coffee Tree
- Mosquito Fern
- Nodding Pogonia
- Pitcher's Thistle
- Plymouth Gentian
- Purple Twayblade
- Red Mulbery
- Sand Verbena
- Small-flowered Lipocarpha
- Sweet Pepperbush
- Tyrell's Willow
- Western Blue Flag
- Western Spiderwort

Species: Any indigenous species, subspecies, or geographically separate population.
Extinct: A species indigenous to Canada that is no longer known to exist anywhere.
Extirpated: A species no longer existing in the wild in Canada but occurring elsewhere.
Endangered: A species threatened with imminent extirpation or extinction throughout all or a significant part of its Canadian range.
Threatened: A species likely to become endangered in Canada if the factors affecting its vulnerability are not reversed.
Status determinations included in this list are determined by the Committee on the Status of Endangered Wildlife in Canada (COSEWIC).

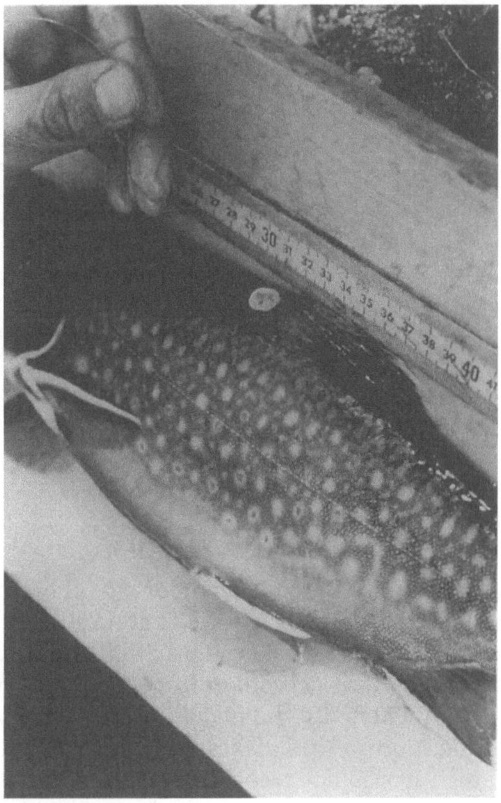

Figure 11.1. Male brook trout. The differences in coloration between the male brook trout (spotted sides) and the male aurora trout (Fig. 11.2) prompted early classification of the aurora trout as a separate species. It is now believed to be a rare strain or race of brook trout. (Photo by V. Liimatainen.)

Figure 11.2. Male aurora trout. (Photo by E. Snucins.)

The classification of the aurora trout has been a source of controversy. It was originally classified as a distinct species (*Salvelinus timagamiensis*) (Henn and Rinkenbach 1925) until a closer affiliation with the brook trout subsequently found favor. On the basis of differences in behavior, coloration, and other characteristics, a subspecies classification was proposed (Sale 1967; Qadri 1968; Parker and Brousseau 1988). Recent work suggests that the genetic differences are not sufficient to justify subspecies status and that the aurora

FIGURE 11.3. Female aurora trout. (Photo by E. Snucins.)

trout is simply a unique strain or race of brook trout (McGlade 1981; Grewe et al. 1990).

The aurora trout lakes lie within the area affected by acid deposition from the Sudbury metal smelters. By the middle of the century, acidification of these lakes was occurring, although it was not recognized at the time. In 1951, the Ontario government began to monitor the aurora trout populations. Angling was no longer permitted on the lakes, but by the late 1950s, the populations had noticeably declined, and by 1967, the aurora trout had disappeared from its home range. The demise of these populations coincided with the acidification of the lakes to near pH 5.0 (Keller 1978), the threshold for brook trout survival (Beggs and Gunn 1986).

Fortunately, before the aurora trout completely disappeared, fertilized eggs were collected from both native lakes, and a hatchery brood stock was established (Fig. 11.5). The work of Paul Graf and colleagues at Hills Lake Provincial Hatchery, their efforts at spawn collection, and the discovery in 1958 of a successful artificial rearing method saved the aurora trout from extinction. The lineage of all aurora

FIGURE 11.4. Aerial view showing the rugged Precambrian Shield terrain surrounding the home range of the aurora trout. This rare strain of brook trout likely developed when Whirligig and Whitepine lakes were isolated from surrounding waterbodies after the last continental glaciers retreated. (Photo by E. Snucins.)

FIGURE 11.5. When it was recognized that the aurora trout populations were declining, fisheries managers collected spawn to establish and maintain a hatchery stock. The aurora trout was subsequently extirpated from its home range until water quality improvements made it possible to re-introduce the fish to Whirligig and Whitepine lakes. (Photo by Ontario Ministry of Natural Resources.)

trout in existence today can be traced to the 1958 spawn collection, when 3644 eggs were collected from one Whitepine Lake and two Whirligig Lake females. The eggs from each female were mixed with the sperm from two males. Thus, the founding population size was only nine individuals and may have been as few as six if all males did not contribute to fertilization. The stock has been artificially maintained in the hatchery ever since.

By the late 1980s, the prospects of maintaining the captive aurora trout population in the hatchery became worrisome. Concern arose over the potentially deleterious effects of generations of domestication on the fitness of the stock (Franklin 1980; Hynes et al. 1981; Lacy 1992). Because selection pressures in the hatchery differ from those in the wild, the acquisition of characteristics that promote success in the hatchery can occur at the expense of other characteristics that are required for survival and reproductive success in the wild. The small founding population size of six to nine individuals may have limited the assortment of genes within the captive population, and genetic diversity could have been further reduced after a few generations in the hatchery. Also, genetic drift caused by nonrepresentative sampling within a small gene pool can alter gene frequencies, and in some species, inbreeding can result in lower viability and reduced fecundity.

Rehabilitation of the Native Lakes

Given the failure of the many attempts made since the 1950s to establish reproducing populations of aurora trout in non-native lakes, fisheries managers in Ontario decided that the best chance of success would be to return these fish to their native lakes. However, the water quality in the native lakes was still too acidic to allow for the survival of the aurora trout. Therefore, both Whirligig Lake (pH 4.8) and its headwater Little Whitepine Lake (pH 5.6) were treated with 21 tonnes of powdered limestone in October 1989; this increased the pH of both lakes to 6.5 (Fig. 11.6). In May 1990, 950 aurora trout hatchery brood stock (aged 2–5 years) were introduced into Whirligig Lake.

During late October of the same year, biologists assessed spawning behavior of introduced fish in the limed lake. A group of about 40 fish was observed congregated at a near-shore groundwater upwelling site. The fish were sexually mature and in good condition, having experienced a threefold increase in weight during the 5 months that they had resided in the lake. However, no spawning was observed and a search for young fish during the spring of 1991 was unsuccessful. It seemed no reproduction had occurred.

FIGURE 11.6. Ontario government helicopter slings powdered limestone into a northeastern Ontario lake. Lime treatment was used to raise the pH levels of lakes in the home range of the aurora trout so that the extirpated aurora trout population could be re-established in the wild. (Photo by E. Snucins.)

Biologists believed that the failure to spawn may have been due to a low water table in 1990 and the consequent absence of high-quality groundwater upwelling sites, the typical spawning area of brook trout. Therefore, in 1991, two artificial upwellings were constructed at the location where the fish had congregated the previous October. Water from a small inlet stream was piped to two wooden boxes, each filled with limestone and granitic gravel, and lined with perforated pipe along the bottom. Water percolated up through the gravel, simulating a groundwater upwelling. During October, 11 adults were captured and injected with salmon pituitary extract to induce maturation. Again, despite these extraordinary efforts, no spawning was observed.

The results were disappointing to those working on the restoration project, and doubts arose over the reproductive ability of the introduced fish. Perhaps after many generations in the hatchery, the aurora trout was no longer able to reproduce in the wild. Much time, effort, and money had been spent, possibly to no avail. But worries quickly vanished the following year when two young aurora trout were observed by a diver swimming along the shoreline (Fig. 11.7).

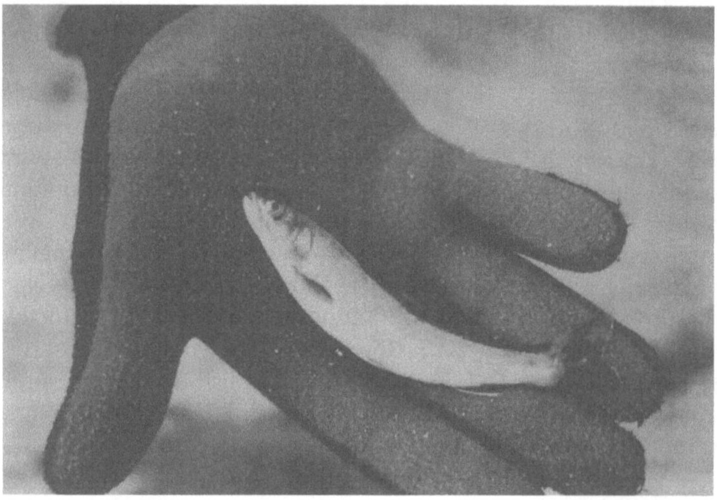

FIGURE 11.7. After nearly 2 years of searching, biologists discovered the offspring of aurora trout in Whirligig Lake in 1992, positive proof that the re-introduced species can reproduce in the wild. (Photo by E. Snucins.)

FIGURE 11.8. Changes in the pH of Whirligig Lake before and after treatment with limestone.

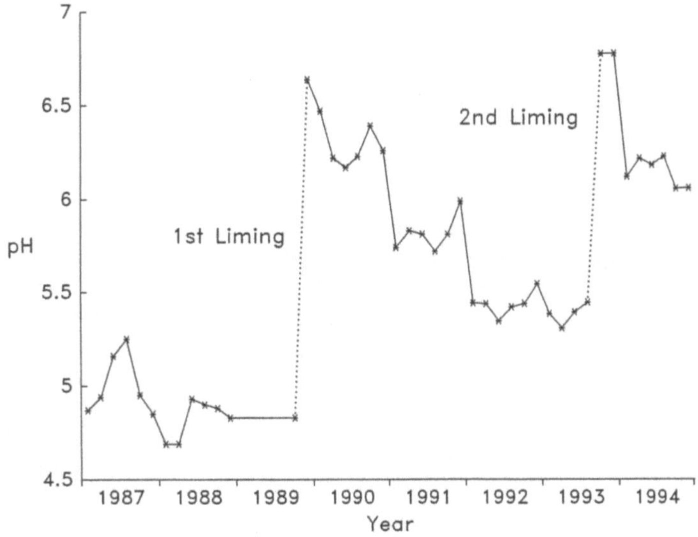

Two near-shore nests, or redds, were found in 1992, but it was not until 1993 that the primary spawning sites were discovered. Most redds were constructed at depths of 3–4 m, and the lake's tea-colored water had hidden this deep spawning habitat from surface observation. The age distribution of juvenile fish in the population indicated that successful spawning had occurred every year since the fish were re-introduced.

The discovery that the aurora trout was still capable of reproducing in the wild was very encouraging, but the realization that the lake was re-acidifying and would soon be too acidic for fish survival was soon to follow. During 1992, the pH of Whirligig Lake fell to 5.4 (Fig. 11.8). Much of the acidic input seemed to be coming from a nearby wetland. During September 1992, 32 tonnes of agricultural limestone was applied to the wetland in an attempt to improve the lake's water quality, but this treatment was not immediately effective and it was necessary to lime the lake itself in September 1993. This succeeded in raising the lake pH to 6.8.

Because Whitepine Lake, the second native lake of the aurora trout, receives the runoff from Whirligig Lake, its water may also have benefited from the liming. The pH of Whitepine Lake rose from 4.9 in 1990 to 5.1 in 1993. In response to these improved conditions, aurora trout were re-introduced to Whitepine Lake in the spring of 1994 as the next step in the restoration plan.

Summary

Our observations of good reproduction indicate that the return of the aurora trout to its native waters has, at least initially, been successful, but the long-term viability of the re-introduced population remains unknown. It is possible that the fitness of the stock for life in the wild has been reduced through inadvertent selection in the hatchery or because genes critical for survival were lost during the genetic bottleneck. However, if the original genetic material was not critically altered, the prospects for long-term persistence of the population in Whirligig Lake are probably good.

The continued survival of the aurora trout in Whirligig Lake depends on maintaining good water quality. Metal smelter emission reductions that began in 1994 will help reduce acid loading and may prolong the time to re-acidification. Water quality monitoring will need to continue, and if current pollution controls are not sufficient to prevent re-acidification, the lake will need to be relimed.

The genetic diversity represented by individual fish stocks uniquely adapted to local condi-

Box 11.2. Convention on Biological Diversity

The Convention on Biological Diversity was signed at the United Nations Conference on Environment and Development in Rio de Janeiro, Brazil (June 3–14, 1992). Under the convention, steps will be taken to protect endangered species and their habitats. All signatory nations agreed to adopt regulations to conserve biological resources, including establishing a system of protected areas. Most of the world's biological diversity exists in the tropical developing countries. To promote conservation in these economically impoverished areas, the developed countries will help supply the required financial resources and technical expertise. Also, to more equitably distribute the financial benefits of genetic resources and thus encourage their preservation, the convention promotes biological diversity as a financial asset that can generate income for the country of origin, in particular the local indigenous peoples.

tions is an important resource for environmental restoration and management. The initial success of the aurora trout restoration program demonstrates that it may be possible to preserve at least some locally adapted gene pools despite alteration of their natural habitat. However, such efforts are costly and not every endangered stock or species can receive such intensive care. Although some success was achieved in saving the aurora trout, the main lesson learned was that habitat protection pays far greater dividends than single-species restoration efforts. The preservation of natural ecosystems with all their species is by far the most effective means of conserving biodiversity (Box 11.2), and it would be dangerous to focus on preserving individual stocks or species without addressing the root causes (habitat alteration by smelter emissions in the case of the aurora trout) that threaten their existence.

Acknowledgments. Many individuals are responsible for the success of the aurora trout restoration program. The stock was saved from certain extinction by the foresight and initiative of the Hills Lake Hatchery staff and their associates in the Department of Lands and Forests. Since 1983, a management committee, composed of Ministry of Natural Resources biologists and hatchery personnel, has directed rehabilitation efforts. The restoration work is financed and conducted by the Ontario Ministry of Natural Resources. Between 1992 and 1994, additional funding was obtained from the Endangered Species Recovery Fund, cosponsored by World Wildlife Fund, Canada and the Canadian Wildlife Service of Environment Canada. Water quality assessments, field support, and additional funding are provided by the Ontario Ministry of the Environment and Energy, Sudbury.

References

Beggs, G.L., and J.M. Gunn. 1986. Response of lake trout (*Salvelinus namaycush*) and brook trout (*Salvelinus fontinalis*) to surface water acidification in Ontario. Water Air Soil Pollut. 30:711–718.

Franklin, I.R. 1980. Evolutionary change in small populations, pp. 135–149. *In* M.E. Soule and B.A. Wilcox (eds.). Conservation Biology: An Evolutionary-Ecological Perspective. Sinauer Associates, Sunderland, MA.

Grewe, P.M., N. Billington, and P.D.N. Hebert. 1990. Phylogenetic relationships among members of *Salvelinus* inferred from mitochondrial DNA divergence. Can. J. Fish. Aquat. Sci. 47:984–991.

Henn, A.W., and W.H. Rinkenbach. 1925. Description of the aurora trout (*Salvelinus timagamiensis*), a new species from Ontario. Ann. Carnegie Museum 16:131–141.

Hynes, J.D., E.H. Brown, Jr., J.H. Helle, N. Ryman, and D.A. Webster. 1981. Guidelines for the culture of fish stocks for resource management. Can. J. Fish. Aquat. Sci. 38:1867–1876.

Kaufman, L. 1992. Catastrophic change in species-rich freshwater ecosystems. The lessons of Lake Victoria. Bioscience 42:846–858.

Keller, W. 1978. Limnological Observations on the Aurora Trout Lakes. Technical Report. Ontario Ministry of the Environment, Sudbury.

Lacy, R.C. 1992. The effects of inbreeding on isolated populations: are minimum viable population sizes predictable? pp. 277–296. *In* P.L. Fiedler and S.K. Jain (eds.). Conservation

Biology: The Theory and Practice of Nature Conservation, Preservation, and Management. Chapman and Hall, New York.

Matuszek, J.E., D.L. Wales, and J.M. Gunn. 1992. Estimated impacts of SO_2 emissions from Sudbury smelters on Ontario's sportfish populations. Can. J. Fish. Aquat. Sci. 49(Suppl. 1):87–94.

May, R.M. 1990. How many species? Phil. Trans. R. Soc. Lond. 330b:293–304.

McGlade, J.M. 1981. Genotypic and phenotypic variation in the brook trout, *Salvelinus fontinalis* (Mitchell). Doctoral thesis, University of Guelph, Guelph, Canada.

McNeely, J.A., K.R. Miller, M.V. Reid, R.A. Mittermeier, and T.B. Werner. 1990. Conserving the World's Biological Diversity. IUCN, Gland, Switzerland; WRI,CI,WWF-US, and the World Bank, Washington, DC.

Moyle, P.B., and R.A. Leidy. 1992. Loss of biodiversity in aquatic ecosystems: evidence from fish faunas, pp. 127–169. *In* P.L. Fiedler and S.K. Jain (eds.). Conservation Biology: The Theory and Practice of Nature Conservation, Preservation, and Management. Chapman and Hall, New York.

Neary, B.P., P.J. Dillon, J.R. Munro, and B.J. Clark. 1990. The Acidification of Ontario Lakes: An Assessment of Their Sensitivity and Current Status with Respect to Biological Damage. Technical Report. Ontario Ministry of the Environment, Dorset, Ontario.

Nehlson, W., J.E. Williams, and J.A. Lichatowich. 1991. Pacific salmon at the crossroads: stocks at risk from California, Oregon, Idaho, and Washington. Fisheries 16(2):4–21.

Parker, B.J., and C. Brousseau. 1988. Status of the aurora trout, *Salvelinus fontinalis timagamiensis*, a distinct stock endemic to Canada. Can. Field-Nat. 102 (1):87–91.

Qadri, S.U. 1968. Morphology and taxonomy of the aurora char, *Salvelinus fontinalis timagamiensis*. Natl. Museums Can. Contrib. Zool. 5:1–18.

Raup, D.M. 1986. Biological extinction in earth history. Science 231:1528–1533.

Sale, P.F. 1967. A re-examination of the taxonomic position of the aurora trout. Can. J. Zool. 45:215–225.

Simpson, G.G. 1952. How many species? Evolution 6:342.

Wilson, E.O. 1984. Biophilia. Harvard University Press, Cambridge, MA.

12

Partnerships for Wildlife Restoration: Peregrine Falcons

Chris G. Blomme and Karen M. Laws

Conservation agencies throughout the world are working to protect and restore endangered species of wildlife. Some of the better known restoration programs are those involving species such as the whooping crane (*Grus americana*) and California condor (*Gymnogyps californianus*), but there are also many less publicized efforts in aid of a wide variety of species. Unfortunately, many of these restoration programs, especially those involving species that have reached critically low numbers, prove to be extremely difficult and often unsuccessful (Halliday 1978; Griffith et al. 1989). However, the need for these programs continues to rise under increasing pressure of habitat loss, overexploitation, introduction of exotic species, and environmental pollution.

Effective wildlife restoration programs often require a large commitment of technical and financial support. Formation of partnerships between conservation agencies and interested corporate sponsors is one means of providing this support. Such partnerships have become more common in recent years. In this chapter, we describe how such a partnership was formed to conduct the successful re-introduction of the eastern subspecies of peregrine falcon (*Falco peregrinus anatum*) to the Sudbury area.

Species Description

The peregrine falcon is one of the world's fastest birds, attaining diving speeds of up to 300 km/h, their powerful talons act as bludgeons, knocking their prey out of the sky (Fig. 12.1). There are three recognized subspecies of peregrine falcon in Canada and the United States. The western peregrine falcon (*F. p. pealei*) breeds along the British Columbia coastline and offshore islands, north through to the Aleutians. The Arctic subspecies (*F. p. tundrius*) breeds in areas above the treeline, along the coastlines of Hudson Bay and in northwestern Canada (Godfrey 1966). The eastern subspecies (*F. p. anatum*), although never very common before the use of pesticides, was found in many regions throughout Canada and the United States until it was extirpated from much of its range from the end of World War II to the present.

The decline of peregrine falcons in eastern North America was monitored and recorded by several researchers (Hickey 1942; Cade and Fyfe 1970; Fyfe et al. 1976). The development and large-scale use of the insecticide DDT from 1947 to the late 1960s is thought to be the main cause for the loss of peregrine populations in North America, as well as Europe (Hickey 1969). However, it was not until several years after the introduction of DDT that the deleterious side effects and the prolonged persistence of the metabolite DDE (dichlorodiphenyldichloroethylene) in the environment became known (Ratcliffe 1967; Hickey and Anderson 1968; Cade et al. 1971; Peakall

FIGURE 12.1. Streamlined body with narrow pointed wings and tail allow the peregrine falcon to move swiftly after prey. (Photo by C. Blomme.)

1976). Rachel Carson's book *Silent Spring*, published in 1962, brought this problem to the attention of the general public, an action that many consider instrumental in initiating the "environmental movement."

Thinning and breakage of egg shells, chick and adult death, and aberrant breeding behavior in adults are some of the adverse effects of long-term exposure to DDT and other chlorinated hydrocarbons (Ratcliffe 1970; Cade et al. 1971; Newton and Bogan 1974). Other species have also been affected (Box 12.1). Although DDT was banned in Canada in 1972 and in much of the United States in 1974, its metabolites are still found throughout the environment (Luoma 1991). Unfortunately, DDT is still used in some central and South American countries such as Mexico and is therefore available as a source of contamination for migratory species including peregrines (Fyfe et al. 1991).

North American Peregrine Recovery Program

Once it was recognized that the eastern subspecies *anatum* was extirpated from much of eastern North America, researchers in Canada (Fyfe 1976, 1988) and the United States (Cade et al. 1988) initiated breeding programs and tested a number of techniques for re-introduction. The Canadian government, through the Canadian Wildlife Service, has been the main contributor to the re-introduction program in Canada (Erickson et al. 1988). For example, the service established a program at Camp Wainwright in Alberta where captive surrogate parents are used to successfully rear chicks for shipment to release sites throughout Canada.

Although the captive breeding and release program has provided large numbers of birds for re-introduction, the program has been criticized by some researchers. For example, the difficulty in locating released birds and confirming that they are successfully reproducing is to some a sign that the program is not working (Cade et al. 1988; Kiff 1988; Nisbet 1988). To date, more than 400 birds have been released in Ontario, and only seven active nest sites have been identified.

Project Peregrine Sudbury

The decision to re-introduce peregrine falcons to the Sudbury area was made for a variety of reasons. First, the return of this native species to the area had been a dream of many local biologists and naturalists since peregrine releases began in North America. Two release

Box 12.1.

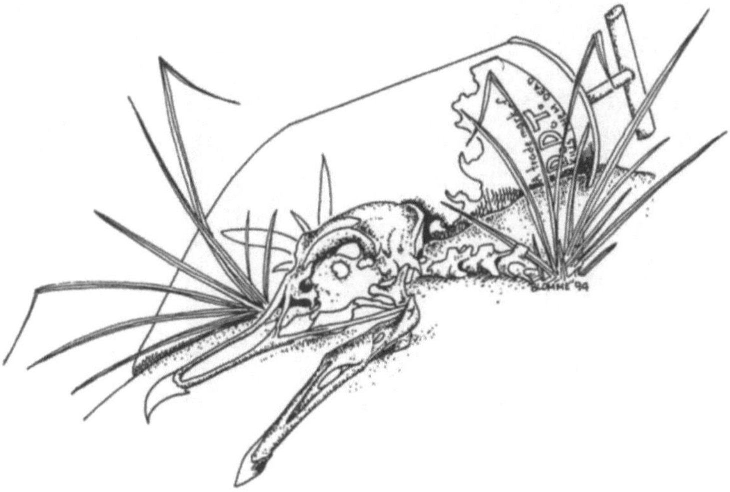

The peregrine falcon has received much public attention and support since the effects of organochlorines were first suspected. However, it is not the only species that suffered severe population declines as a result of these chemicals. The osprey (*Pandion haliaetus*) and bald eagle (*Haliaeetus leucocephalus*) became endangered, whereas the herring gull (*Larus argentatus*), double-crested cormorant (*Phalacrocorax auritus*), and several other fish-eating birds were affected at various locations.

Herring gulls nest on islands and shorelines throughout the Great Lakes area. They usually lay three eggs in a clutch and average one to two offspring from the colony per season. In 1969, the colony on Scotch Bonnet Island in Lake Ontario revealed an alarming young/adult ratio of 1:8 instead of the expected 1:2 ratio. Further studies showed that herring gulls were accumulating high concentrations of organochlorines such as DDT (DDE), dieldrin, PCB (polychlorinated biphenyls), and dioxin in its tissues. These compounds resulted in behavioral and reproductive problems, causing low recruitment of young.

Since the discovery at Scotch Bonnet Island, the herring gull has become a very important part of the biomonitoring program for the Great Lakes. Egg samples from several herring gull colonies now provide an annual measure of the presence of organochoride compounds. Most contaminants have shown a marked decrease since the early 1970s. Dieldrin has shown only a gradual decrease, and localized populations of gulls in highly industrialized areas of Lake Erie and Lake Ontario still show physical and reproductive problems associated with chemical contaminants. The biomonitoring of contaminants in colonial birds such as the herring gull has therefore become an essential source of information about the condition of the environment in which we live.

efforts had been made previously in northern Ontario, but neither was near the Sudbury area. Second, the steady improvement in the diversity of plant and animal life in the industrially damaged area was thought adequate to support a population of peregrines. A total of 266 species of birds are known to occur in the area (Whitelaw 1989), with several being suitable prey. Third, Falconbridge Limited, one of the mining companies originally responsible

Figure 12.2. Large hack boxes located on the east end of the student residence at Laurentian University overlook the city of Sudbury. Large painted numbers on the sides aided volunteers in describing locations of their observations. (Photo by K.P. Morrison.)

for some of the environmental degradation, considered the project an ideal opportunity to illustrate their commitment to improving and restoring industrially damaged ecosystems. The fact that the species of interest was a falcon, the corporate logo of the company, made support for this particular program especially appropriate. Finally, the project was a chance for local government agencies to show that multiagency partnerships could be used to participate effectively in natural resources management and restoration efforts.

Partnership

The path from talking about a release to actually making one happen requires the efforts of many dedicated individuals, contribution of money, and administrative support. The process began with the involvement of several dynamic and determined individuals who promoted the program as a partnership between government, industry, and academia and capitalized on the heightened interests in environmental restoration in the Sudbury area. However, what began as a partnership between agencies (Ontario Ministry of Natural Resources, Falconbridge Limited, Noranda Inc., Laurentian University, World Wildlife Fund) soon expanded to involve the general public. Several hundred volunteers helped monitor the movement of birds, especially in the few days after releases.

Urban Release—Sudbury (1990–1991)

The release of peregrines into an area can occur by augmentation, cross-fostering, or hacking (Burnham et al. 1978; Sherrod et al. 1981). The first two techniques depend on having established breeding pairs in the region. Therefore, only the hacking technique was available to us.

The roof top of one of the student residences at Laurentian University in Sudbury was chosen for the hack box site (Fig. 12.2) (see Idle 1990) for details of construction). The site overlooked a woodland and nearby lake and met the following criteria: (1) an unobstructed vista to permit peregrines to orient themselves to surrounding landscapes during the hacking phase; (2) protection from vandalism and predation; (3) suitable locations for observers to watch released birds; and (4) adequate natural food supply and habitat for young falcons to hunt.

Birds arrived by air freight from Edmonton each summer. A media presentation at the time of arrival gave members of the press the opportunity to obtain photographs of the young birds (Fig. 12.3). The Sudbury peregrine story was

FIGURE 12.3. Warren Holmes (Vice President and General Manager, Falconbridge Ltd.) and Karen Laws (District Biologist, Ministry of Natural Resources-Sudbury) launch the first year of Project Peregrine. The co-operative venture was the first of its kind in Ontario. (Photo by M. Roche.)

FIGURE 12.4. Peregrine chicks are removed from their shipping boxes and transferred to the hack after the bands are verified. Note the sharp talons and beak used for capturing and tearing prey. (Photo by C. Blomme.)

televised on the Canadian national news in 1990. Media personnel enthusiastically covered the peregrine release projects each summer.

Each bird was examined and its band number recorded before being placed in a hack box (Fig. 12.4). Standard, individually numbered, eight-digit aluminum Canadian Wildlife Service bands had been previously placed on one leg of each bird. On the other leg, a bright red aluminum band individually labeled with ei-

FIGURE 12.5. These 4-week-old chicks have just been placed in the hack box. The interior provides perches, a feeding station, gravel, a hide, and a vista of the surrounding landscape. (Photo by C. Blomme.)

ther two numbers, two letters, or a combination (3E, A2) were present. These bands were used to identify individual birds at a distance using binoculars or a spotting scope. Birds were distributed to the various boxes according to age. The birds' ages at arrival ranged from 29 to 35 days, with an average of 32 days for 1990 and 1991.

Life in the Hack Box

During the 2-week hacking period, the young falcons, called eyasses, were monitored 24 hours a day using concealed windows and remote cameras (Fig. 12.5). Initially, the birds were fed small pieces of quail, but as the falcons grew, whole quail were delivered through feeding tubes. Considerable care was taken to avoid contact between the care givers and the birds to prevent domestication.

Wing flapping behavior increased as development progressed. As release time approached, some birds would cling to the front aluminum bars, flap their wings vigorously, and begin elevating off the floor. Sexual dimorphism was evident, with females being larger than males of equivalent age.

Release Day

Sherrod et al. (1981) recommend that birds should be 42–45 days of age when released. Sudbury's birds were 42–48 days of age on release in 1990 and 41–46 days of age in 1991. Two releases occurred each year, with totals of 15 and 17 birds released in 1990 and 1991, respectively. Each release day depended on weather conditions and the developmental stage of the chicks. Ideal conditions for release were clear days with a light breeze providing slight updrafts to assist with first flights. On the morning of the release, food was placed outside the hack so that birds would initially associate food with the hack site. A water basin with fresh water was placed near the hack boxes for bathing.

FIGURE 12.6. This 49-day-old peregrine falcon perched near an observer during its third day of freedom. The red band on the left leg (band number 08) indicates it is a male from Sudbury's release. (Photo by C. Blomme.)

Observers with portable radios maintained a continuous watch over the birds for the first 5 days after release, a critical period during which the birds attempt their first flights, learn to manoeuver, and land. Many volunteers were needed for these intense observations. In 1990, 70 volunteers and, in 1991, 75 volunteers from a variety of backgrounds and disciplines contributed to this co-operative venture.

Our first release day was July 11, 1990. To the delight of gathered representatives of the partners, television and other media crews, and many interested members of the general public, all eight birds flew on the first day, three of them within the first hour (Fig. 12.6). The second release in 1990 and the two releases in 1991 were equally successful, with most birds flying on the first day. The longest delay was female "HR" (46 days old), who took her first flight 5 days post-release.

Training and experience were obviously essential for the young peregrines to develop their tremendous aerobatic skills. For example, one bird took off and circled the residence, but when it approached for landing on the roof, it pulled up at the last second, crashing into the wall, and fortunately, slid safely to the ground. These types of incidents were noted on several "first" flights and were a constant concern for the observers watching the birds learn to fly and land.

Peregrine visual capabilities are acute. After release, many birds would be flying after dark (Courtin 1991; Hillis 1992). Hunting behavior was noted within the first 5 days of release. Perched birds would peck at and eat insects and spiders, and flying birds would chase monarch butterflies (*Danaus plexippus*), sulphurs (*Colias* sp.), and hornets. As flying skills improved, the young peregrines concentrated more on pursuing birds as prey. American goldfinch (*Carduelis tristis*), starling (*Sturnus vulgaris*), and tree swallow (*Iridoprocne bicolor*) were chased. Even a large great blue heron (*Ardea herodias*) was harassed by the young peregrines. These chases appeared to be precursors to actual capturing of prey at a later age. Confirmed kills were difficult to substantiate or observe. A rock dove (*Columbia livia*) was the first confirmed kill by a peregrine at 16 days after the first release. One peregrine was also observed eating the remains of a sharp-shinned hawk (*Accipiter striatus*).

Long periods of observations by volunteers and project staff resulted in many confirmed sightings of individual birds well after release. In 1990, the last bird observation (band number DU) was recorded on September 12, 6 weeks after release. The average age of the 15 birds at the time they were last observed was 76 days. As many as 11 falcons were seen on August 17, when the youngest and oldest birds would be 67 and 83 days of age, respectively. Birds began their southern movements between August 10 and September 12, 1990. In 1991, the average age of the 17 birds when last confirmed was 77 days. Three birds were seen on September 9.

Wilderness Release—Killarney Provincial Park (1992–1993)

Peregrine falcons reach maturity between 2 and 3 years of age. At this time, a breeding territory would usually be established and defended by a male. If a nest happened to be located close to the hack box site, a territorial male may chase newly released peregrines. Therefore, in the third year of the program the hack boxes were moved to a new site, located at Hawk Ridge (elevation, 310 m) in a wilderness park, approximately 60 km south of Sudbury. Steep cliffs in the park provided potential nesting sites; natural food was plentiful; risk of predation and human interference was low; and access for staff was challenging but possible. Again, more than 70 volunteers from the park campgrounds and Sudbury community offered assistance at the new site.

Many challenging logistical problems had to be overcome at this remote site. Four hack boxes had to be secured to the edge of the cliff with stainless-steel cables (Fig. 12.7), and concealed observation stations had to be established at either end of the ridge to monitor birds during the releases. A base camp was established at the bottom of the cliff for project staff. A helicopter was needed to transport the crates containing the falcon chicks to the site.

The birds were fed and monitored daily for 2 weeks after arrival. Black bear (*Ursus americanus*) were common in the area and were a concern due to the presence of dead quail. Misto-van deodorizing chemical was applied on the ground surrounding the hack boxes and proved to be a deterrent to bears approaching the hack sites.

During the releases, volunteers in canoes observed the peregrines from the lake surface at the base of the cliff. Climbing gear and safety equipment were needed for safe approach to the hack boxes to remove frontal barriers. Feeding boards were placed on the roof of two hack boxes, and later, only one feeding station was maintained for the birds of both releases.

Release of the 59 young peregrines at the remote location went smoothly, with no groundings observed. All birds flew within 1 hour of barrier removal. This was noticeably different from the urban releases and was attributed to the older ages of the released birds. Birds were released at between 47 and 52 days of age at Hawk Ridge. The extra hack time was intended to ensure stronger development of young falcons to reduce the chance of groundings and slow fliers, as observed in the urban releases.

Flight development and hunting techniques developed similarly to the urban released birds but were more spectacular to observe over the hills and forests of this wilderness park. Birds dispersed from the area by the third week of August. The average time to dispersal for the urban and wilderness releases was similar at 35 days.

North American Monitoring

Monitoring of the success of this endangered species recovery project would not be possible

FIGURE 12.7. The location of the hack boxes on Hawk Cliff above George Lake in Killarney Provincial Park was very different from the urban release at Sudbury. The vista revealed little human development. (Photo by C. Blomme.)

without an intricate network of communications across North America. Observations by avid naturalists, conservationists, and other interested people provide reports of band sightings to the U.S. Department of Fish and Wildlife or the Canadian Wildlife Service. In 1990, one of Sudbury's birds, female "DY," was identified and photographed in Terre Haute, Indiana, on September 12, 35 days and 1000 km away from her Sudbury origin. Female Z4, released July 23, 1993, was observed in good condition in Cape May, New Jersey, on September 27.

Since beginning in 1990, our project has released 91 falcons to their historical range (Fig. 12.8). Fifteen sightings have been reported, 6 of which were mortalities. Two birds were shot, one while a person was protecting his pigeons; one bird hit a glass building in Los Angeles, California; one bird was found as a carcass; and two birds were hit by cars while in pursuit of prey. One bird fell into a sewage treatment site and was subsequently rehabilitated and released in Florida. Two males and a female returned to Sudbury 1 year after release. A female released in Sudbury in 1990 was observed with a mate at the Toronto waterfront in October 1993. One of the returning birds has been the subject of a detailed follow-up study that is described in Box 12.2. As of June 1994, there have been no other confirmed sightings of the rest of the birds, but we assume that many are alive.

Benefits of the Partnership

This re-introduction of peregrine falcons could not have been as successful without the partnership approach. The birds were the principal beneficiary of the partnership, but the partners also benefited. For example, there were immediate and tangible benefits for partners such as tax credits and positive publicity for sponsoring industries, as well as less tangible but still very useful information exchange through personal involvement, a benefit that

FIGURE 12.8. An adult peregrine falcon looks very different from first- and second-year birds. This male is one of the original breeders from Camp Wainwright, Alberta. Note the prominent black facial crest and the buffy clear chest. (Photo by C. Blomme.)

can greatly assist in future cooperative efforts. The direct exchange of environmental and natural history information, particularly with the volunteers and other members of the general public, was one of the most rewarding parts of this program.

One should not overstate the ease with which such partnerships can be formed or suggest that they will always be successful. Successful partnerships require considerable commitment, flexibility, awareness of the goals and needs of all involved, frequent and extensive communication, humility, and an appreciation of the contributions and sometimes the limitations of each member. No doubt, many partnerships can be expensive, inefficient, and difficult to maintain, but to our knowledge there has been little objective analysis of the factors (e.g., number of partners, level of financial contribution, du-

Box 12.2. Female AD's Story

On May 10, 1991, female AD was observed in Sudbury on top of the building near the site where she was released a year earlier. In September of that year, she was regularly observed in downtown Detroit where she paired and mated with an unidentified subadult male in the spring of 1992. These birds nested and laid three eggs on a ledge beneath a fire escape on the thirty-third floor of an office building. The eggs proved non-viable. Close monitoring by Detroit peregrine biologists resulted in a fostering operation. Two 10-day-old chicks, one male and one female, were exchanged for the eggs. The two eyasses were immediately adopted and reared successfully. After fledging, the fostered young female established residence in downtown Detroit. Her brother was observed flying toward Windsor, Ontario, near the end of summer 1992.

The Detroit peregrine staff have renamed our Sudbury female "Judy" and her mate "Pop." In 1993, Judy and Pop overwintered in downtown Detroit. Mating occurred in mid-March. Four eggs were laid, with two healthy chicks hatching in mid-May. This was the first recorded wild hatching of peregrine falcons in the Detroit, Michigan, area. Both chicks had fledged by the end of August 1993 and were residing in the Detroit area. In 1994, Judy and Pop again produced two healthy chicks.

ration of project, administrative structure) that contribute to success or failure. Given the need for partnership in environmental or ecological restoration, such analysis should be encouraged.

Acknowledgments. We thank the partners and all the volunteers for their interest and contributions. A. Bradshaw, J. Gunn, J.D. Shorthouse, and M. Wiseman provided many helpful suggestions and review comments.

References

Burnham, W.A., J. Craig, J.H. Enderson, and W.R. Heinrich. 1978. Artificial increase in reproduction of wild peregrine falcons. J. Wildlife Management 42:625–628.

Cade, T.J., J.H. Enderson, C.G. Thelander, and C.M. White (eds.). 1988. Peregrine Falcon Populations: Their Management and Recovery. The Peregrine Fund, Boise, ID.

Cade, T.J., and R. Fyfe (eds.). 1970. The North American peregrine survey, 1970. Can. Field-Naturalist 84:231–245.

Cade, T.J., J.L. Lincer, C.M. White, D.G. Roseneau, and L.G. Swartz. 1971. DDE residues and eggshell changes in Alaskan falcons and hawks. Science 172:945–957.

Carson, R. 1962. Silent Spring. Houghton Mifflin, Boston.

Courtin, C. 1991. Project Peregrine/Projet Pèlerin-Sudbury 1991. Summary report. Laurentian University Press, Sudbury.

Erickson, G., R. Fyfe, R. Bromley, G.L. Holroyd, D. Mossop, B. Munro, R. Nero, C. Shank, and T. Wiens. 1988. *Anatum* Peregrine Falcon Recovery Plan. Canadian Wildlife Service, Environment Canada, Ottawa.

Fyfe, R.W. 1976. Rationale and success of the Canadian Wildlife Service Peregrine Breeding Project. Can. Field-Naturalist 90:308–319.

Fyfe, R.W. 1988. The Canadian peregrine falcon recovery program, 1967–1985, pp. 773–778. *In* T.J. Cade, J.H. Enderson, C.G. Thelander, and C.M. White (eds.). Peregrine Falcon Populations: Their Management and Recovery. The Peregrine Fund, Boise, ID.

Fyfe, R.W., U. Banasch, V. Benavides, N.H. de Benavides, A.Luscombe, and J. Sanchez. 1991. Organochlorine residues in potential prey of peregrine falcons, *Falco peregrinus*, Latin America. Can. Field-Naturalist 104:285–292.

Fyfe, R.W., S.A. Temple, and T.J.Cade. 1976. The 1975 North American peregrine falcon survey. Can. Field-Naturalist 84:228–273.

Godfrey, W.E. 1966. The Birds of Canada. Queen's Printer, Ottawa.

Griffith, B., J.M. Scott, J.W. Carpenter, and C. Reed. 1989. Translocation as a species conservation tool: status and strategy. Science 245: 477–480.

Halliday, T. 1978. Vanishing Birds: Their Natural History and Conservation. Holt, Rinehart and Winston, NY.

Hickey, J.J. 1942. Eastern population of the duck hawk. Auk 59:176–204.

Hickey, J.J. (ed.). 1969. Peregrine Falcon Populations: Their Biology and Decline. University of Wisconsin Press, Madison.

Hickey, J.J., and D.W. Anderson. 1968. Chlorinated hydrocarbons and eggshell changes in raptorial and fish-eating birds. Science 162: 271–273.

Hillis, T. 1992. Project Peregrine/Projet Pèlerin-Wilderness Release, Killarney Provincial Park 1992. Summary report. Laurentian University Press, Sudbury.

Idle, P.D. 1990. Project Peregrine/Projet Pèlerin-Sudbury 1990. Summary report. Laurentian University Press, Sudbury.

Kiff, L.F. 1988. Commentary—changes in the status of the peregrine in North America: an overview, pp. 123–140. *In* T.J. Cade, J.H. Enderson, C.M. Thelander, and C.M. White (eds.). Peregrine Falcon Populations: Their Management and Recovery. The Peregrine Fund, Boise, ID.

Luoma, J.R. 1991. The deadly legacy of DDT. Wildlife Conservation 94:26–35.

Newton, I., and J. Bogan. 1974. Organochlorine residues, eggshell thinning, and hatching success in British sparrowhawks. Nature 249:582–583.

Nisbet, I.C.T. 1988. Summary, pp. 851–855. *In* T.J. Cade, J.H. Enderson, C.G. Thelander, and C.M. White (eds.). Peregrine Falcon Populations: Their Management and Recovery. The Peregrine Fund, Boise, ID.

Peakall, D.B. 1976. The peregrine falcon (*Falco peregrinus*) and pesticides. Can. Field-Naturalist 90: 301–307.

Ratcliffe, D.A. 1967. Decrease in eggshell weight in certain birds of prey. Nature 215:208–210.

Ratcliffe, D.A. 1970. Changes attributable to pesticides in egg breakage frequency and eggshell thickness in some British birds. J. Appl. Ecol. 7:67–115.

Sherrod, S.K., W.R. Heinrich, W.A. Burnham, J.H. Barclay, and T.J. Cade. 1981. Hacking: A Method for Releasing Peregrine Falcons and Other Birds of Prey. The Peregrine Fund, Ithaca, NY.

Whitelaw, C. 1989. Seasonal Occurrence of Birds in the Sudbury District. Sudbury. Unpublished report.

Section D

Research Topics in Restoration Ecology

John Cairns, Jr.

Since the beginning of the agricultural revolution, the rate of ecological destruction has markedly exceeded the rate of ecological recovery. Natural ecosystems have suffered perturbations that have resulted in displacement of species, loss of ecological function and structure, and diminution of ecological integrity. The newly developing field of restoration ecology is, in a very real sense, an attempt to accelerate and enhance the natural recovery processes so that the integrity of a damaged ecosystem can be fully or partially restored. Because each ecosystem is the product of a sequence of biological, climatological, and other processes, precise replication of the predamaged condition is highly improbable (Cairns 1989). Therefore, most restoration will be an attempt to re-establish a naturalistic assemblage of organisms compatible with the chemical/physical environment that exists.

Current evidence suggests that ecological restoration will be most effective when carried out at the landscape level (NRC 1992). Restoration efforts are also more likely to be successful and endure once completed if human society is sufficiently environmentally literate to understand both the ecological values and the human values resulting from restoration. Figure D.1 provides an illustrative matrix showing a number of choices that reflect both ecological and societal values.

Design criteria for ecological restoration and maintenance projects should reflect both human behavior and needs as well as the ecological needs of project species and habitat. Estimating these is a formidable undertaking, far beyond the capabilities of a single individual or single discipline. However, if the present rates of ecological destruction and human population continue to increase, for even a few additional decades, ecosystem services per capita will almost certainly be below the level necessary for sustained long-term use. Illustrative ecosystem services are regulation of the atmospheric gas balance, improvement of water

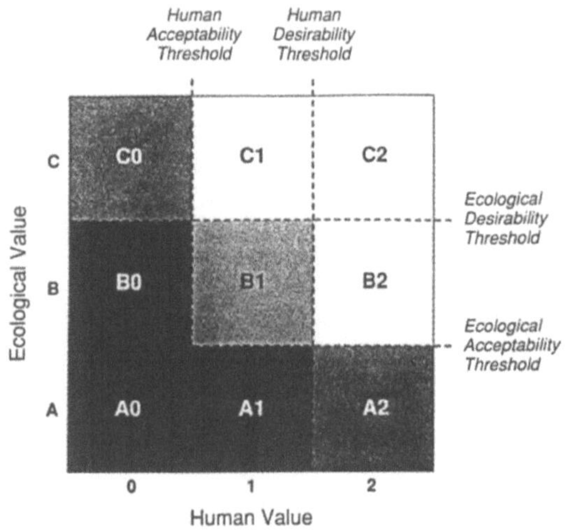

Cell	Ecological Value	Human Value
A0	Unacceptable	Unacceptable
A1	Unacceptable	Acceptable
A2	Unacceptable	Desirable
B0	Acceptable	Unacceptable
B1	Acceptable	Acceptable
B2	Acceptable	Desirable
C0	Desirable	Unacceptable
C1	Desirable	Acceptable
C2	Desirable	Desirable

FIGURE D.1. (From National Research Council 1992.)

quality, provision of models for new drugs and medicines, provision of genetic pools for agricultural crops (especially those that will be necessary if global climate change occurs), and that feeling of well-being so difficult to quantify but which large numbers of people seek on their vacations and free time by associating with natural systems.

Communicating effectively about ecological and societal value systems and reaching compromises will indeed be a challenge. Success at achieving this environmental literacy will occur if ecological values are given the same attention as human values. At present, most humans consider themselves apart from ecosystems rather than a part of ecosystems.

As a number of chapters in this book have shown, eliminating or markedly reducing the source of the problem (e.g., emission controls) may result in remarkable recovery, as was the case for the River Thames in England (Gameson and Wheeler 1977). However, I have personally encountered a number of ecosystems (as remote from my home institute as Hobart, Tasmania, and as close as the surface mining in my home state of Virginia) where ecosystem recovery has not occurred long after the stress had been eliminated or markedly reduced. Clearly, for the latter instances, the ecosystem "medicine" of Rapport (1984) will be necessary if the "patient" is to recover.

FIGURE D.2. (From National Research Council 1992.)

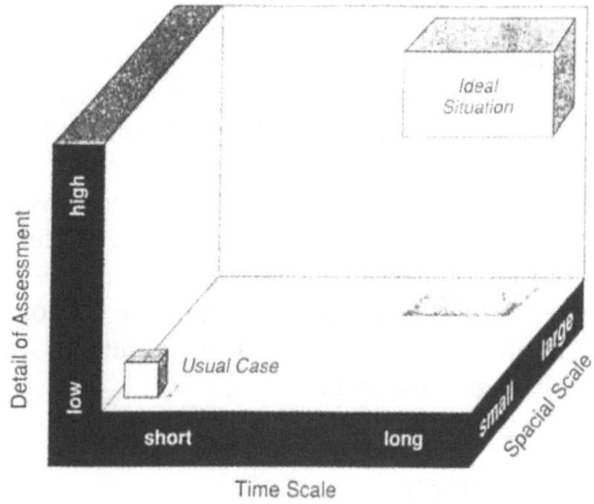

As Kauffman (1993) noted, a new science, the science of complexity, is emerging. This particular science, together with the dawning of the information age, will certainly result in major transformations of the social, economic, and ecological systems in the next century and perhaps at the end of this one. The science of complexity and the information age are inextricably linked, but they are phenomena deserving individual attention. Complex adaptive systems achieve, in a lawlike way, the edge of chaos (Kauffman 1993). This is a consequence of the fact that organisms, economic entities, nations, and ecosystems do not merely evolve but rather coevolve.

One of Kauffman's (1993) hypotheses that strikes me as particularly relevant to restoration ecology is that a coevolving individual does not benefit, but in fact does worse, by calculating too far into the future. My own interpretation of this is that if we are to develop a partnership with nature as an alternative to exploiting natural systems, our coevolution must begin with preventing damage and repairing damage wherever possible. Included in this "wherever possible" are economic considerations that may use opportunity–cost analysis rather than the traditional cost–benefit analysis. This entails repairing ecosystems so that money spent in such activities will generate the greatest social good rather than fixing blame for the destruction and assigning the costs of repair to the "guilty." Figure D.2 illustrates this point. If Kauffman's speculation is correct, society could move gradually from the box in the lower left-hand corner of the figure (usual case) to the box in the upper right-hand corner (ideal situation). Trying to leap immediately to the ideal situation would almost certainly be counterproductive.

The good news about ecological restoration is that, despite the fact that it is a new field (as scientific fields go), practically all attempts at ecological restoration result in significant improvement in ecological integrity and condition. In fact, the field is too new to have a long history that will determine whether the systems return to predisturbance condition or some approximation thereof or, more important, whether they ultimately become self-maintaining. Monitoring recovery rates and trajectories at landscape damage sites such as Sudbury provide important opportunities to track this process. However, despite uncertainties about the future condition of "restored ecosystems," the fact that they are vastly better than the damaged condition is comforting. What is less comforting is the fact that restoration may cost an order of magnitude or more than the cost of preservation, something that is true of health and a variety of other factors affecting the human condition.

References

Cairns, J., Jr. 1989. Restoring damaged ecosystems: is predisturbance condition a viable option? Environ. Prof. 11:152–159.

Gameson, A.L.H., and A. Wheeler. 1977. Restoration and recovery of the Thames estuary, pp. 72–101. *In* J. Cairns, Jr., K.L. Dickson, and E.E. Herricks (eds.). Recovery and Restoration of Damaged Ecosystems. University Press of Virginia, Charlottesville, VA.

Kauffman, S.A. 1993. The sciences of complexity and "origins of order." Ann. Earth 11(3):19–26.

National Research Council. 1992. Restoration of Aquatic Ecosystems. National Academy Press, Washington, D.C.

Rapport, D.J. 1984. State of ecosystem medicine, pp. 315–324. *In* V.W. Cairns, P.V. Hodson, and J.O. Nriagu (eds.). Contaminant Effects on Fisheries. John Wiley & Sons, New York.

13

Dynamics of Plant Communities and Soils in Revegetated Ecosystems: A Sudbury Case Study

Keith Winterhalder

Trigger Factor Effect

Colonization of the Sudbury barrens by metal-intolerant plants is inhibited by metal toxicity under highly acidic conditions, exacerbated by enhanced frost action (see Chapter 7). It was once believed that revegetation of the barren soils would only be achieved if they were deeply tilled, bringing the less-contaminated soil to the surface, but Winterhalder (1974) observed that disturbance or minimal treatment of contaminated soils in the Sudbury area with limestone or fertilizer led to rapid colonization by tickle grass (*Agrostis scabra*). Later, it was found that a thin surface sprinkling of ground limestone would lead to establishment of woody plants such as white birch (*Betula papyrifera*), trembling aspen (*Populus tremuloides*), and willows (*Salix* spp.) (Fig. 13.1). Because this "minimal amelioration" (Skaller 1981) appeared to initiate spontaneous colonization by metal-intolerant plants from a local seed source, it was referred to as a "trigger factor" (Winterhalder 1983).

This simple approach works for several reasons. The stony mantle that covers the eroded, glacial till-derived soils of the slopes traps limestone particles and seeds. Detoxification of the surface allows the development of a deep root system, affording greater stability to the seedling and protecting it against drought and frost-heaving. Ultimately, the insulating effect of leaf litter also contributes to the reduction in frost activity.

Limestone Detoxification of Sudbury Soils

The detoxifying effect of limestone on Sudbury's acidic metal-contaminated soils is the net result of several mechanisms:

1. **precipitation of copper and nickel from solution if a sufficiently high pH is achieved**
2. **reduction in the toxicity of aluminum ions as they combine with hydroxyl ions**
3. **enhanced plasma membrane integrity in the cells of the root hairs, improving the plant's ability to selectively exclude toxic ions**
4. **reducing metal ion uptake by mass action effect of calcium and magnesium on the root hairs' exchange complex**
5. **increased availability of soil phosphorus, which acts as a plant nutrient as well as a partial protection against aluminum toxicity**

Figure 13.1. Colonization by birch seedlings of a 1-m^2 plot in a semibarren birch woodland near Falconbridge, photographed on August 10, 1985. The acid- and metal-tolerant moss *Pohlia nutans* had been sprinkled with a light covering of ground dolomitic limestone on September 22, 1979.

pH and Base Dynamics on Limed Soil

Limestone treatments normally increase soil pH by between one and two units. For example, at the site shown in Figure 13.2, the unlimed land dominated by metal-tolerant tufted hairgrass (*Deschampsia caespitosa*) has a soil pH of approximately 4.0, whereas the adjacent limed land supporting a lush growth of grasses, forbs, and woody plants has a pH of approximately 6.0.

Manual methods of limestone application ensure a mosaic of soil pH and base content, engendering both floristic and genetic diversity in the colonizing species, especially with respect to metal tolerance. A decade after liming, a clear correlation exists between soil pH and the species that it supports. The soil under birch and the acid-tolerant moss is in the pH 4.0–5.0 range, whereas pearly everlasting (*Anaphalis margaritacea*, a native wildflower) and the two seeded legumes (Alsike clover [*Trifolium hybridum*] and birdsfoot trefoil [*Lotus corniculatus*]) are associated with soil in the pH

Figure 13.2. Treated-untreated boundary south of Copper Cliff in July 1992, showing tufted hairgrass (*Deschampsia caespitosa*) in the untreated foreground and a lush growth of woody plants in the background, which was limed, fertilized, and seeded 10 years previously.

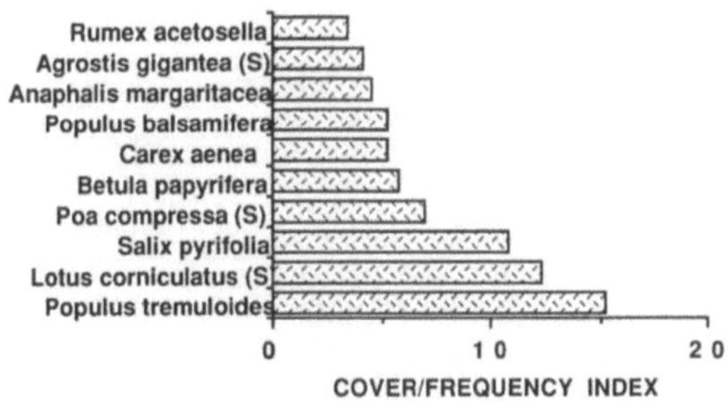

FIGURE 13.3. Ten highest-ranking species on revegetated Sudbury barrens. Species marked with (S) were seeded; others are volunteers. The species are ranked by mean Cover/Frequency Index (Relative Cover + Relative Frequency/2) and based on 1684 quadrats (1 m^2) in 45 transects on land treated 10–14 years previously.

5.0–6.0 range. Calcium and magnesium concentrations (the fraction extractable with dilute acid) show a pattern similar to that for pH, with moss and birch associated with low base cation concentrations and trefoil and clover with high values, pearly everlasting being intermediate. Although these patterns may partially indicate the "preferences" of particular plant species, competition is also an important factor in species distribution. For example, the birch is probably occupying available niches that are not necessarily optimal but are too acid for the other species.

The plants may themselves influence the soil chemistry. Although it might be expected that the neutralizing power of the applied limestone will become exhausted, leading to regression of the plant community, the detoxification of the surface allows for the penetration of roots into a larger volume of soil and facilitates the movement of calcium and magnesium to the surface. This mechanism of soil-base enrichment has been referred to as a "cation pump" by Aber (1987), who pointed out that some tree species are better cation pumps than others, with poplars and spruces being some of the best for base enrichment. However, not all trees produce this effect. The leaf litter from some trees such as pines actually acidifies the forest floor. The current procedure in the Sudbury area, of establishing a vigorous growth of birch and poplar before or concurrently with the planting of pines, therefore seems to be the correct choice.

It has been suggested that as copper and nickel toxicity decreases, calcium and magnesium (as well as phosphorus and nitrogen) might become limiting to plant growth (Lozano and Morrison 1981). Clearly the calcium in the liming material confers plant nutritional benefits, but preliminary experiments have indicated that dolomitic limestone (calcium magnesium carbonate) is more effective than calcitic limestone (calcium carbonate) as a soil ameliorant in certain sites, and it is likely that high calcium levels in soils limed with calcitic limestone can actually induce magnesium deficiency in plants. The possible antagonistic or protective effect of magnesium with respect to nickel toxicity is also under investigation.

Changes in Species Composition after Treatment

In general, a mixture of seeded and volunteer species dominates a site 10 years or so after it has been treated (Fig. 13.3). The principal colonists of amended Sudbury barrens are wind-dispersed, and although there is a small persistent seed bank of tickle grass and white birch, most of the seed source is from the seed rain, which differs from site to site. Figure 13.4 shows an area west of the abandoned smelter at Coniston that was previously barren except for relict white birch and red oak, its grass-legume sward dominated by redtop (*Agrostis gigantea*) 1 year after treatment. Six years later, the grassed area was extensively colonized by white birch. Figure 13.5 shows another barren site, south of the Copper Cliff smelter. Six years after liming and grassing, the dominant woody colonist was trembling aspen.

Plant colonization processes on treated land differ qualitatively as well as quantitatively from natural processes on untreated lands (see Fig. 13.2). In general, the relative importance

Figure 13.4. (*Upper photo*) Grassed site near the abandoned Coniston smelter in 1980 (seeded 1979), showing grass cover dominated by redtop (*Agrostis gigantea*) and scattered "relict" white birch and red oak. (*Lower photo*) The same site in 1986, showing colonization of grassed area by white birch.

of metal-tolerant species is likely to be lower in areas that have been treated. For example, at sites where natural colonization by white birch is occurring, liming allows for the establishment of trembling aspen, but there is no indication that birch growth is also stimulated (Fig. 13.6). In this case, liming has provided a competitive advantage to aspen.

Over the years that follow treatment, the importance of introduced species may rise for a period (see the example of nitrogen-fixing birdsfoot trefoil), but there is a general tendency for introduced species to decrease and for native species to increase (Fig. 13.7). Regardless of treatment differences (e.g., seed mixture, fertilizer, and limestone type and level) at seeded sites, it appears that once the "trigger factor" is applied, other natural forces take over, and a very similar colonization process occurs. However, there are striking differences between sites that have been seeded with the grass-legume mixture and those that are unseeded, with as many as 10 to 20 times fewer woody plant individuals colonizing seeded sites (per unit area).

Because the application of limestone alone is sufficient to initiate plant growth, one could argue that seed and fertilizer application is superfluous, wasteful of resources, and ecologically inappropriate. Seeding does assist in

FIGURE 13.5. (*Upper photo*) North shore of Kelly Lake near the Copper Cliff smelter during the liming procedure in 1983. (*Lower photo*) The same site in 1989, showing colonization of grassed land by trembling aspen.

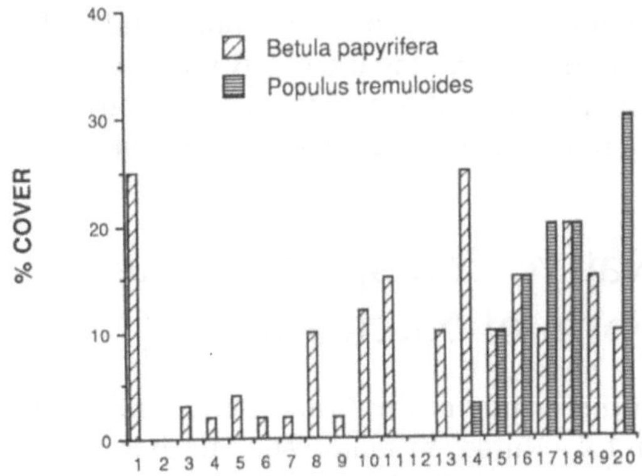

FIGURE 13.6. Percentage cover of white birch (*Betula papyrifera*) and trembling aspen (*Populus tremuloides*) across a liming boundary 14 years after treatment.

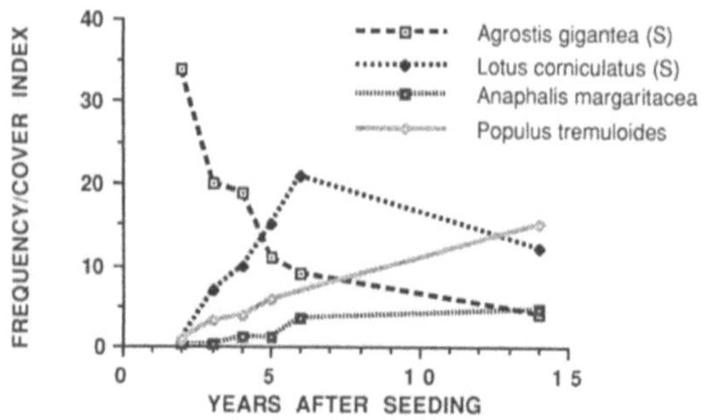

Figure 13.7. Change in relative importance of four plant species over the first 14 years after treatment. S, seeded.

rapidly achieving the aesthetic goals of "regreening" (Chapter 8) and contributes to soil stabilization, microclimate amelioration, and nitrogen fixation, but the agronomic cover also seems to have another beneficial effect. Unseeded areas develop a dense thicket of birch, poplar, and willow undergoing vigorous competition, whereas the woody plants in seeded areas colonize in an open random fashion, leading to good spatial and structural diversity in the stand.

Seeding or planting and natural colonization or "succession" are by no means mutually exclusive, because the seeded or planted vegetation acts as a "nurse crop" to the colonists. Like the pioneer species that initiate natural succession, a nurse crop facilitates colonization by enhancing the microenvironment, reducing evapotranspiration and increasing snow cover. It also plays a critical role in trapping windblown seeds. However, competition, usually from grasses, can form a barrier to colonization and succession on many reclamation sites, as well as competing with planted trees for light and nutrients (Bradshaw and Chadwick 1980) and forming a winter habitat for bark-gnawing rodents. However, in the Sudbury land reclamation program, the use of low seed and fertilizer rates, probably assisted by the stony soil, seems to have eliminated the competition factor.

Native Transplants as a Form of Nucleation

In their study of the Grand Bend sand dunes on Lake Huron, Yarranton and Morrison (1974) noticed that individuals of certain pioneer species such as red cedar (*Juniperus virginiana*) formed nuclei for the initiation of patches of other "persistent" species that spread and eventually coalesced. Miller (1978) suggested that such an approach might be taken in revegetation, in that selected "pioneer" species could be planted in clumps, then the persistent species introduced into the clumps once the pioneers are well established.

A modification of this approach is under investigation on the revegetated Sudbury barrens as a means of introducing understory species characteristic of the targeted plant community. When native species are transplanted from their natural environment in blocks of their own soil, seedlings or propagules of associated species are inevitably introduced with them. For example, species introduced incidentally during the transplantation of soapberry (*Shepherdia canadensis*) and bearberry (*Arctostaphylos uva-ursi*) include common juniper (*Juniperus communis*), white spruce (*Picea glauca*), wild strawberry (*Fragaria virginiana*), wild basil (*Satureja vulgaris*), balsam ragwort (*Senecio pauperculus*), starry false solomon's seal (*Smilacina stellata*), wood-lily (*Lilium philadelphicum*), yellow lady-slipper (*Cypripedium calceolus*), and several species of asters (*Aster* spp.) and goldenrods (*Solidago* spp.). This procedure is likely to be equally beneficial with respect to the introduction of appropriate soil microorganisms.

Acceleration of Succession

The restoration ecologist can test the various hypotheses that attempt to explain succession by the appropriate use of manipulation (Harper 1987). For example, an attempt can be made to "bypass" some of the normal seral stages by soil amelioration (simulating the effects of the previous seral stage according to the relay floristics hypothesis) or by planting species that are usually "later colonists" or have a poor dispersal mechanism. The latter approach is taken in Sudbury's land reclamation program, where the three native pines (jack, red, and white) do not colonize immediately in the absence of a suitable seed bed but are planted on treated land within the first few years after treatment (Beckett and Negusanti 1990). They are planted in small groups to avoid the appearance of a plantation, and it is hoped that they will form a seed source for future colonization.

Although the goal of the revegetation process in Sudbury is a stable, functioning, quasi-natural biotic community (i.e., one that has as large a component as possible of species and biotypes that have evolved under the regional environment), a biotic community identical to the "natural" one may be neither possible nor desirable as an endpoint. Not only is there no uniquely natural association of species in the region, but ecological concepts such as "association" and "climax" are, to quote Whittaker (1977) "abstractions; . . . essentially human creations serving to order, interrelate, and interpret some of the information about natural communities available to us."

Nutrient Cycling and the Potential for Nutrient Limitations

Phosphorus

The phosphorus "working capital" in a soil or in a soil parent material can limit the type of plant community that will ultimately occupy the site (Beadle and Burges 1949). In the short term, however, phosphorus limitation in the Sudbury area has only been shown on some sandy soils, where phosphorus deficiency becomes a secondary limiting factor once metal toxicity is eliminated. Nevertheless, when such sites were limed, fertilized, and sown to Canada bluegrass (*Poa compressa*) in 1974, 20 years later they were able to support clonal patches of trembling aspen and sweet fern (*Comptonia peregrina*), as well as jack pine (*Pinus banksiana*) and red pine (*P. resinosa*) plantations, suggesting that once plants are able to explore the deeper soil, the necessary phosphorus becomes available.

On the stony slopes, the reservoir of phosphorus present in the residual organic matter is sufficient to support plant growth for several years, depending on the degree of erosion that has occurred. Furthermore, limestone application has a beneficial effect in making phosphorus in acid soils available to plants, both through the release of phosphate from inorganic complexation (Brady 1974) and through the enhanced mineralization of organic matter (Fransen 1991). Furthermore, there is also a small reservoir of phosphorus in the unweathered glacial till pebbles that characterize these soils, and the activities of vesicular-arbuscular mycorrhizae, which aid in phosphorus nutrition (Chapter 16), are also stimulated by liming (Blundon 1976).

Plant and soil analysis indicate that a significant buildup of phosphorus has occurred in the revegetated system. The largest buildup is in the aboveground biomass, but with some increase also in root systems and soil. Species play distinct roles in the phosphorus cycle. For example, in Alsike clover, most of the phosphorus is located in the soil and the plant roots, whereas phosphorus buildup in birch is mainly in the shoot system, with a smaller buildup in soil phosphorus.

Nitrogen

Nitrogen deficiency is often the dominant limiting factor in land reclamation (Bradshaw and Chadwick 1980), but on the Sudbury barrens, the residual organic matter satisfies nitrogen needs for the first few years after treatment. Nevertheless, once the residual nitrogen is exhausted, biological nitrogen fixation must become part of the revegetation formula if a maintenance-free system is to be restored.

The mature native forest of the Sudbury area contains no leguminous species, nor do

FIGURE 13.8. Black locust growing on gullied clay soils east of the abandoned Coniston smelter.

leguminous species play a role in succession after disturbance. The only common nitrogen-fixer in upland sites is sweet fern, which plays a role in jack pine succession. In the Sudbury area, relict stands of sweet fern occur on the barrens and often colonize sites that have been limed but not necessarily seeded. Because volunteer establishment of sweet fern on restored sites is intermittent, acid-tolerant, cold-hardy, and easily obtainable, nitrogen-fixing species such as birdsfoot trefoil and Alsike clover are incorporated into the seed mixture.

On difficult sites, such as the silty clay "badlands" where the direct seeding approach does not work, the small leguminous tree black locust (*Robinia pseudo-acacia*) is planted (Beckett and Negusanti 1990) (Fig. 13.8). In its native habitat, black locust is a vigorous pioneer colonist of disturbed sites in the southeastern deciduous forest. Its tendency toward a growth decrease and possible mortality after 10–20 years (Boring and Swank 1984) suggests that it might play a useful role in soil stabilization and soil humus and nitrogen buildup before giving way to native woody species rather than persisting as an aggressive weed. Also, it will be interesting to observe what effect black locust will have on community dynamics, because there are reports that this species provides a favorable environment for colonizing by other woody species (e.g., Ashby et al. 1980).

Cycling and Fate of Potentially Toxic Metals

A potential negative effect of revegetation might be the translocation of metals by plants from the soil into the terrestrial food chain. Certainly, the copper and nickel contents of the herbaceous vegetation first established on reclaimed land are approximately 10 times as high as Freedman and Hutchinson's (1981) "normal" values (Winterhalder et al. 1984). In the case of tree species, Beckett and Negusanti (1990) compared 3-year-old jack pine needles from a revegetated site with those from a control site 50 km from Sudbury and found elevated levels of aluminum, copper, and nickel in the former. Nevertheless, liming itself reduces the potential metal uptake by plants. Winterhalder et al. (1984) have shown that with metal-tolerant strains of tufted hairgrass, liming brings about a significant decrease in leaf content of aluminum, copper, and nickel (aluminum > nickel > copper).

Although the detoxifying effect of liming on soil is a complex one, the reduced availability of metals after liming can still be demonstrated in chemical terms alone. Total metals in the soil have shown a reduction on revegetated land over the 12-year period after liming as a result of uptake of metal into plants and loss to the general environment. Some changes have also taken place in untreated soils, and a reduction in water-soluble copper and nickel has been found in barren soils collected west of the inactive Coniston smelter, presumably due to a combination of pH amelioration, leaching, and erosion.

General Discussion

Cairns (1979) proposed three management options for the reclamation of mined land: (1) doing nothing, (2) restoring to the original condition, or (3) reclaiming to an ecologically improved and socially acceptable state. The burgeoning "restorationist" movement has the second option as its commendable goal, but as Cairns pointed out, this is rarely attainable, and the mutual acceptance of option 3 by ecologists and industry is the one most likely to lead to a nonadversarial and productive working relationship. The Sudbury approach appears to be a compromise of the sort proposed by Cairns, in that the goal is a quasi-natural functioning ecosystem. Nevertheless, wherever possible, barriers to the ultimate development of the ecosystem toward the "climax" should be removed (e.g., by providing nuclei of understory species and microbiota). In a later paper, Cairns (1983) split his second option into "rehabilitation," in which the reclamation moves in the general direction of restoration, and "alternative ecosystems." Clearly, the Sudbury experience falls into the rehabilitation category.

In the Sudbury Land Reclamation Program (Winterhalder 1985, 1987, 1988), the use of minimal amelioration and minimal seeding rates makes for a lean, diverse physical environment, with sparse initial cover, which is very suitable for colonization by a diversity of species and probably by a diversity of genetic variants of some of the species. It also appears to achieve the twin goals of optimal cover and optimal diversity, considered by Davis et al. (1985) to be incompatible.

Based on a relatively short time frame of 15 years, the Sudbury story appears to be an exception to Bradshaw's (1987b) statement that "in the case of metal-contaminated sites . . . , the nature of the toxicity is such that direct treatment is not completely satisfactory." Bradshaw goes on to suggest that one should look to metal-tolerant ecotypes as the answer, but the Sudbury experiment strongly suggests that in the case of metal-contaminated soils (if not in the case of mine wastes), once the initial barrier to plant growth is overcome, the vegetation itself will continue to ameliorate the soil by the production of insulating leaf litter and metal-chelating humus, by transporting bases to the surface, by fixing nitrogen, and by modifying the microclimate.

Ecologists can learn a great deal from the observation of ecosystem degradation and restoration (Cairns 1981; Bradshaw 1987a). The dynamics of the Sudbury landscape will provide research opportunities for decades to come, especially the chance to compare parallel changes in treated and untreated land.

Acknowledgments. John Cairns, John Gunn, Robert Hedin, and Tom Peters kindly provided review comments.

References

Aber, J.D. 1987. Restored forests and identification of critical factors in species–site interactions, pp. 241–250. *In* W.R. Jordan, M.E. Gilpin, and J.D. Aber (eds.). Restoration Ecology. Cambridge University Press, Cambridge.

Ashby, W.C., C. Kolar, and N.F. Rodgers. 1980. Results of 30-year old plantations on surface mines in the Central States, pp. 99–107. *In* Proceedings of Trees for Reclamation in the Eastern United States Symposium, Lexington, Kentucky, October 27–29. General Technical Report NE-61. USDA Forest Service, Broomal, PA.

Beadle, N.C.W., and A. Burges. 1949. Working capital in a plant community. Aust. J. Sci. 11:207–208.

Beckett, P.J., and J. Negusanti. 1990. Using land reclamation practices to improve tree condition in the Sudbury smelting area, Ontario, Canada, pp. 307–320. *In* J. Skousen et al. (eds.). Proceedings of 1990 Mining and Reclamation Conference and Exhibition. West Virginia University, Morgantown.

Blundon, R.A. 1976. Vesicular-arbuscular mycorrhizae in industrially denuded soils. B.Sc. thesis, Laurentian University, Sudbury, Ontario.

Boring, L.R., and W.T. Swank. 1984. The role of black locust (*Robinia pseudo-acacia*) in forest succession. J. Ecol. 72(3):749–766.

Bradshaw, A.D. 1987a. The reclamation of derelict land and the ecology of ecosystems, pp. 23–29. *In* W.R. Jordan, M.E. Gilpin, and J.D. Aber (eds.). Restoration Ecology. Cambridge University Press, Cambridge.

Bradshaw, A.D. 1987b. Restoration: an acid test for ecology, pp. 54–74. *In* W.R. Jordan, M.E. Gilpin, and J.D. Aber (eds.). Restoration Ecology. Cambridge University Press, Cambridge.

Bradshaw, A.D., and M.J. Chadwick. 1980. The Restoration of Land. University of California Press, Berkeley.

Brady, N.C. 1974. The Nature and Properties of Soils. 8th ed. Macmillan, New York.

Cairns, J. 1979. Ecological considerations in reclaiming surface mined lands. Min. Environ. 1:83–89.

Cairns, J. 1981. Restoration and management: an ecologist's perspective. Restoration Management Notes 1(1):6–8.

Cairns, J. 1983. Management options for rehabilitation and enhancement of surface-mined ecosystems. Miner. Environ. 5:32–38.

Davis, B.N.K., K.H. Lakhani, M.C. Brown, and D.G. Park. 1985. Early seral communities in a limestone quarry—an experimental study of treatment effects on cover and richness of vegetation. J. Appl. Ecol. 22(2):473–490.

Fransen, K. 1991. The effects of land reclamation liming practices on phosphate fixation and release in acid, metal-contaminated Sudbury-region soils. B.Sc. thesis, Laurentian University, Sudbury, Ontario.

Freedman, B., and T.C. Hutchinson. 1981. Sources of metal and elemental contamination of terrestrial environments, pp. 35–94. *In* N.W. Lepp (ed.). Effect of Heavy Metal Pollution on Plants. Vol. 1. Effects of Trace Metals on Plant Function. Applied Science Publishers, London.

Harper, J.L. 1987. The heuristic value of ecological restoration, pp. 23–29. *In* W.R. Jordan, M.E. Gilpin, and J.D. Aber (eds.). Restoration Ecology. Cambridge University Press, Cambridge.

Lozano, F.C., and I.K. Morrison. 1981. Disruption of hardwood nutrition by sulfur dioxide, nickel and copper near Sudbury, Canada. J. Environ. Qual. 10(2):198–204.

Miller, G. 1978. A method of establishing native vegetation on disturbed sites, consistent with the theory of nucleation, pp. 322–327. *In* S.B. Lowe (ed.). Proceedings of the Third Annual Meeting, Canadian Land Reclamation Association, Laurentian University, May 29–June 1, 1978. CLRA, Guelph, Ontario.

Skaller, P.M. 1981. Vegetation management by minimal interference: working with succession. Landscape Planning 8:149–174.

Whittaker, R.H. 1977. Recent evolution of ecological concepts in relation to the eastern forests of North America, pp. 340–358. *In* History of American Ecology. Arno Press, New York.

Winterhalder, K. 1974. Reclamation studies on industrial barrens in the Sudbury area. *In* Proceedings of the Fourth Annual Workshop, Ontario Cover Crop Committee, University of Guelph, Ontario, December 1974.

Winterhalder, K. 1983. Limestone application as a trigger factor in the revegetation of acid, metal-contaminated soils of the Sudbury area, pp. 201–212. *In* Proceedings of the Canadian Land Reclamation Association, University of Waterloo, Ontario, August 1983. CLRA, Guelph, Ontario.

Winterhalder, K. 1985. The use of manual surface seeding, liming and fertilization in the reclamation of acid, metal-contaminated land in the Sudbury, Ontario mining and smelting region of Canada, pp. 196–204. *In* D. Williams and S.E. Fisher (eds.). Proceedings of the 1985 National Meeting, American Society for Surface Mining and Reclamation, Denver, Colorado, October 8–10, 1985. ASSMR, Princeton, WV.

Winterhalder, K. 1987. The Sudbury Regional Land Reclamation Program—an ecologist's perspective, pp. 81–89. *In* P.J. Beckett (ed.). Proceedings of the Twelfth Annual Meeting, Canadian Land Reclamation Association, Laurentian University, June 7–11, 1987. CLRA, Guelph, Ontario.

Winterhalder, K. 1988. Trigger factors initiating natural revegetation processes on barren, acid, metal-contamined soils near Sudbury, Ontario smelters, pp. 118–124. *In* Mine Drainage and Surface Mine Reclamation. Vol. 2. Information Circular 9184. Dept of the Interior. U.S. Bureau of Mines, Pittsburgh, PA.

Winterhalder, K., P.J. Beckett, and M.R. Todd. 1984. Metal dynamics in a revegetated ecosystem at Sudbury, Ontario, pp. 499–504. *In* Proceedings of the First Annual Conference on Environmental Contamination, London, July 1984. CEP Consultants, Edinburgh.

Yarranton, G.A., and R.G. Morrison. 1974. Spatial dynamics of a primary succession: nucleation. J. Ecol. 62(2):417–428.

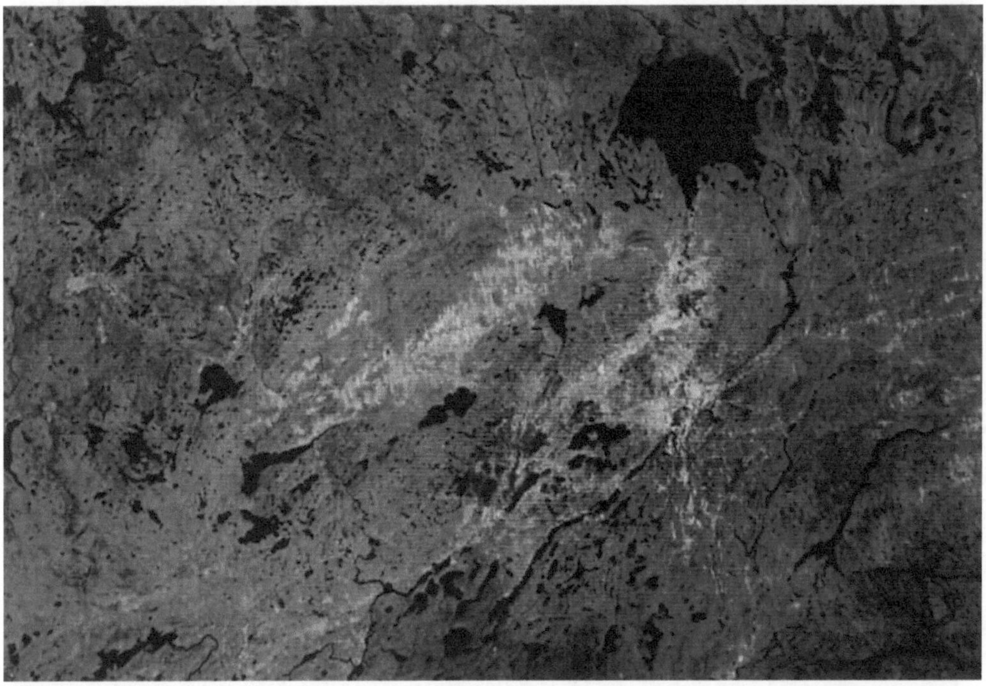

PLATE 1. LANDSAT image of Sudbury's industrial barrens (blue-gray area). Satellite image recorded July 25, 1987.

PLATE 2. A 1979 picture of the 2286-m long O'Donnell roast yard that operated from 1916 to 1929. (Photo by W. McIlveen.)

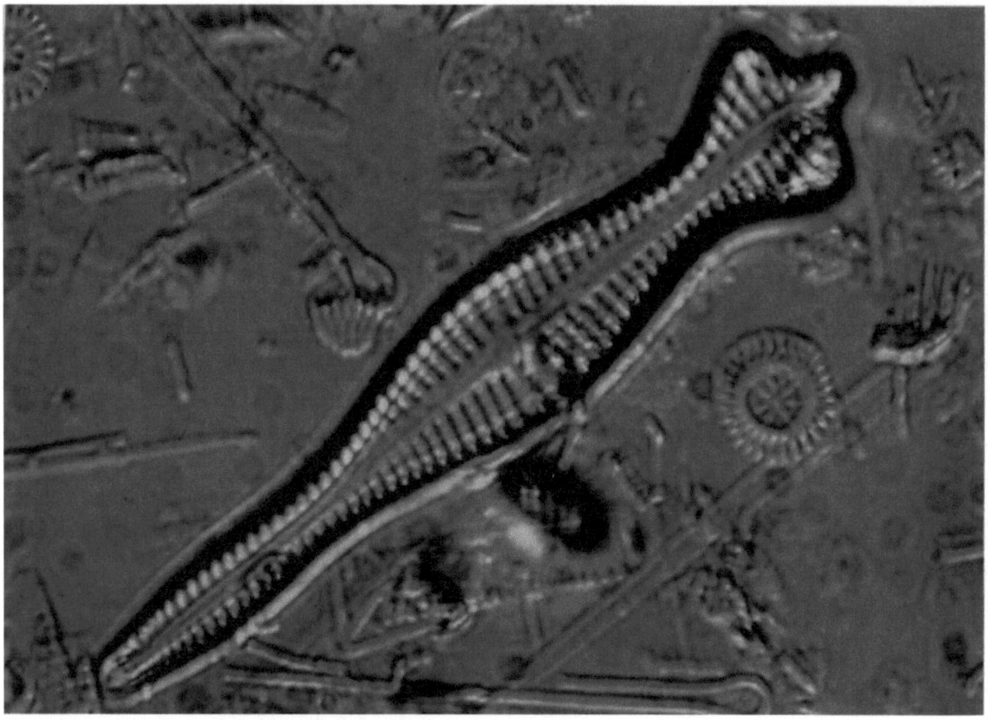

PLATE 3. Microfossils of diatoms from lake sediments. (Photo courtesy of J. Smol.)

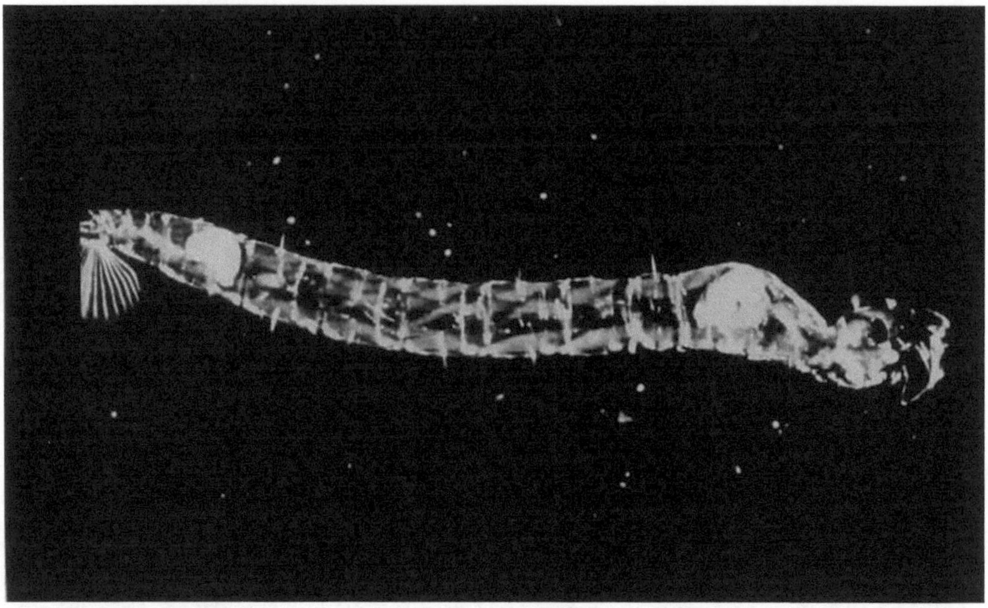

PLATE 4. *Chaoborus* larvae. (Photo courtesy of A.Uutala.)

PLATE 5. Sulfur dioxide injury to white pine. (Photo by P. McGovern.)

PLATE 6. Sulfur dioxide injury to white birch. (Photo by J. Negusanti.)

PLATE 7. Coppiced birch forest. (Photo by K. Winterhalder.)

PLATE 8. A fruticose lichen, an indicator of improving air quality. (Photo by J. Gunn.)

PLATES 9–11. Revegetation of the Cambrian Heights site in Sudbury. (Photos by K. Winterhalder.)

July 29, 1981

May 20, 1982

July 6, 1988

PLATE 12. Aurora trout lakes 110 km north of Sudbury. (Photo by E. Snucins.)

PLATE 13. Pouring slag. (Photo by B. Chambers.)

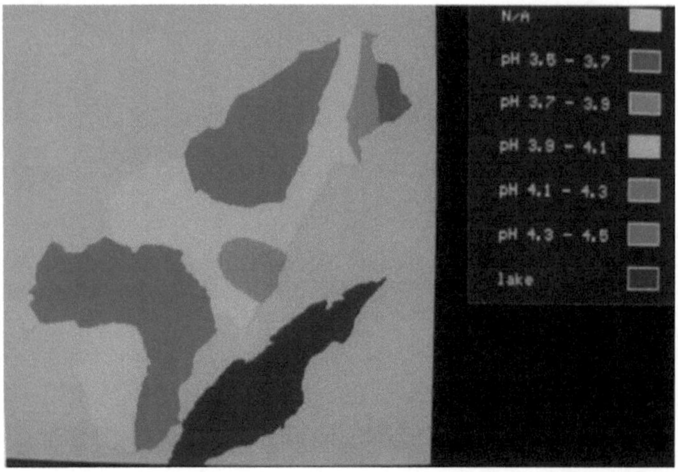

PLATE 14. Soil pH for two sub-waterheads at Daisy Lake. (Photo by A. Gallie.)

PLATE 15. Male aurora trout in spawning colors. (Photo by E. Snucins.)

PLATE 16. Tailings disposal. (Photo by K. Winterhalder.)

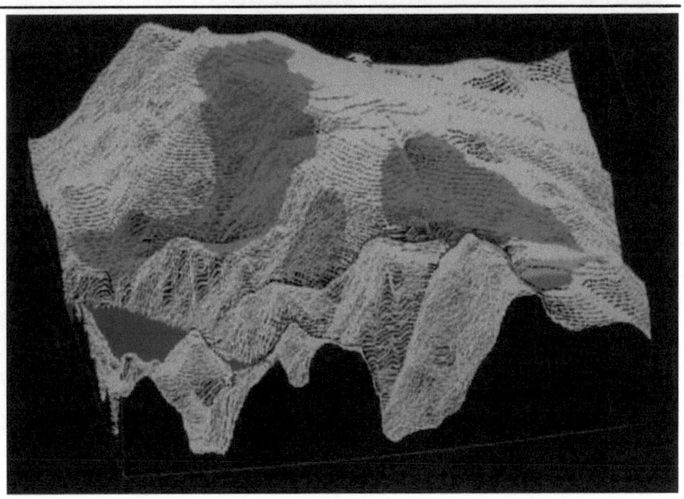

PLATE 17. Soil pH draped over topography at Daisy Lake subwatersheds. (Photo by A. Gallie.)

PLATE 18. Aerial view of downtown Sudbury.

14

Lake Sediments: Sources or Sinks of Industrially Mobilized Elements?

Nelson Belzile and J. Robert Morris

The once rather pessimistic outlook on the environmental impact of trace elements dispersed from Sudbury's smelters is gradually being replaced by a more optimistic view. The optimism has been prompted by several clear indications that the recent period of declining smelter emissions has resulted in both biological (Gunn and Keller 1990; Keller et al. 1992a) and chemical (Keller et al. 1992b) improvements in local lakes.

Lake sediments can become the ultimate site of deposition for many of the trace elements introduced into the environment, but sediments cannot be considered as a stable and non-reactive milieu. Potent chemical and biochemical reactions and transformations occur in the sediments, making the biogeochemical behavior of deposited trace elements highly dynamic. The relevant processes include microbially mediated decomposition of organic matter, precipitation and dissolution of minerals, sorption and desorption of trace elements on living and non-living particles, and the diffusion of dissolved constituents along concentration gradients (Fig. 14.1).

In this context, it is relevant to ask whether processes in the lake sediments influence the rate or extent of whole-lake recovery and whether the sediment environment itself exhibits either beneficial or undesirable responses. In particular, if sediments release potentially harmful chemicals back to the overlying waters, this might slow the rate of lake recovery. There are strong indications that in sulfur-rich sediments containing little or no oxygen, the precipitation of sulfide minerals can greatly limit the solubility of many trace elements (Carignan and Nriagu 1985). Although, in contrast, oxygen-rich sediments generally show undersaturation with respect to trace element solid phases (Tessier 1992), trace element dissolution may, in fact, be insignificant because oxygen-rich sediments commonly have a high content of iron and manganese oxyhydroxides, and these insoluble compounds exhibit a great capacity for trace metal sorption.

In this chapter, we present details on some of the processes controlling the solubility and hence the mobility of metals and other trace elements in lake sediments near Sudbury and discuss the consequences of lowered sulfur emission rates and in-lake sulfur reduction on sediment pH and trace element geochemistry.

Chemical Element Profiles in Lake Sediments

Facilities for smelting copper and nickel ores are everywhere known as a major source of atmospheric emissions of iron, sulfur, and the trace elements antimony, arsenic, cadmium, cobalt, copper, lead, nickel, selenium, and zinc (Nriagu and Pacyna 1988). What happens to

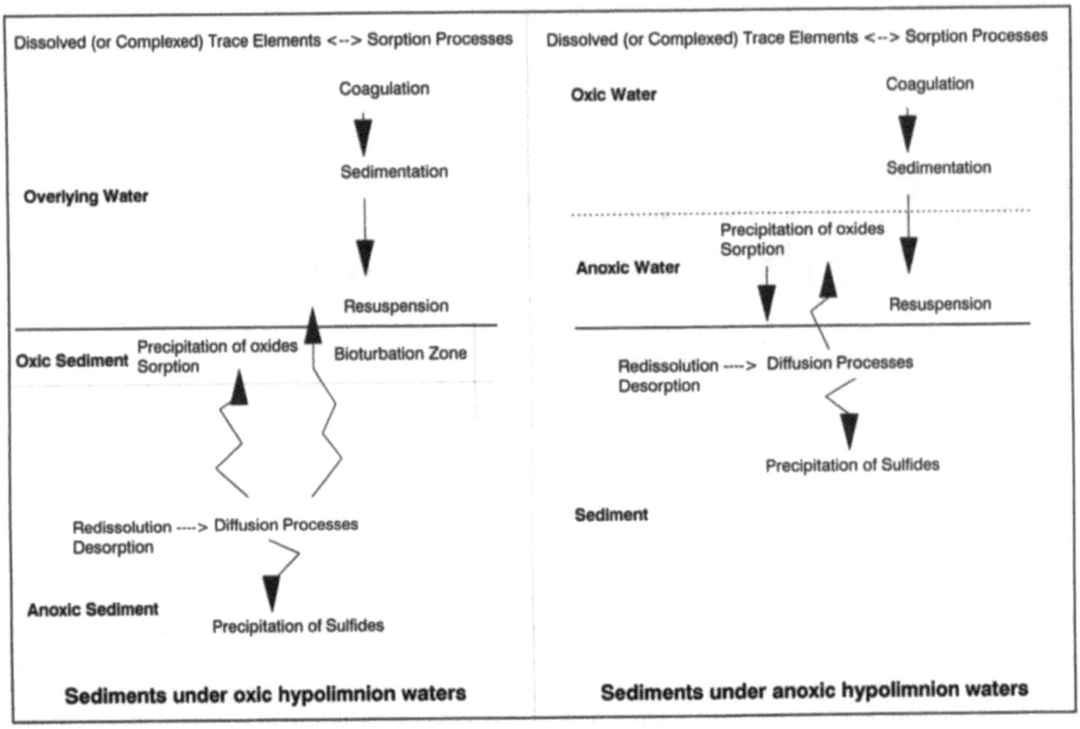

FIGURE 14.1. Schematic representation of the processes controlling the behavior of trace elements in sediments.

these elements after they fall into or are carried into lakes in the Sudbury area?

It is apparent from many investigations that significant quantities of trace elements have been deposited in the sediments of Sudbury area lakes, and sulfur, copper, and nickel enrichment can still be detected more than 60 km away (Semkin and Kramer 1976; Nriagu et al. 1982). Enrichment factors, defined as the ratio of total element concentration in the surficial layer to the average concentration in the deeper or precolonial layers, are generally lower in the acidic than the non-acidic lakes near Sudbury, reflecting the higher solubility of trace elements at lower pH (Carignan and Nriagu 1985; Tessier 1992). Figure 14.2 shows total concentrations of several elements as a function of depth below the sediment surface observed in McFarlane Lake, a mesotrophic near-neutral lake in Sudbury, during the early 1980s (Nriagu et al. 1982; Nriagu 1983; Nriagu and Coker 1983; Nriagu and Wong 1983). More recent studies (Dillon and Smith 1984; Nriagu and Rao 1987) show a sharp decline in the metal content of the most recently deposited sediment in Clearwater, Silver, and McFarlane lakes.

The decline in metal levels in the upper layer (i.e., the most recent sediments) could be a reflection of lowered smelter emission rates (Nriagu and Rao 1987). However, the fine-scale distribution patterns of trace elements in the surface sediments must be interpreted with caution because trace elements levels may be influenced by chemical transformations and element remobilization that occurred after deposition (Carignan and Nriagu 1985; Carignan and Tessier 1985; Belzile and Tessier 1990). The degree of oxidation or reduction of the sediment surface and the position of the sulfate-reducing zone seem to be particularly important factors. The location of the sulfate-reducing zone within the sediment profile controls the location of metal sulfide precipitation, and in sulfur-enriched lakes, such precipitates are undoubtedly important. In Sudbury

FIGURE 14.2. Concentration profiles of arsenic, selenium, nickel, copper, and sulfur in sediments of McFarlane Lake showing the surface enrichment attributed to emissions from smelters (from Nriagu et al. 1982; Nriagu 1983; Nriagu and Coker 1983; Nriagu and Wong 1983).

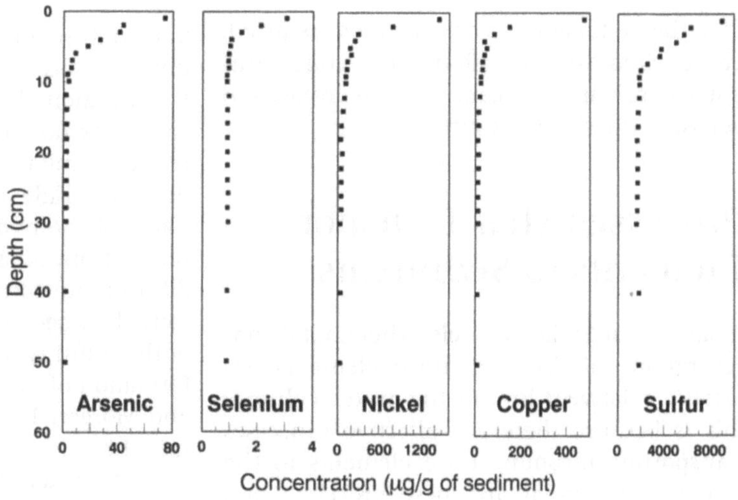

area lakes, the zone of maximum sulfur concentration may sometimes occur immediately beneath the sediment surface, but it has frequently been found below the 5-cm depth (Fig. 14.3) (Morris, *unpublished data*). Lowered trace element concentrations could also result from dilution effects associated with the sediment surface. Potential diluting materials are freshly sedimented organic material, especially in productive lakes, and iron and manganese oxyhydroxides, particularly in unproductive lakes. In unproductive lakes, with little organic matter settling to the bottom, organic decomposition may consume so little oxygen

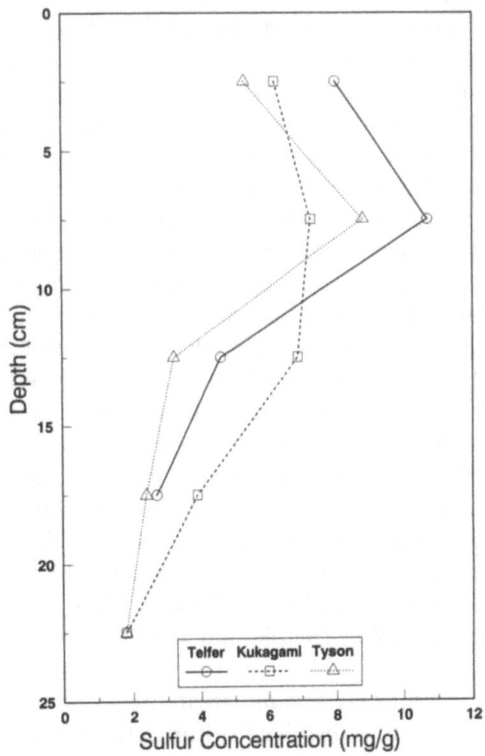

FIGURE 14.3. Depth-distributions of total sulfur in three lakes near Sudbury (from Morris, *unpublished data*).

that the sediment surface remains oxidized. These circumstances allow high concentrations of iron and manganese oxyhydroxides to persist at the sediment surface.

Processes That Transport Elements to Sediments

In lakes near Sudbury, as elsewhere, total concentrations of dissolved trace elements are usually relatively low in the water column; this is because there are effective processes transporting incoming trace elements to the lake bottom. What are these processes and how do they function?

The sedimentation of biogenic particulates is an important mechanism by which trace elements are removed from the water column (Sigg 1985; Sigg et al. 1987). Various planktonic organisms take up trace elements from the water by physiological processes. More important, chemical complexation may bind trace elements to dead plankton, to particles of decaying plant material, to fecal pellets of planktonic animals and protozoans, and to humic particles, and all these tend to sink to the lake bottom and thus remove trace elements from the water column. Another important mechanism moving trace elements from solution to the sediments involves the more or less continuous scavenging by iron and manganese oxyhydroxides. Iron and manganese are very abundant in most soils. Because these two metals tend to precipitate rapidly in the presence of oxygen, they typically enter lakes dissolved in anoxic groundwater seepages or directly as oxyhydroxides. Once in the lake, dissolved forms usually encounter dissolved oxygen and a near-neutral pH, and insoluble oxyhydroxides begin to form (Stumm and Morgan 1981). Because precipitated iron and manganese oxyhydroxides readily adsorb or form complexes with many dissolved ions, trace elements are firstly scavenged from the water column and then are carried to the bottom when the precipitates sink (Belzile and Tessier 1990; Tessier 1992).

Sulfur may accumulate in the sediments because more dissolved sulfur ions, such as sulfate, diffuse into the sediment than diffuse out. This occurs because anaerobic layers in the sediment harbor species of bacteria that can break down or oxidize organic matter by reducing sulfate and other sulfur ions to hydrogen sulfide (Stumm and Morgan 1981). Some of the hydrogen sulfide generated may return from the sediment to the lake water by diffusing upward or in rising bubbles, but because hydrogen sulfide is highly reactive, much of the sulfur will be fixed in solids through the formation of highly insoluble sulfides (Stumm and Morgan 1981). In acidic Clearwater Lake (pH 4.8), concentration profiles of dissolved copper, nickel, and zinc near the sediment–water interface suggest that these metals were also diffusing into the sediments along a concentration gradient and then precipitating (Fig. 14.4). These findings contradict other field studies (Sigg et al. 1987) that have emphasized the removal of trace elements from the water column through binding to sinking particles, and it may be that the diffusion of trace metals into sediments is only significant in very acidic lakes.

Chemical Transformations within the Sediments

What circumstances or processes influence the mobility and distribution of trace elements after they have become part of the sediments? Here, more detailed focus must be put on factors that control the stability of iron and manganese oxyhydroxides and the formation of metal sulfides.

Importance of Iron Oxyhydroxides

At or near their surface, typical lake sediments become anoxic, reducing environments because of oxygen consumption during the microbial decomposition of organic matter. In circumstances in which stronger oxidants such as dissolved oxygen, nitrate, or manganese oxides have already been depleted, the iron in

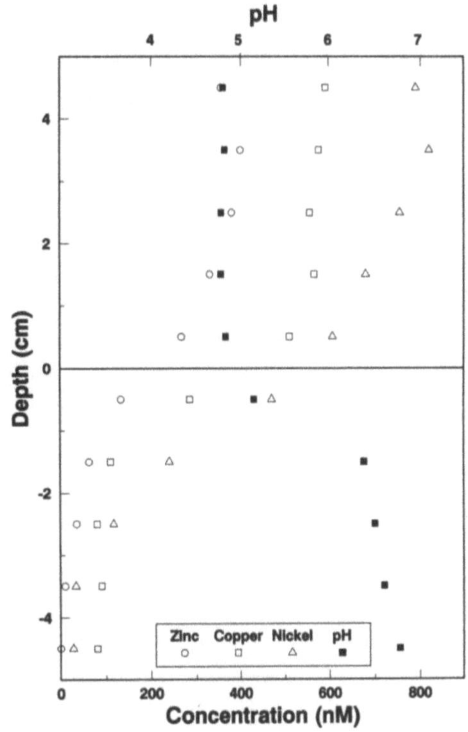

FIGURE 14.4. Depth-distributions of dissolved zinc, copper, nickel, and pH values in porewaters and overlying waters at a littoral station of Clearwater Lake. *Horizontal line* represents the sediment–water interface (from Carignan and Nriagu 1985).

oxyhydroxides converts from the ferric oxidation state Fe(III) to the ferrous state Fe(II), and the former iron oxyhydroxides dissolve. Consequently, trace elements complexed with iron oxyhydroxides are released into solution and are then immediately free to move away by diffusion.

Arsenic provides an excellent example of the close coupling between a trace element's chemical behavior and that of iron (Belzile and Tessier 1990). In Figure 14.5, each profile of dissolved iron and dissolved arsenic shows a peak located a few centimeters below the sediment–water interface; both elements were released to solution by the reductive dissolution of buried iron oxyhydroxides. In summer or winter lake stagnation periods, oxygen depletion at the sediment surface and the consequent upward migration of the oxidation-reduction transition zone into the water column can also lead to seasonal episodes of iron and trace element dissolution (see Fig. 14.1). Because the peaks in the concentration of dissolved iron Fe(II) and dissolved arsenic are situated below the sediment surface (Fig. 14.5), two sink propagating mechanisms are implied. Above the peaks, oxygen was apparently diffusing downward from the lake water, oxidizing upward diffusing Fe(II), and enriching the surficial sediments in iron oxyhydroxides. Simultaneously, upward diffusing arsenic was being removed from solution in this zone by iron oxyhydroxide sorption (Belzile and Tessier 1990). Below the peaks, downward diffusing dissolved Fe(II) and arsenic were probably precipitated by reactions involving hydrogen sulfide. Sufficient hydrogen sulfide can be generated because Sudbury's lakes are sulfur-enriched.

The above hypotheses were confirmed when Teflon sheets inserted into the sediments collected iron oxyhydroxide precipitates near the sediment surface, and iron sulfide precipitates a few centimeters deeper (Belzile et al. 1989). The decline in dissolved arsenic below its peak may have resulted

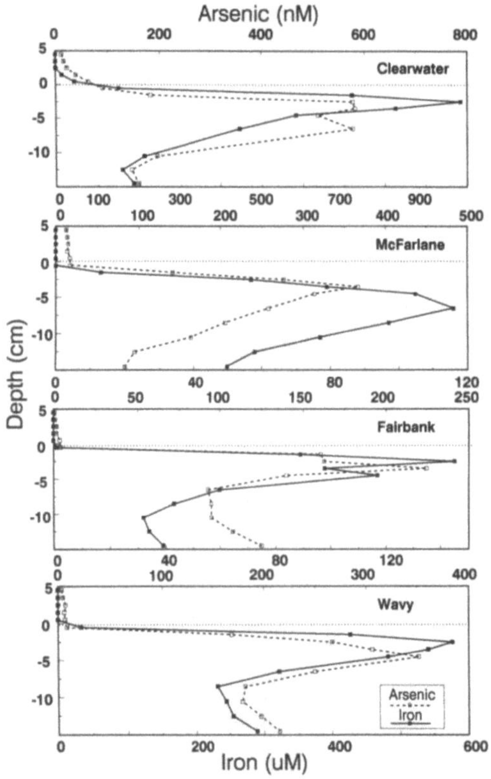

FIGURE 14.5. Depth-distributions of total dissolved arsenic and iron in porewaters and overlying waters of four lakes near Sudbury. *Horizontal dotted lines* represent the sediment-water interface (from Belzile and Tessier 1990).

from precipitation of arsenic sulfides (As_2S_3 or FeAsS) and arsenic adsorption onto iron sulfide; these reactions have been reported frequently elsewhere (Aggett and O'Brien 1985; Belzile and Lebel 1986; Edenborn et al. 1986; Belzile 1988).

If the sediment's surface is sufficiently reduced, dissolved ferrous iron and trace elements may be released directly into the overlying anoxic lake water, and the sediment may become a source rather than a sink for dissolved trace elements. This is shown for arsenic in Figure 14.5 at a site in Clearwater Lake. Two factors tend to minimize the significance of such releases. First, oxygenated lake water will usually occur at some level higher in the water column. Here, oxygen will limit the upward diffusion of ferrous iron by again precipitating iron oxyhydroxides (see Fig. 14.1), which will bind with trace elements returning them again toward the lake bottom. Second, in sulfur-enriched Sudbury lakes, the hydrogen sulfide available near the surface of anoxic sediments may be sufficient to react with most of the dissolving arsenic, ferrous iron, and trace metal ions and may render them insoluble before they have time to diffuse appreciably into the lake water.

Precipitation of Sulfides—Saturation Indices

Porewater profiles of dissolved copper, nickel, and zinc in Clearwater Lake showed distinct losses in metal solubility that were associated with the transition to anoxia at 2 cm below the sediment–water interface (see Fig. 14.4). This correlation with anoxia suggested that the low dissolved metal concentrations may have resulted from precipitation of copper, nickel, and zinc sulfides. Saturation index calculations (Box 14.1) for the zone observed to be low in dissolved metals indeed showed that saturation of the porewaters was to be expected with respect to zinc sulfide and nickel

Box 14.1.

Saturation index (SI) is defined as log (IAP/Ksp), where Ksp is the solubility-product constant. The ion activity product (IAP) is equal to the product of the specific activity (*a*) of the different ions forming a precipitate or solid. A simplified way of calculating the IAP for iron sulfide (FeS) would be

$$\text{IAP FeS} = (a_{Fe}2+)\ (a_S2-)$$

But because of the very weak second dissociation constant of H_2S, the HS^- ion is more relevant to this calculation than the S^{2-} ion considered above. A better way of calculating the IAP would therefore be

$$\text{IAP FeS} = (aFe^{2+})\ (a_{HS}-/a_H+)$$

Negative values of IAP indicate undersaturation and no precipitation. The precipitate should be forming if the IAP value is either zero, indicating saturation, or greater than zero, indicating supersaturation.

sulfide and indicated oversaturation with respect to copper sulfide (Carignan and Nriagu 1985; Carignan and Tessier 1985). It was concluded, therefore, that sediments more than 2 cm below the sediment–water interface are a sink for trace metals in Sudbury area lakes. These sediments would only cease to be a sink in the event that they ceased to be anaerobic. Such a situation occurred when Lake Laurentian was drained and reflooded in 1982. Subsequent increases in the concentrations of metals in water and macrophyte tissues were observed (Keller 1984).

Similar patterns of trace element insolubility have also been found with arsenic and iron, and they occurred in lakes representing a range of pH levels (Fig. 14.5). Saturation index calculations and analysis of black deposits on Teflon collectors (Belzile et al. 1989) indicate that dissolved iron concentrations can also be controlled in anoxic porewaters by precipitation of a sulfide such as iron sulfide. Finally, an increase in pH has been observed in Sudbury area sediments at the levels of intense sulfate reduction (see Fig. 14.4), especially where the sediment surface is anoxic and high in porosity (Fig. 14.6). This phenomenon is consistent with the hypothesis that the reduction of iron oxyhydroxides, the reduction of sulfates to sulfides, the precipitation of iron sulfide, and the generation of alkalinity occur concurrently.

$$4Fe(OH)_3(s) + 4SO_42- + 9CH_2O \rightarrow 4FeS(s) + CO_2 + 8HCO_3- + 11H_2O \quad (1)$$

Metal Sorption on Oxide Surfaces—The Influence of pH

In contrast with anoxic porewaters, where precipitation of sulfides seems to control the concentrations and solubilities of dissolved trace element, oxic porewaters and overlying waters do not show saturation with known solid phases of trace elements. More specifically, arsenic (Belzile and Tessier 1990), cadmium, copper, lead, nickel, and zinc (Tessier et al., 1994) should not form any oxide, hydroxide, or carbonate precipitates either within or above oxic lake sediments, because the relevant saturation indices have negative values. Because iron and sometimes manganese oxyhydroxides do precipitate in and above oxic sediments and because these compounds have the ability to adsorb trace elements, the evidence for adsorption should be examined in more detail. According to the surface complexation theory for low-density adsorption (Tessier et al., 1994), the propensity of any trace metal to adsorb onto iron oxyhydroxide can be represented by an apparent equilibrium constant (K_M), which can be determined by

$$K_M = \frac{N_{Fe}\ {}^*K_M}{[H^+]^{m+1}} = \frac{\{Fe\text{–}O\text{–}M\}}{\{Fe\text{–}Ox\}[M^{z+}]} \quad (2)$$

where {Fe-O-M} is the equilibrium concentration of the adsorbed trace metal, {Fe-Ox} is the concentration of the iron fixed in iron oxyhydroxides, $[M^{z+}]$ is the equilibrium concentration of the trace metal ion remaining free in solution, and N_{Fe} is the number of adsorbing sites per mole of iron oxyhydroxides.

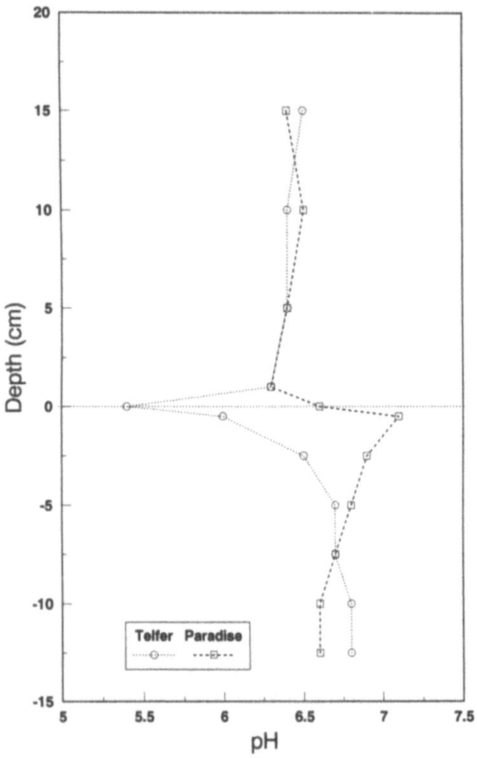

FIGURE 14.6. Profiles of pH values measured at the sediment interface of two lakes located at approximately 50 km from Sudbury. Paradise Lake showed a negative redox potential at 2.5 cm in the sediments that had also a low bulk density (high water content) and a high excess of sulfur. Telfer Lake showed a positive redox potential value at 2.5 cm, and sediments had a high bulk density and a smaller excess of sulfur than Paradise Lake. *Horizontal dotted line* represents the sediment–water interface (from Morris, *unpublished data*).

The concentration of both iron oxyhydroxides {Fe-Ox} and adsorbed trace metal {Fe-O-M} can be determined by analyzing partial chemical extracts from samples of oxic sediments or by inserting Teflon collectors in oxic sediments and subsequently dissolving and analyzing the iron oxyhydroxide-trace metal complexes that deposit on them. The concentrations of the dissolved free metal ions [M^{z+}] is calculated from the total dissolved trace metal and the inorganic ligands concentrations in the overlying waters, using porewater peepers (Carignan 1984).

On theoretical grounds, the adsorption of trace metals by iron oxyhydroxides should increase with ambient pH (Equation 2). To test this prediction in the natural environment, the apparent equilibrium constant for the adsorption of zinc on iron oxyhydroxides (K_{Zn}) was determined for oxic sediments from 41 lakes spread over an area of 350,000 km^2 in Ontario and Quebec. When these K_{Zn} values are regressed on lake pH (Fig. 14.7), the expected trend is observed, and differences in pH explained about 90% (r^2 = .89) of the variation in K_{Zn} (Tessier et al. 1989). In the same geographic area, similar linear regression models having slopes close to 1 were also obtained for the adsorption of cadmium (r^2 = .80; n = 26), copper (r^2 = .75; n = 39), lead (r^2 = .81; n = 7), and nickel (r^2 = .87; n =29) (Tessier 1992).

Substantial reductions in sulfur emissions from Sudbury area smelters have allowed lake pH values to increase, and sulfate reduction has increased sediment pH in many local lakes. Trace metal solubilities should therefore be decreasing in most Sudbury area lakes because sorption to iron oxyhydroxides has undoubtedly been increasing. In the past 20 years, the pH of Clearwater Lake has increased from 4.3 (Dillon et al. 1986) to 5.0 (Belzile, *unpublished data*). Assuming a constant ratio of adsorbed metal to iron oxyhydroxides, Equation 2 indicates that the solubility of zinc has declined by more than seven times during the past two decades, and the situation should be similar for other trace metals.

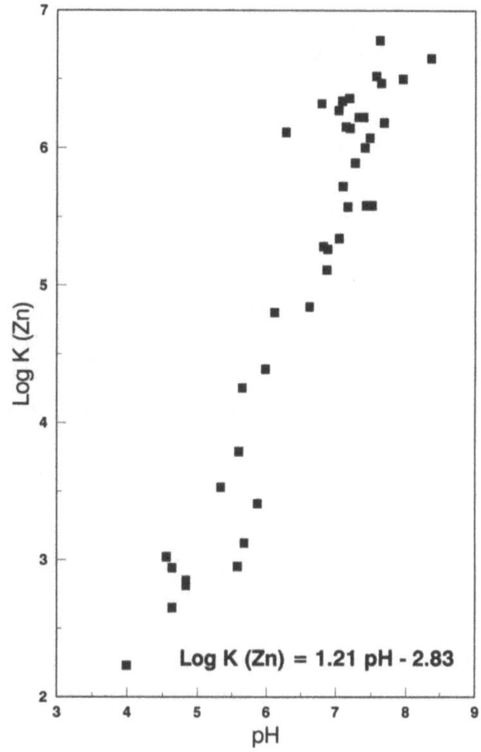

FIGURE 14.7. Relationship between the apparent equilibrium constant for the adsorption of zinc on natural iron oxyhydroxides (expressed here as the log of the ratio of adsorbed zinc to iron oxyhydroxides and dissolved zinc) and pH obtained for 41 lakes (from Tessier et al. 1989).

Management Implications

In some cases, the natural recovery of lake systems is not possible in a short period of time, and remedial measures (NRC 1992; Rhodes and Wiley 1993) have to be taken to mitigate the effects of acidification and contamination. One approach that has been commonly used involves the addition of a base (such as $CaCO_3$), resulting in increased pH and acid-neutralizing capacity and reduced trace elements concentrations. However, hydrologic inputs of acidic water from precipitation, stream inflow, and groundwater can cause reacidification and an eventual deterioration in water and sediment quality (Driscoll et al. 1989). The coverage of contaminated sediments or their removal by dredging represents other approaches for which the feasibility and the cost have to be considered (Box 14.2). The dredging option requires subsequent treatment and/or disposal to reduce or eliminate the toxicity of contaminants.

Box 14.2. Ore Grade Sediments

Kelly Lake is located in the immediate vicinity of the Inco smelter. There are dumps of tailings nearby, and for many years, it has received mining and smelter effluents. Also, the lake received high atmospheric inputs of metals and trace elements, resulting in significant sediment accumulation of metals such as copper (5200 $\mu g{\cdot}g^{-1}$) and nickel (10,560 $\mu g{\cdot}g^{-1}$) but there are also significant amounts of noble metals such as platinum (1.8 $\mu g{\cdot}g^{-1}$), palladium (0.2 $\mu g{\cdot}g^{-1}$), gold (0.3 $\mu g{\cdot}g^{-1}$), and iridium (0.03 $\mu g{\cdot}g^{-1}$) in the lake sediments (Crocket and Teruta 1976). Most of the early processing methods were largely ineffective at extracting the noble metals. Considering the high value of noble metals, it might now be economically feasible to dredge and process Kelly Lake sediments to extract those metals. The challenge is to find an environmentally sound means of disposal of the extracted sediments.

Summary

Knowledge of the sedimentary processes controlling the behavior of trace elements is now good but still remains incomplete. Sediments remain a complex environment, and the contrasting behavior of some trace elements may indicate that opposite mechanisms (sedimentation/resuspension; diffusion to/from sediments; precipitation/dissolution; sorption/desorption; complexation/decomplexation) may often, in fact, operate concurrently. It is nevertheless clear that reductions of sulfur emissions greatly improved surface water quality and, consequently, sediment quality. The continuous pH increase contributes to decrease the solubility (and mobility) of trace elements through sorption processes on surficial oxic sediments. Precipitation of trace elements as sulfides that probably occur in anoxic sediments may ultimately constitute a major sink for toxic trace elements.

References

Aggett, J., and G.A. O'Brien. 1985. Detailed model for the mobility of arsenic in lacustrine sediments based on measurements in Lake Ohakuri. Environ. Sci. Technol. 19:231–238.

Belzile, N. 1988. The fate of arsenic in sediments of the Laurentian Trough. Geochim. Cosmochim. Acta 52:2293–2302.

Belzile, N., R.R. De Vitre, and A. Tessier. 1989. In situ collection of diagenetic iron and manganese oxyhydroxides from natural sediments. Nature (Lond.) 340:376–377.

Belzile, N., and J. Lebel. 1986. Capture of arsenic by pyrite in near-shore sediments. Chem. Geol. 54: 279–281.

Belzile, N., and A.Tessier. 1990. Interactions between arsenic and iron oxyhydroxides in lacustrine sediments. Geochim. Cosmochim. Acta 54:103–109.

Carignan, R. 1984. Interstitial water sampling by dialysis: methodological notes. Limnol. Oceanogr. 14:454–458.

Carignan, R., and J.O. Nriagu. 1985. Trace metal deposition and mobility in the sediments of two lakes near Sudbury, Ontario. Geochim. Cosmochim. Acta 49:1753–1764.

Carignan, R., and A. Tessier. 1985. Zinc deposition in acid lakes: the role of diffusion. Science 228: 1524–1526.

Crocket, J.H., and Y. Teruta. 1976. Pt, Pd, Au and Ir content of Kelley Lake bottom sediments. Can. Miner. 14:58–61.

Dillon, P.J., R.A. Reid, and R. Girard. 1986. Changes in the chemistry of lakes near Sudbury, Ontario following reductions of SO_2 emissions. Water Air Soil Pollut. 31:59–65.

Dillon, P.J., and P.J. Smith. 1984. Trace metal and nutrient accumulation in the sediments of lakes near Sudbury, Ontario, pp. 375–426. *In* J.O. Nriagu (ed.). Environmental Impact of Smelters. John Wiley and Sons, New York.

Driscoll, C.T., W.A. Ayling, G.F. Fordham, and L.M. Oliver. 1989. Chemical response of lakes treated with $CaCO_3$ to reacidification. Can. J. Fish. Aquat. Sci. 46:258–267.

Edenborn, H.M., N. Belzile, A. Mucci, J. Lebel, and N. Silverberg. 1986. Observations on the diagenetic behaviour of arsenic in a deep coastal sediment. Biogeochemistry 2:359–376.

Gunn, J.M., and W. Keller. 1990. Biological recovery of an acid lake after reductions in industrial emissions of sulfur. Nature (Lond.) 345:431–433.

Keller, N. 1984. Changes in the chemical composition of water, sediment and aquatic plants related to the reflooding of a drained lake. B.Sc. thesis, Laurentian University, Sudbury, Ontario.

Keller, W., J.M. Gunn, and N.D. Yan. 1992a. Evidence of biological recovery in acid-stressed lakes near Sudbury. Environ. Pollut. 78:79–85.

Keller, W., J.R. Pitblado, and J. Carbone. 1992b. Chemical responses of acidic lakes in the Sudbury, Ontario, area to reduced smelter emissions, 1981–1989. Can. J. Fish. Aquat. Sci. 49(Suppl. 1):25–32.

National Research Council (U.S.A.). 1992. Restoration of Aquatic Systems: Science, Technology, and Public Policy. National Academy Press, Washington, DC.

Nriagu, J.O. 1983. Arsenic enrichment in lakes near the smelters at Sudbury, Ontario. Geochim. Cosmochim. Acta 47:1523–1526.

Nriagu, J.O., and R.D. Coker. 1983. Sulphur in sediments chronicles past changes in lake acidification. Nature (Lond.) 303:692–694.

Nriagu, J.O., and J.M. Pacyna. 1988. Quantitative assessment of worldwide contamination of air, water, and soils by trace metals. Nature (Lond.) 333:134–139.

Nriagu, J.O., and S.S. Rao. 1987. Response of lake sediments to changes in trace metal emission from the smelters at Sudbury, Ontario. Environ. Pollut. 44:211–218.

Nriagu, J.O., and H.K.T. Wong. 1983. Selenium pollution of lakes near the smelters at Sudbury, Ontario. Nature (Lond.) 310:55–57.

Nriagu, J.O., H.K.T. Wong, and R.D. Coker. 1982. Deposition and chemistry of pollutants metals in lakes around smelters at Sudbury, Ontario. Environ. Sci. Technol. 16:551–560.

Rhodes, S.L., and K.B. Wiley. 1993. Great Lakes toxic sediments and climate change. Global Environ. Change 3:292–305.

Semkin, R.G., and J.R. Kramer. 1976. Sediment geochemistry of Sudbury area lakes. Can. Miner. 14:73–90.

Sigg, L. 1985. Metal transfer mechanisms in lakes: The role of settling particles, pp. 285–310. *In* W. Stumm (ed.). Chemical Processes in Lakes. John Wiley and Sons, New York.

Sigg, L., M. Sturm, and D. Kistler. 1987. Vertical transport of heavy metals by settling particles in Lake Zurich. Limnol. Oceanogr. 32: 112–130.

Stumm, W., and J.J. Morgan. 1981. Aquatic Chemistry. 2nd Ed. John Wiley and Sons, New York.

Tessier, A. 1992. Sorption of trace elements on natural particles in oxic sediments, pp. 426–453. *In* J. Buffle and H.P. van Leeuven (eds.). Sampling and Characterization of Environmental Particles. Lewis, Chelsea, MI.

Tessier, A., R. Carignan, and N. Belzile. 1994. Processes occurring at the sediment–water interface: emphasis on trace elements, pp. 139–175. *In* J. Buffle and R.R. De Vitre (eds.). Chemical and Biological Regulation of Aquatic Systems. Lewis, Chelsea, MI.

Tessier, A., R. Carignan, B. Dubreuil, and F. Rapin. 1989. Partitioning of zinc between the water column and the oxic sediments in lakes. Geochim. Cosmochim. Acta 53:1511–1522.

15

Liming of Sudbury Lakes: Lessons for Recovery of Aquatic Biota from Acidification

Norman D. Yan, W. (Bill) Keller, and John M. Gunn

When the pH of lakes falls to less than 6.0, many plant and animal species suffer appreciable damage (see Fig. 5.2). Many species disappear (Schindler et al. 1989). In the 1980s, there were about 19,000 lakes in Ontario with a pH less than 6 (Neary et al. 1990). Roughly one-third of these lakes are near Sudbury. They were acidified by long-term emissions of sulfur dioxide from local smelters (see Chapter 3).

The solution to this problem is to generate less acid at the source (i.e., to reduce emissions of sulfur dioxide to the atmosphere). Neutralizing the acidity at the receptor by adding base to or "liming" entire lakes is not a solution to the acid rain problem for Ontario. It is impractical because of the large number and remoteness of the damaged lakes and the need for retreatment if inputs of acid remain elevated. Also, liming addresses only one of the problems attributable to an acidified atmosphere, the acidification of lakes.

Liming is considered a more effective part of the overall solution to the acid rain problem in other parts of the world where the problem is more advanced (Box 15.1). Those interested in the engineering and scientific aspects of liming as a management tool are encouraged to consult the recent books of Dickson (1988), Olem (1991), Olem et al. (1991), and Brocksen et al. (1992). However, this chapter has other purposes.

Although it is not a general solution to the acid rain problem, liming is warranted in some circumstances. It may be the only way to protect unique species or habitats threatened by acid rain. Chapter 11 provides an excellent example—the liming of the native habitat of the aurora trout, a unique color variant of the brook trout, *Salvelinus fontinalis*. Liming experiments can also contribute to our understanding of the factors that regulate the recovery of biota from acidification. This is particularly important at the moment because of the enormous magnitude of the programs designed to reduce sulfur dioxide emissions across North America over the next two decades (e.g., NAPAP 1993). This chapter provides a brief review of long-term liming experiments conducted in the Sudbury area, highlighting what they can teach us about the potential for the biota of acidified lakes to recover if water quality improves in response to lowered rates of acid input.

Given this objective, can liming experiments really be used in this larger context? What can the addition of base to a few polluted Sudbury lakes teach us about the future of the thousands of Canadian acidified lakes? For several reasons, liming experiments can teach us a great deal. First, liming produces water quality changes in lakes that are similar, although not identical, to those that accompany reductions in acid input (i.e., dramatic increases in lake

Box 15.1.

Lake or watershed liming studies have been carried out in many areas of the world, including Canada, the United States, Scotland, Wales, Norway, and Sweden (Olem et al. 1991); however, the largest operational liming program is in Sweden. Sweden had about 16,000 acidified lakes in the mid-1980s, most of them privately owned and accessible. Given the scale of the damage and with much pressure for solutions, it was decided to proceed with an operational liming program until the effective control of acidic deposition was achieved. Various methods are used to apply limestone directly to lakes, and in some cases, limestone is also applied to surrounding wetlands and watersheds or to streams. After lakes are initially neutralized, the strategy is to retreat them before they reacidify enough to cause damage to the biological communities that have become re-established. The Swedish government now subsidizes the repeated liming of about 6000 lakes and about 10,000 km of running water, a program that uses more than 150 million kg of limestone annually.

pH and alkalinity and decreases in toxic metals) (Yan and Dillon 1984). Second, the Sudbury liming experiments were conducted in lakes that varied widely in acidity. Because the biota of several of the limed lakes are representative of other Ontario lakes with similar acidity (Yan et al., under review), the experimental results should be broadly applicable. Finally, biological changes in limed and naturally recovering lakes have proved to be similar in the few cases in which such comparisons were possible (Keller et al. 1992a) (see Chapter 5).

Sudbury Liming Experiments

Eight whole-lake liming experiments have been conducted over the past two decades in Ontario. Seven of these lakes were in the large area (see Fig. 3.2) affected by the Sudbury smelter emissions. Between 1973 and 1976, staff of the Ontario Ministries of the Environment and Natural Resources limed Middle, Hannah, and Lohi lakes—three severely acidified, metal-contaminated lakes located close to the smelters—and Nelson Lake, an intermediately acidic, lake trout (*Salvelinus namaycush*) lake, located about 30 km from Sudbury. The objective of these experiments was to determine how the severity of damage would influence the rate of recovery of the lakes after water quality was improved (Scheider and Dillon 1976; Yan et al. 1977; Dillon et al. 1979; Yan and Dillon 1984; Gunn et al. 1988; Yan et al., under review). In the early 1980s, two new liming experiments were initiated by the Ontario Ministries of Natural Resources and Environment, with a focus on lake trout fisheries. Bowland Lake, 70 km from Sudbury, was limed in 1983 and restocked with lake trout to determine if this former lake trout lake could support a self-sustaining fishery after water quality and food web structure improved (Molot et al. 1986; Gunn et al. 1990; Jackson et al. 1990; Keller et al. 1990a,b; Molot et al. 1990a,b; Keller et al. 1992b). Trout Lake, near Parry Sound, was limed in 1984 to determine if a lake trout fishery that was threatened by acidification could be protected by liming (Howell et al. 1991). Finally, Whirligig Lake, one of the two lakes that comprise the entire native habitat of the aurora trout was limed in 1989, along with an upstream reference lake, Little Whitepine Lake, before restoring this unique trout into its native habitat (Snucins et al., in press) (see Chapter 11). Whirligig lake, 107 km north of Sudbury, was later relimed in 1993.

The eight experimental lakes spanned the full range of damage known in the Sudbury area. At one extreme were the acidic, metal-contaminated, fishless Hannah, Middle, and Lohi lakes, which had impoverished species assemblages at every level of their food webs (Dillon et al. 1979). At the other extreme were the slightly damaged Nelson and Trout lakes, which still supported fisheries at the time of liming, despite degraded water quality.

Calcium hydroxide and/or calcium carbonate were the neutralizing agents selected in all experiments. These materials are inexpensive, readily available, and for the latter agent, safe and easy to handle. They provide neutralizing substances that are normally important in the acid base chemistry of lakes. Finally, excellent dosage and treatment duration models exist for these materials (Sverdrup and Bjerle 1982; Sverdrup et al. 1984). In all cases, the base was added to the surface of the experimental lakes by boat or aircraft as a dry powder or a fine aqueous slurry (Fig. 15.1).

The investigators in each study wished to raise lake pH to near 7 and provide some residual buffering capacity. The dosages of base required to achieve these targets were calculated in various ways as the engineering of liming advanced between 1973, when Middle and Lohi lakes were limed, and 1989, when Whirligig and Little Whitepine lakes were limed. However, because the two most important parameters in the dosage calculations are the lake's volume and acidity, there is a strong negative relationship between application rate, expressed volumetrically, and the preliming pH of the study lakes (Fig. 15.2).

A detailed discussion of the changes in water quality that followed liming is beyond the scope of this review. In all cases, liming increased alkalinity and pH and decreased metal levels (e.g., Yan and Dillon 1984). The duration of effect varied dramatically from lake to lake and was influenced by the lakes' flushing rates, by input rates of acid, by rates of internal acid generation or consumption, and by any additional applications of limestone to the watersheds. For example, the pH of Nelson Lake, with its flushing rate of 10 years, is still greater than 6, 16 years after the addition of base (Fig. 15.3). By contrast, Lohi and Bowland lakes quickly re-acidified to pH less than 6, corresponding with their short flushing rates of 1 and 2.5 years, respectively. Like Nelson Lake, the pH of Middle Lake also remains greater than 6, 15 years after the additions of base, despite its short flushing rate of 1.5 years. In this case, the longevity of the treatment is mainly attributable to the liming of Hannah Lake in 1975, the lake upstream of Middle Lake, and to the liming of the catchments of Middle and Hannah lakes by the municipality of Sudbury in the early 1980s as part of land reclamation efforts (see Chapter 8).

Lessons of Liming

Viewed collectively, the liming experiments provide several general observations about biological recovery that follows reductions in the acidity of lakes:

1. **rapid large elevations in pH are detrimental to the aquatic biota of acid lakes in the short term**
2. **rapid small increases in pH do no harm in the short term**
3. **biological communities can recover from excess acidity**
4. **rate of recovery of species is related in part to the severity of stress before liming**
5. **rate of recovery of species is also related to their fecundity and dispersal ability**
6. **several features of community recovery are directly attributable to improvements in water quality, but**

Figure 15.1. Various lime application procedures used in Sudbury lakes: (**A**) Liming Hannah Lake in 1973 by boat and slurry pump. (Photo by Ontario Ministry of Environment.) (**B**) Liming Bowland Lake by fixed-wing aircraft in 1983. (Photo by B. A. R. Environmental.) (**C**) Liming Whirligig Lake by hand from a boat in 1989 (Photo by E. Snucins) and (**D**) by helicopter in 1993 (Photo by W. Keller.).

7. **much of the recovery is only indirectly related to improvements in water quality, depending instead on a rebuilt food web**

Each of these observations is discussed in turn.

Liming represents a stress to those organisms that have adapted to acid conditions. Therefore, it should come as no surprise that rapid large increases in pH are detrimental to some of the biota of acid lakes in the short term (observation 1). Yan and Dillon (1984) noted that after they raised the pH of Middle, Hannah, and Lohi lakes from 4.4 to near 7, algal biomass was depressed by an order of magnitude for several months. Further, they and Yan et al. (under review) noted that acid-tolerant zooplankton were decimated by the additions of base, leaving an extremely pecu-

FIGURE 15.2. Scattergram of the dose of base initially applied to each lake versus the preliming pH. Note that the mass of material added per unit volume decreased with increases in pH.

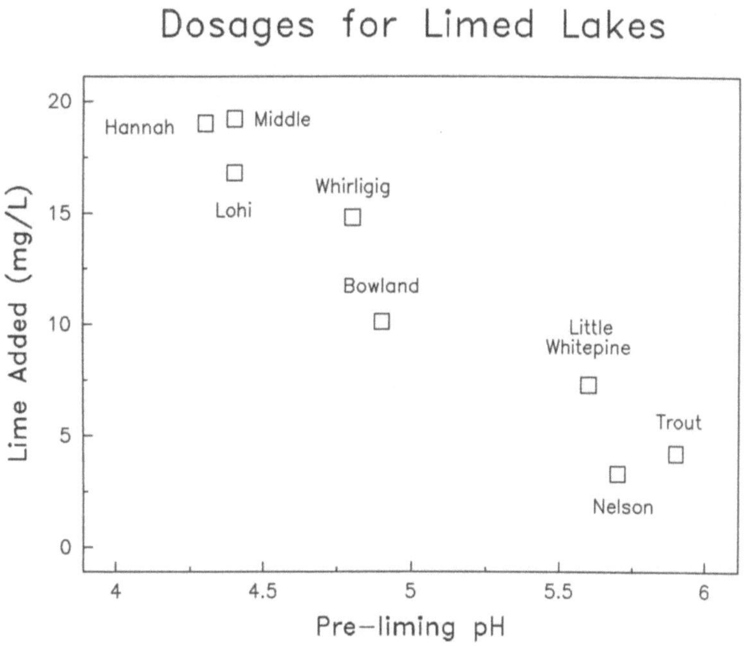

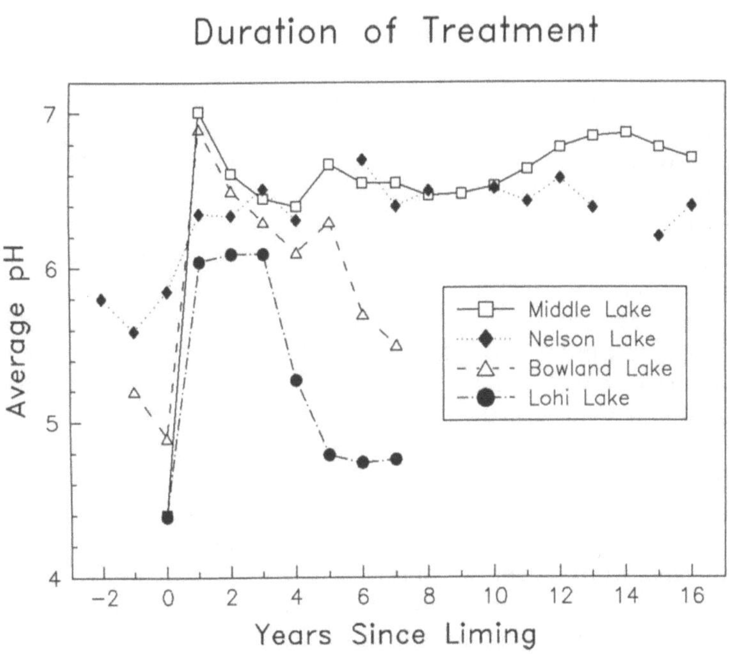

FIGURE 15.3. Changes in pH (annual average) of limed lakes before and after liming.

liar zooplankton community composed of littoral zone opportunists to dominate the lakes for almost a decade.

In contrast, smaller additions of base that produce smaller increases in pH do not harm aquatic communities in the short term (observation 2). Yan et al. (1977) raised the pH of Nelson Lake from 5.7 to 6.3 in 1975. This did not change plankton standing stocks or community structure. Molot et al. (1986) raised the pH of Trout Lake from 5.9 to 6.5 in 1984 with no discernable negative impacts on nutri-

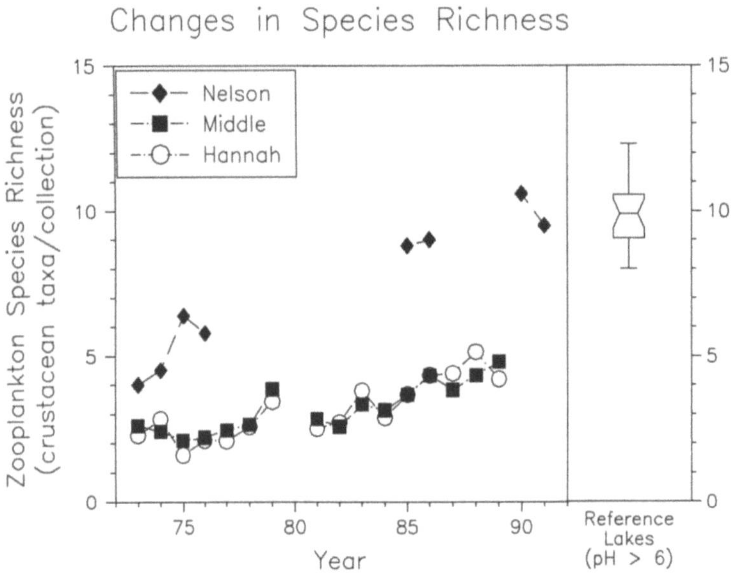

FIGURE 15.4. Recovery of zooplankton species richness in Middle, Hannah, and Nelson lakes in comparison with the expected richness in non-acidified lakes. The lakes were limed in 1973, 1975, and 1975, respectively, and have been non-acidic ever since. The notched box plot summarizes richness in the 22 reference lakes with pH greater than 6. The median richness is located at the narrowest part of the notch. The box encompasses 50% of the observations; the whiskers 99% of them.

ent regimes, phytoplankton, zooplankton, or fish (Howell et al. 1991). Finally, Yan and Dillon (1984) noted that the first liming of Lohi Lake decimated the phytoplankton, but subsequent additions that produced smaller changes in pH did not harm the community.

In the long term, the liming experiments clearly demonstrate that many aquatic biota can recover from acidity (observation 3). For example, in Whirligig Lake, re-introduced aurora trout are surviving and reproducing in their ancestral habitat (see Chapter 11). In Bowland Lake, stocked lake trout reproduced (Gunn et al. 1990), and communities of phytoplankton (Molot et al. 1990a), littoral algae (Jackson et al. 1990), zooplankton (Keller et al. 1992b), and some bottom-dwelling invertebrates (Keller et al. 1990b) changed in ways indicative of recovery.

The rate of recovery of aquatic biota in the liming experiments was related in part to the severity of preliming acidity (observation 4). In Nelson Lake, for example, the plankton communities returned to those characteristic of non-acidic lakes within 10 years of the addition of base. However, the zooplankton communities of the more severely stressed Middle and Hannah lakes remain impoverished in comparison with non-acidic reference lakes 16 years after the addition of base (Fig. 15.4).

The severity of acidity influences rates of recovery of biota for two reasons. First, fish exert a strong influence on the size structure and taxonomic composition of aquatic food webs. Hence, recovery of the ecosystem depends in part on the re-establishment of a normal fish community (Henrikson et al. 1985). This may take a very long time in severely acidified lakes, or it may never occur without the help of management agencies. Second, recovery must be delayed in severely acidified lakes, because many more species will have disappeared during the process of acidification (Yan and Welbourn 1990). In such cases, recolonization may be the rate-limiting step for recovery.

In fact, the liming experiments do provide evidence that the rate of recovery of different groups of biota is related to their fecundity and dispersal ability (observation 5). Educated guesses of the relative magnitude of these two parameters for various groups of aquatic and semiaquatic organisms allow estimation of the relative recovery rates of organisms from local extinction (Fig. 15.5). For example, aquatic insects with winged adult life stages should recover relatively quickly from acidification both because of the vast numbers of offspring they produce and because adult insects often fly tens of kilometers within a few days of emergence (Baldwin et al. 1975). Similarly,

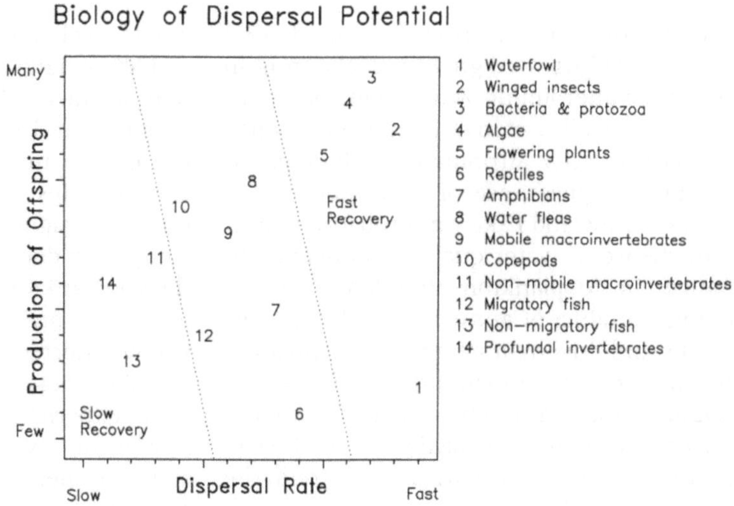

FIGURE 15.5. Relative recovery potential of biota estimated from their fecundity and dispersal ability.

phytoplankton, littoral algae, bacteria, and protozoa are prolific, reproduce frequently, and disperse through the air (Maguire 1963; Parker 1970). These organisms should recover quickly as long as there are source pools in the vicinity (Cairns 1991). By contrast, non-migratory fish and profundal invertebrates such as the opossum shrimp, *Mysis relicta,* an important acid-sensitive macroinvertebrate predator (Nero and Schindler 1983), should recover very slowly from local extinction because they disperse so slowly. The distribution of the opossum shrimp in Ontario has not changed since the retreat of the glaciers (Dadswell 1974).

Although Figure 15.5 is consistent with colonization theory (Mooney and Drake 1989), the dispersal rates are so poorly known that the figure should be regarded mainly as an hypothesis generator. Nevertheless, data from the limed lakes support the ideas. In Middle and Lohi lakes, bacterial abundance increased and community composition normalized almost immediately after liming (Scheider and Dillon 1976). Phytoplankton community composition recovered many attributes of non-acidic lakes within a few years of liming (Yan and Dillon 1984). By contrast, although there are promising signs of recovery, zooplankton community composition remains unusual in the lakes 15 years after the addition of base (see Fig. 15.4). Similarly in Bowland Lake, filamentous algae responded immediately to liming (Jackson et al. 1990), rotifers responded within 1 year, but crustacean zooplankton communities did not resemble those of non-acidic lakes 4 years after additions of base (Keller et al. 1992b).

Some features of biological recovery in the experimental lakes were directly attributable to improvements in water quality (observation 6). Examples include the increase in abundance of the opossum shrimp in Trout Lake (Howell et al. 1991), the recruitment of stocked bass (*Micropterus dolomeiui*) in Nelson Lake (Gunn et al. 1988) and lake trout in Bowland Lake (Gunn et al. 1990), the reappearance of acid-sensitive plankton species in Bowland Lake (Molot et al. 1990a; Keller et al. 1992b), and the recolonization of Middle and Hannah lakes by the acid-sensitive (Keller et al. 1990c) water flea, *Daphnia galeata mendotae* (Yan et al., under review). Because changes such as these depend only on the restoration of water quality they augur well for the remainder of Ontario's acid lakes, as long as seed populations of the biota are available.

Other community changes in the limed lakes are not directly attributable to water quality improvements, rather they are indirect effects of changes in food webs, which may themselves be directly related to additions of base (observation 7). For example, Howell et al. (1991) thought that increased growth of lake trout in Trout Lake after liming was attributable to increased availability of invertebrate prey, which was itself directly attributable to liming. Although the appearance of new zooplankton species in Bowland Lake was probably a direct consequence of water quality improvements, changes in zooplankton abundance were a complex product of annual variability in vertebrate and invertebrate predation pressure and the water temperature (Keller et al. 1992b). These sorts of changes, which are dependent on complex interactions in food webs, are more difficult to generalize beyond the experimental lakes.

Conclusions

The recovery of organisms from stress is influenced by many factors. These include the severity and duration of stress, the condition of the habitat after removal of the stress, the presence of refuges, the availability of colonists, their productivity and dispersal ability, and barriers to their dispersal (Cairns 1990; Niemi et al. 1990; Detenbeck et al. 1992). Management agencies can also directly increase rates of recovery by re-introducing locally extinct species or manipulating damaged habitats. With such complexity, it is not surprising that restoration ecology is a science in its infancy. At the moment, rates or patterns of aquatic community recovery cannot be predicted with certainty, because of the complexity of the science, the paucity of theoretical frameworks, and the scarcity of experiments designed to test hypothetical regulators of recovery from long-term stressors such as acidification (Niemi et al. 1990). Studies of recovery of experimentally acidified lakes, such as those of Lake 223 in northwestern Ontario (see Box 5.1), will be extremely useful to this field in years to come if they are continued (Schindler et al. 1991).

The Sudbury liming experiments can contribute to the emerging discipline of restoration ecology in several practical ways. First, the liming experiments have provided approaches to setting recovery targets. For example, Keller et al. (1992b) and Yan et al. (under review) used reference data sets from non-acidified lakes to determine the normal temporal and spatial variability in zooplankton communities characteristic of "healthy" communities. Second, liming experiments provide insights about required durations of study. Two decades have not been enough for the zooplankton of Middle and Hannah lakes to recover but were adequate for the less severely affected community of Nelson Lake (see Fig. 15.4). Third, as previously discussed, these experiments provide insight about the role of the severity of the stress and dispersal ability of species for biological recovery. These are two of the hypothesized regulators of recovery. Finally, the experiments can identify predictable aspects of recovery. Those changes that we can attribute directly to improving water quality are probably most predictable, and they occur at the top and bottom of food webs. At the bottom, bacteria, phytoplankton, and littoral algae respond rapidly and directly to water quality improvements alone. At the top, the recruitment of piscivores (from relict adult or restocked populations) also responds directly to the improved water quality in spawning habitats. Changes in the middle of the food web are more difficult to predict because they are influenced by both direct and indirect regulators of recovery. Unfortunately, our understanding of the acid-sensitivity of most species is too poor (Locke 1991) and food web linkages are too complex to predict whether direct or indirect effects will predominate in the recovery of most biota in individual lakes.

The management of lakes in North America owes a great deal to whole-lake manipulation experiments. They have provided many insights into the impacts of pollutants on ecosystem function that smaller-scale experiments could not provide, and they have provided crucial tests of competing hypotheses with enormous management implications (e.g., Schindler 1990). It is expected that atmospheric emissions of sulfur dioxide will decline substantially in North America during the next 20 years, as Canadian and American governments attempt to reduce the acidity of our atmosphere. As we endeavor to predict the benefits of these programs, the Sudbury liming experiments of the 1970s and 1980s will provide some of the best models of the recovery of lakes from acidification.

References

Baldwin, W.F., A.S. West, and J. Gomery. 1975. Dispersal pattern of black flies (Diptera: Simuliidae) tagged with 32P. Can. Entomol. 107:113–118.

Brocksen, R.W., M.D. Marcus, and H. Olem. 1992. Practical Guide to Managing Acidic Surface Waters and Their Fisheries. Lewis Publications, Boca Raton, FL.

Cairns, J., Jr. 1990. Lack of a theoretical basis for predicting rate and pathways of recovery. Environ. Manage. 14:517–526.

Cairns, J., Jr. 1991. Probable consequences of a cosmopolitan distribution. Specul. Sci. Technol. 14: 41–50.

Dadswell, M.J. 1974. Distribution, ecology, and postglacial dispersal of certain crustaceans and fishes in eastern North America. Nat. Museum Can. Publ. Zool. 11.

Detenbeck, N.E., P.W. DeVore, G.J. Niemi, and A. Lima. 1992. Recovery of temperate-stream fish communities from disturbance: a review of case studies and synthesis of theory. Environ. Manage. 16:33–53.

Dickson, W. (ed.). 1988. Liming of Lake Gårdsjön—An Acidified Lake in SW Sweden. Report 3426. National Swedish Environmental Protection Board, Solna, Sweden.

Dillon, P.J., N.D. Yan, W.A. Scheider, and N. Conroy. 1979. Acidic lakes in Ontario, Canada: characterization, extent and responses to base and nutrient additions. Arch. Hydrobiol. Beih. 13:317–336.

Gunn, J.M., J.G. Hamilton, G.M. Booth, C.D. Wren, G.L. Beggs, H.J. Rietveld, and J.R. Munro. 1990. Survival, growth and reproduction of lake trout (*Salvelinus namaycush*) and yellow perch (*Perca flavescens*) after neutralization of an acidic lake near Sudbury, Ontario. Can. J. Fish. Aquat. Sci. 47:446–453.

Gunn, J.M., M.J. McMurtry, J.M. Casselman, W. Keller, and M.J. Powell. 1988. Changes in the fish community of a limed lake near Sudbury, Ontario: effects of chemical neutralization, or reduced atmospheric deposition of acids? Water Air Soil Pollut. 41:113–136.

Henrikson, L., H.G. Nyman, H.G. Oscarson, and J.A.E. Stenson. 1985. Changes in the zooplankton community after lime treatment of an acidified lake. Verh. Int. Verein. Limnol. 22:3008–3013.

Howell, E.T., G. Coker, G.M. Booth, W. Keller, B. Neary, K.H. Nicholls, F.D. Tomassini, N. Yan, J.M. Gunn, and H. Rietveld. 1991. Ecosystem responses of a pH 5.9 lake trout lake to whole lake liming, pp. 61–95. *In* H. Olem, R.K. Schreiber, R.W. Brocksen, and D.P. Porcella (eds.). International Lake and Watershed Liming Practices. The Terrene Institute, Washington, DC.

Jackson, M.B., E.M. Vandermeer, N. Lester, J.A. Booth, and L. Molot. 1990. Effects of neutralization and early reacidification on filamentous algae and macrophytes in Bowland Lake. Can. J. Fish. Aquat. Sci. 47:432–439.

Keller, W., D.P. Dodge, and G.M. Booth. 1990a. Experimental lake neutralization program: overview of neutralization studies in Ontario. Can. J. Fish. Aquat. Sci. 47:410–411.

Keller, W., J.M. Gunn, and N.D. Yan. 1992a. Evidence of biological recovery of acid stressed lakes near Sudbury, Canada. Environ. Pollut. 78:79–85.

Keller, W., L.A. Molot, R.W. Griffiths, and N.D. Yan. 1990b. Changes in the zoobenthos community of acidified Bowland Lake after whole-lake neutralization and lake trout (*Salvelinus namaycush*) reintroduction. Can. J. Fish. Aquat. Sci. 47:440–445.

Keller, W., N.D. Yan, K.E. Holtze, and J.R. Pitblado. 1990c. Inferred effects of lake acidification on *Daphnia galeata mendotae*. Environ. Sci. Technol. 24:1259–1261.

Keller, W., N.D. Yan, T. Howell, L.A. Molot, and W.D. Taylor. 1992b. Changes in zooplankton during the experimental neutralization and early re-acidification of Bowland Lake, near Sudbury, Ontario. Can. J. Fish. Aquat. Sci. 49(Suppl. 1):52–62.

Locke, A. 1991. Zooplankton responses to acidification—a review of laboratory bioassays. Water Air Soil Pollut. 60:135–148.

Maguire, B., Jr. 1963. The passive dispersal of small aquatic organisms and their colonization of isolated bodies of water. Ecol. Monogr. 33:161–185.

Molot, L., L. Heintsch, and K.H. Nicholls. 1990a. Response of phytoplankton in acidic lakes in Ontario to whole-lake neutralization. Can. J. Fish. Aquat. Sci. 47:422–431.

Molot, L.A., P.J. Dillon, and G.M. Booth. 1990b. Whole-lake and nearshore water chemistry in Bowland Lake, before and after treatment with $CaCO_3$. Can. J. Fish. Aquat. Sci. 47:412–421.

Molot, L.A., J.G. Hamilton, and G.M. Booth. 1986. Neutralization of acidic lakes: short-term dissolution of dry and slurried calcite. Water Res. 20: 757–761.

Mooney, H., and J.A. Drake. 1989. Biological invasions: a SCOPE program overview, pp. 491–506. *In* J. A. Drake (ed.). Biological Invasions: A Global Perspective. J. Wiley & Sons, Chichester, UK.

National Acid Precipitation Assessment Program (NAPAP). 1993. 1992 Report to Congress. NAPAP, Washington, DC.

Neary, B.P., P.J. Dillon, J.R. Munro, and B.J. Clark. 1990. The Acidification of Ontario Lakes: An Assessment of Their Sensitivity and Current Status with Respect to Biological Damage. Technical Report. Ontario Ministry of Environment, Dorset, Ontario.

Nero, R.W., and D.W. Schindler. 1983. Decline of *Mysis relicta* during the acidification of Lake 223. Can. J. Fish. Aquat. Sci. 40:1905–1911.

Niemi, G.J., P. DeVore, N. Detenbeck, D. Taylor, A. Lima, J. Pastor, J.D. Yount, and R.J. Naiman. 1990. Overview of case studies on recovery of aquatic systems from disturbance. Environ. Manage. 14:571–588.

Olem, H. 1991. Liming Acidic Surface Waters. Lewis Publications, Boca Raton, FL.

Olem, H., R.K. Schreiber, R.W. Brocksen, and D.P. Porcella. 1991. International Lake and Watershed Liming Practices. The Terrene Institute, Washington, DC.

Parker, B.C. 1970. Life in the sky. Nat. Hist. 79:54–59.

Scheider, W., and P.J. Dillon. 1976. Neutralization and fertilization of acidified lakes near Sudbury, Ontario, pp. 93–100. *In* Proceedings of the 11th Canadian Symposium on Water Pollution Research in Canada.

Schindler, D.W. 1990. Experimental perturbations of whole lakes as tests of hypotheses concerning ecosystem structure and function. Oikos 57:25–41.

Schindler, D.W., T.M. Frost, K.H. Mills, P.S.S. Chang, I.J. Davies, L. Findlay, D.F. Malley, J.A. Shearer, M.A. Turner, P.G. Garrison, C.J. Watras, K. Webster, J.M. Gunn, P.L. Brezonik, and W.A. Swenson. 1991. Comparisons between experimentally- and atmospherically-acidified lakes during stress and recovery. Proc. R. Soc. Edin. 97B:193–227.

Schindler, D.W., S.E.M. Kasian, and R.H. Hesslein. 1989. Biological impoverishment in lakes of the midwestern and northeastern United States from acid rain. Environ. Sci. Technol. 23:573–580.

Snucins, E.J., J.M. Gunn, and W. Keller. In press. Restoration of the aurora trout to their acid-damaged native lakes. Cons. Biol.

Sverdrup, H., and I. Bjerle. 1982. Dissolution of calcite and other related minerals in acidic aqueous solution in a pH-stat. Vatten 38:59–73.

Sverdrup, H., R. Rasmussen, and I. Bjerle. 1984. A simple model for the reacidification of limed lakes, taking the simultaneous deactivation and dissolution of calcite in the sediments into account. Chemica Scripta 24:53–66.

Yan, N.D., and P.J. Dillon. 1984. Experimental neutralization of lakes near Sudbury, Ontario, pp. 417–456. *In* J. Nriagu (ed.). Environmental Impacts of Smelters. John Wiley & Sons, New York.

Yan, N.D., W. Keller, K.M. Somers, T.W. Pawson, and R. Girard. Submitted. The recovery of zooplankton from acidification: comparing manipulated and reference lakes.

Yan, N.D., W.A. Scheider, and P.J. Dillon. 1977. Chemical and biological changes in Nelson Lake, Ontario following experimental elevation of lake pH, pp. 213–231. Proceedings of the 12th Canadian Symposium on Water Pollution Research in Canada.

Yan, N.D., and P.M. Welbourn. 1990. The impoverishment of aquatic communities by smelter activities near Sudbury, Canada, pp. 477–494. *In* G.M. Woodwell (ed.). The Earth in Transition: Patterns and Processes of Biotic Impoverishment. Cambridge University Press, Cambridge.

16

Trends in Waterfowl Populations: Evidence of Recovery from Acidification

Donald K. McNicol, R. Kenyon Ross, Mark L. Mallory, and Lise A. Brisebois

The loss and degradation of aquatic habitat has serious implications for waterfowl species breeding in the northern forests of eastern Canada, an area whose importance to continental populations is only beginning to be appreciated (NAWMP 1986). Acid rain poses a serious threat to waterfowl that rely on healthy lakes and wetlands for nest sites, protection of young, and most important, food. Both ducks and loons require high-protein, mineral-rich foods (small fish or invertebrates) for the production of eggs and rapid growth of young (Reinecke and Owen 1980). As lakes acidify, the accompanying biological changes disrupt normal prey communities. Changes in the composition, abundance, and nutritional value of prey can reduce the quality of breeding habitat (McNicol and Wayland 1992) and affect reproduction (Scheuhammer 1991). Consequently, reversal of acidification would benefit a significant waterfowl resource. An estimated 192,000 pairs of ducks and loons breed on the exposed Precambrian Shield of central and northeastern Ontario, where surface waters have a limited capacity to neutralize acidic inputs (McNicol et al. 1990).

Efforts to reduce sulfur dioxide emissions are aimed at protecting sensitive aquatic habitats. Where improvements to water quality are expected, little is known about the response of biological communities, because most research has documented aspects of decline. A notable exception is recent evidence of biological changes that have accompanied chemical improvements of acidic lakes near Sudbury after reductions in local smelter emissions (Keller et al. 1992a). Such evidence confirms that damaging effects of acidification can be reversed; already, some biotic communities in recovering lakes are becoming more typical of those found in non-acidic lakes. Sudbury area lakes provide a unique opportunity to monitor biological recovery across a broad range of initial chemical conditions.

Waterfowl are being studied as indicators and integrators of the biological responses to reduced acid deposition and chemical recovery observed in Sudbury area lakes. As fish and invertebrate prey return to lakes, the quality of breeding habitat for some species is expected to improve. This chapter examines trends in waterfowl populations breeding in the Sudbury area as possible evidence of reversibility of acidification. Relationships between the distribution, density, and productivity of waterfowl and various habitat parameters including pH are examined. Finally, present knowledge is used to predict future responses of waterfowl to the environmental effects of additional abatement programs.

Waterfowl and Habitat Parameters

Waterfowl that commonly breed near Sudbury are shown in Figure 16.1 and include, for this exercise, the Common Loon (*Gavia immer*), as well as the following ducks: Common Merganser (*Mergus merganser*), Common Goldeneye (*Bucephala clangula*), Hooded Merganser (*Lophodytes cucullatus*), American Black Duck (*Anas rubripes*), Ring-necked Duck (*Aythya collaris*), Mallard (*Anas platyrhynchos*), and Wood Duck (*Aix sponsa*). Aspects of their breeding biology are summarized in Table 16.1.

Studies of waterfowl breeding populations, habitat quality, and aquatic food webs were conducted in the Sudbury area between 1983 and 1993. This study area covers approximately 17,000 km^2, generally corresponding to the area southwest and northeast of Sudbury (Fig. 16.2), defined by Neary et al. (1990) as the zone influenced by past sulfur emissions from Sudbury smelters (see Chapter 3). Surveys of breeding populations were undertaken in 4-km^2 plots located throughout this zone between 1985 and 1993 (Ross 1987) (Fig. 16.2). Numbers of wetlands and estimates of open water area were determined for each plot (Fig. 16.3). Data from 363 lakes sampled in the early 1980s (Neary et al. 1990) were used to map lake pH distributions in the area; survey plots were then assigned a pH class.

The ecology of waterfowl breeding in the acid-stressed area north of Lake Wanapitei has been studied since 1983. A total of 174 lakes and wetlands was surveyed or sampled to varying degrees in the Wanapitei study area (see Fig. 16.2). These lakes varied substantially in pH (3.9–7.6) and fish and invertebrate populations (Bendell and McNicol 1987, in press). The relative production of waterfowl breeding at Wanapitei was compared among pH classes based on the most recent pH reading, usually 1993.

Simple correlation was used to examine relationships between waterfowl populations and habitat parameters. For plots within the Sudbury study area, no significant correlations were evident among habitat parameters (Table 16.2). Average numbers of wetlands were higher in central plots compared with the northeast and far southwest, where fewer but larger lakes predominate (Fig. 16.3). North of Sudbury, high wetland densities coincide with high open water areas, a situation that might arise if large numbers of small lakes were found in close proximity. Large areas northeast and southwest of Sudbury contained surface waters with pH less than 6 in the early 1980s (Fig. 16.3).

After emission reductions in the 1970s, acidic lakes distant from Sudbury showed marked increases in pH and alkalinity and declines in sulfate between 1980 and 1987 (Keller et al. 1992b) (see Chapter 5). Similar trends were evident in small lakes monitored for waterfowl near Wanapitei (McNicol and Mallory 1994). However, further improvements to water quality have not been observed since 1987, nor are any expected until additional abatement measures are implemented.

The invertebrate communities in these lakes are strongly influenced by the presence or absence of fish (McNicol and Wayland 1992). Approximately 25% of lakes sampled at Wanapitei in 1993 continue to have pH less than 5 and are fishless. This abundance of small (<8 ha), fishless, acidic lakes is a unique feature of the Wanapitei area. In the absence of fish predation, these small acid lakes contain large populations of a few types of acid-tolerant invertebrate taxa. Larger lakes (>8 ha), ranging in pH between 5 and 6.3, often support acid-tolerant fish species, such as yellow perch (*Perca flavescens*), which, coupled with acid stress, reduce invertebrate prey abundance and diversity (Bendell and McNicol 1987, in press).

Figure 16.1. Waterfowl species that commonly breed in the Sudbury area: (**A**) Common Loon, (**B**) Common Merganser, (**C**) Common Goldeneye, (**D**) Hooded Merganser, (**E**) Black Duck, (**F**) Ring-necked Duck, (**G**) Mallard, (**H**) Wood Duck. (Photos for A, B, E–H by Michael W.P. Runtz; photos for C and D by Jim Flynn.)

TABLE 16.1. Habitat requirements of waterfowl species found in the Sudbury area.

Species	Diet	Nesting habitat	Brood-rearing habitat
Common Loon	Small to medium-sized fish	Ground (prefers islands) near water	Lakes; medium-to-large, meso- to oligotrophic
Wood Duck	Adults—plant matter Ducklings—invertebrates	Tree cavity	Wetlands; small, meso- to eutrophic with marshy cover
Black Duck	Adults—plant matter Ducklings—invertebrates	Ground in heavy cover	Wetlands; small, meso- to eutrophic with shoreline cover
Mallard	Adults—plant matter Ducklings—invertebrates	Ground in heavy cover	Wetlands; small, eutrophic with marshy cover
Ring-necked Duck	Adults—plant matter Ducklings—invertebrates	Ground in heavy cover near water	Wetlands; small, meso- to eutrophic with shoreline cover
Common Goldeneye	Mostly invertebrates	Tree cavity	Lakes; small-to-medium, oligotrophic
Hooded Merganser	Small fish and invertebrates	Tree cavity	Wetlands; small, meso- to oligotrophic with shoreline cover
Common Merganser	Small to medium-sized fish	Tree cavity	Rivers and medium-to-large lakes

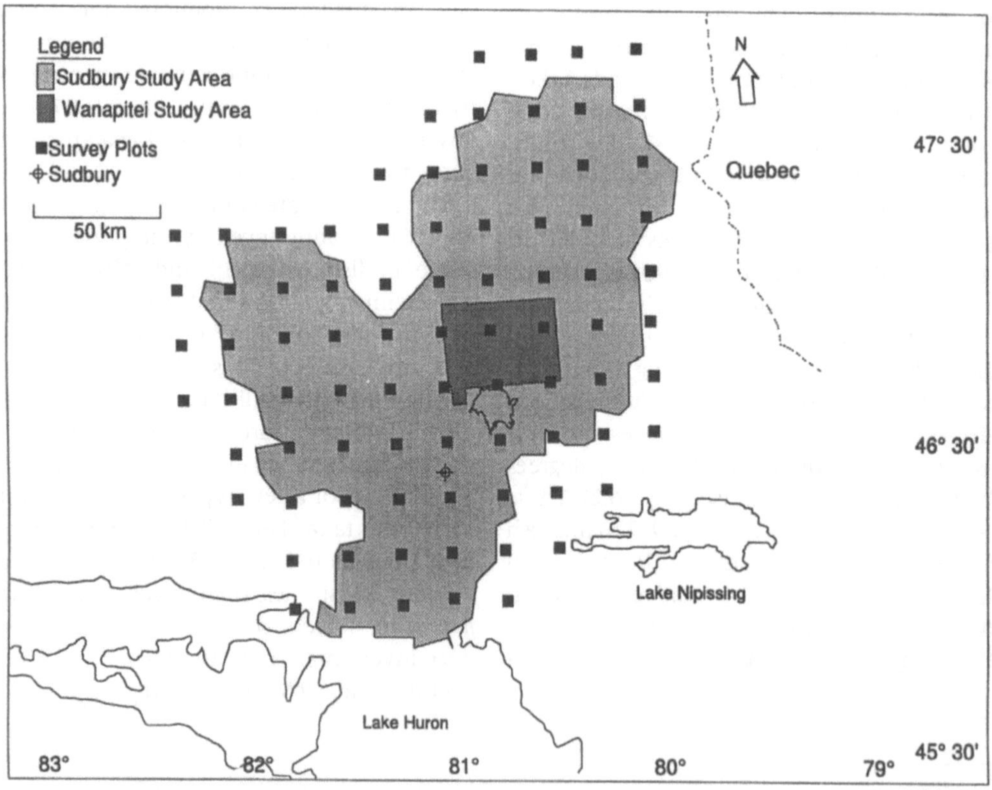

FIGURE 16.2. Map of northeastern Ontario showing outlines of the Sudbury study area (from Neary et al. 1990) and the Wanapitei study area (from McNicol et al. 1987a). Waterfowl and habitat survey plots (2 × 2 km) that fall inside the Sudbury area ($N = 57$) or within 20 km ($N = 26$) are also identified.

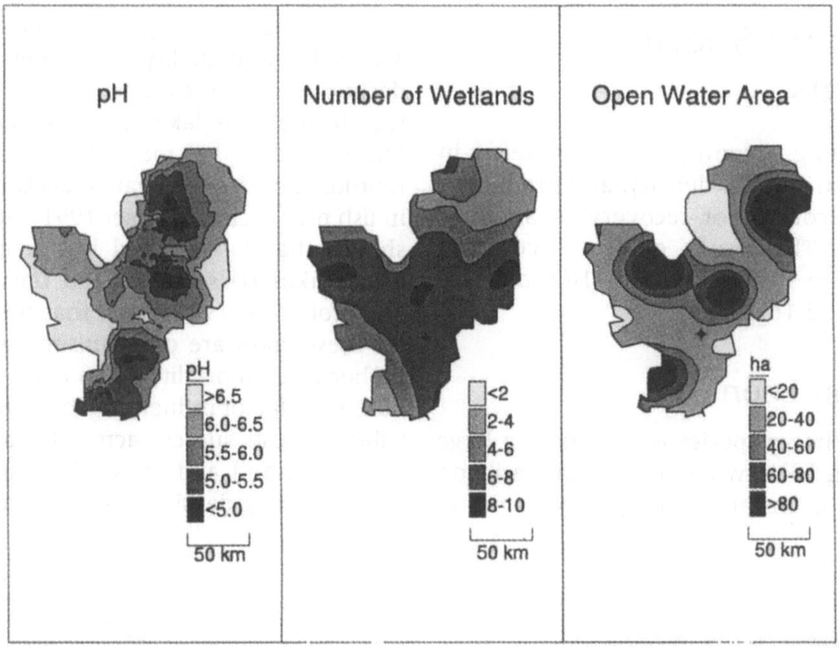

FIGURE 16.3. Maps of the distribution of pH, number of wetlands, and open water area (derived from 83 4-km² plots) in the Sudbury area. The pH map was derived from 363 lakes (from Neary et al. 1990). Maps were constructed using the POTMAP use of SPANS (Intera Tydac 1991). Sudbury is shown by ⊕.

TABLE 16.2. Relationships between the densities of eight waterfowl species breeding in the Sudbury area and habitat parameters, based on 5-year average densities for 57 plots, expressed as Spearman rank correlation coefficients.

	Habitat parameter		
	pH	No. wetlands	Open Water Area (ha)
No. wetlands	0.04		
Open Water Area	0.10	0.19	
Common Loon	0.26*	0.37**	0.78**
Common Merganser	−0.08	−0.11	0.31*
Common Goldeneye	−0.29**	0.16	0.11
Hooded Merganser	0.07	0.61**	0.20
Black Duck	−0.06	0.45**	0.06
Ring-necked Duck	−0.20	0.45**	0.06
Mallard	−0.02	0.18	−0.15
Wood Duck	−0.11	0.08	−0.03

* $P < .05$; ** $P < .01$.

Waterfowl Species Accounts

The following accounts discuss patterns in species' distribution, density, and productivity in the context of recovery of Sudbury area lakes. The results of the surveys are illustrated in Figures 16.3 to 16.6 and Tables 16.1 and 16.2.

Common Loon

This well-known species is common on large oligotrophic lakes, where it feeds primarily on small fish. Loons and other piscivores are at risk from acidic precipitation due to reduced availability and quality of fish prey as lake pH declines. Because they are long-lived and return to the same lake each year, loons breeding on acidic lakes may experience impaired reproduction from elevated levels of mercury in fish prey (Scheuhammer 1991). Studies have shown that loons breed less frequently on acidic lakes (DesGranges and Darveau 1985; McNicol et al. 1987a) and that nest attempts on these lakes are often unsuccessful due to higher chick mortality (Alvo et al. 1988). The dominant factor influencing loon breeding distribution and success across Ontario is lake area (Wayland and McNicol 1990). At Sudbury, breeding densities correlate strongly with

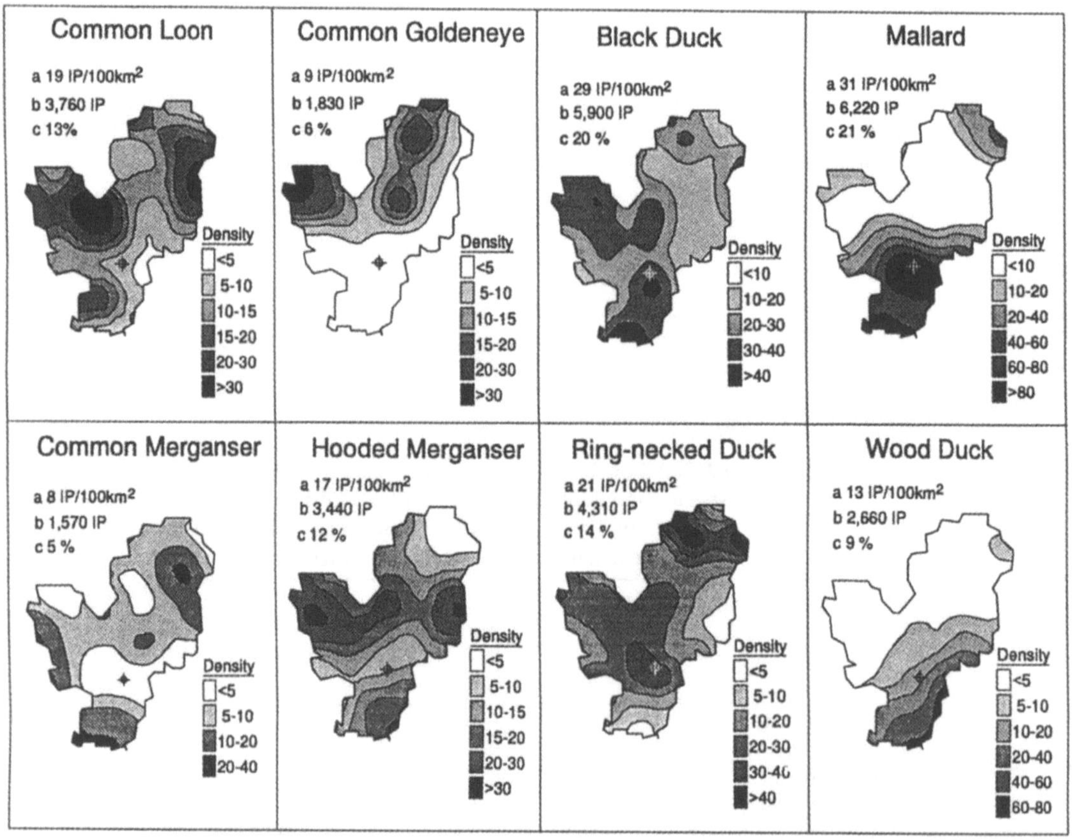

FIGURE 16.4. Maps of the distribution and density (indicated pairs per 100 km^2) of eight waterfowl species breeding in the Sudbury area, based on 5-year average densities for 83 survey plots (from Ross 1987). The location and number of pairs were determined using helicopter surveys conducted during nest initiation in May 1985, 1987–89, and 1993. Maps were constructed using the POTMAP use of SPANS (Intera Tydac 1991). Estimates of (*a*) mean density, (*b*) total number of indicated breeding pairs (*IP*), and (*c*) percentage of total waterfowl count are included for each species. Sudbury is shown by ⊕.

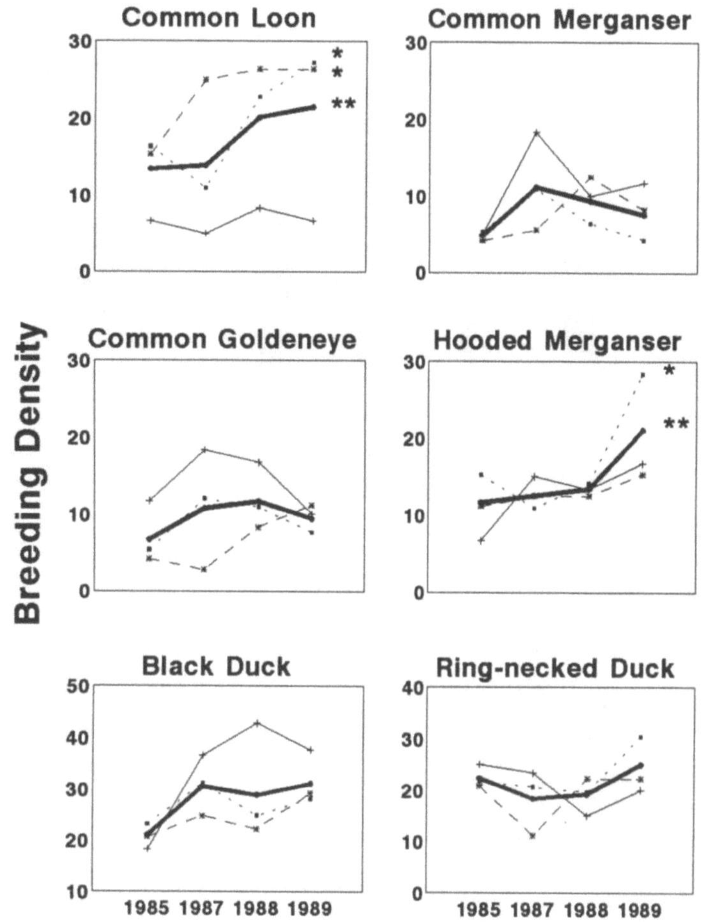

FIGURE 16.5. Trends in the average density (indicated pairs per 100 km²) of six waterfowl species breeding in 56 plots within the Sudbury study area surveyed in 4 years between 1985 and 1989 (*thick line*). Also presented are trends for plots assigned to broad pH classes: less than 5.5 (+ — +; $N = 15$), 5.5–6.3 (* — *; $N = 18$), greater than 6.3 (■ ···· ■; $N = 23$). Statistical tests are based on non-parametric trend analyses (* – $P < .05$; ** – $P < .01$) from Ross et al. 1984).

open water area, reflecting its preference for larger water bodies, but also correlate with pH. Loons are abundant in the northwest (Lake Onaping area), far northeast (Lady Evelyn area), and southwest (Lake Panache area), where average pHs are generally greater than 5.5. Loons have a very localized distribution in the immediate vicinity of Sudbury, where appropriate nesting lakes are rare. Northeast of Sudbury, near Lake Chiniguchi, suitable breeding habitat is available (i.e., large lakes), but few loons are recorded. Extremely acidic lakes (pH <5.5), typical of the general area, are used infrequently by loons.

Common Loon numbers have increased near Sudbury in recent years (Fig. 16.5). Although average breeding densities in acidified areas (pH <5.5) have not changed and remain low, populations in medium (5.5–6.3) and high (>6.3) pH areas have increased substantially. Presumably, increases in local breeding populations are related to recovery in water quality and biological communities (see Chapter 5), especially to the west and far north of Sudbury, where pHs in the early 1980s were greater than 5.5. As the fish prey base in large lakes improves and habitat suited for nesting and raising chicks becomes more available, increased immigration, coupled with enhanced productivity, has ultimately led to the increased recruitment to local breeding populations witnessed recently.

Common Merganser

In summer, this fish-eating species favors clear lakes, rivers, and streams of the Precambrian Shield. As with loons, mergansers display an

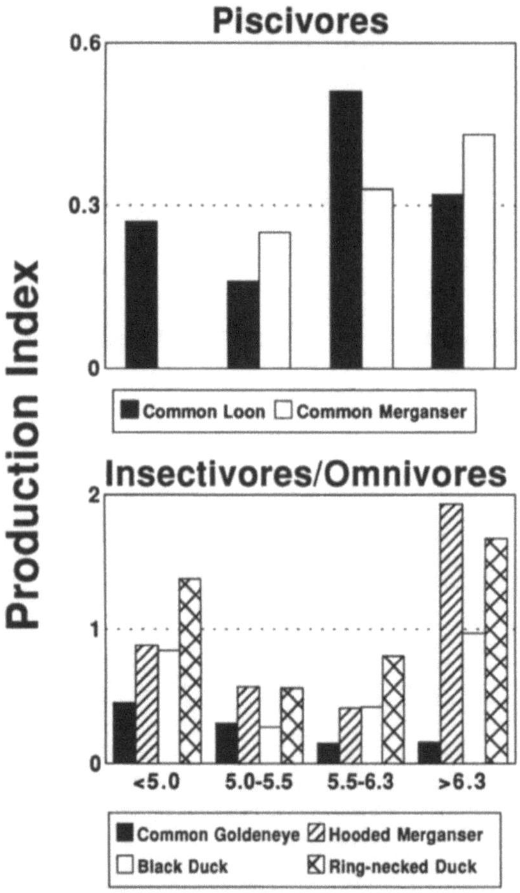

FIGURE 16.6. Comparisons of relative production of young (% class 2 broods/% indicated pairs) for piscivores (Common Loon, Common Merganser) and insectivores (Common Goldeneye, Hooded Merganser) or omnivores (Black Duck, Ring-necked Duck) using 174 lakes in the Wanapitei study area. Indicated pair data were collected in 1983, 1985–86, and 1993, and brood data were collected in 1983–87 and 1993. Data were then pooled across years per lake, and lakes were scored as used or unused by each species. Number of lakes in each pH class: less than 5 ($N = 43$), 5–5.5 ($N = 39$), 5.5–6.3 ($N = 46$), greater than 6.3 ($N = 46$).

obvious preference for lakes greater than pH 6 that contain fish (McNicol et al. 1990). However, these two fish-eating species differ in their breeding habits; loons raise only one or two young, which are restricted to their nesting lake, whereas Common Mergansers attempt to raise many ducklings that are highly mobile and can move to lakes with a greater food supply.

Fewer than 1600 pairs of Common Mergansers nest in the Sudbury area, with especially low breeding densities (less than five pairs per 100 km^2) recorded near Sudbury, where fish populations continue to be stressed (see Chapter 5). Although production in the acid-stressed Wanapitei area was strongly related to pH, no population trends were evident among any pH classes. Breeding densities were only correlated with open water area, confirming their preference for larger water bodies (McNicol et al. 1987a). Breeding distributions were influenced by two main factors: (1) preference to breed near very large inland lakes (Lady Evelyn, Temagami, Onaping, Wanapitei, and Panache) and near Georgian Bay, and (2) preference for large river systems, such as the Spanish. Both patterns support the tendency of broods to move progressively downstream to brood-rearing habitat after hatch.

Common Goldeneye

The Common Goldeneye is one of the few species that can exploit acidified environments associated with industrial regions (Gilyazov 1993). Early studies indicated that goldeneyes compete with fish for similar invertebrate prey (Eriksson 1979), which influences their distribution (McNicol et al. 1987a) and partially explains the negative correlation with pH ob-

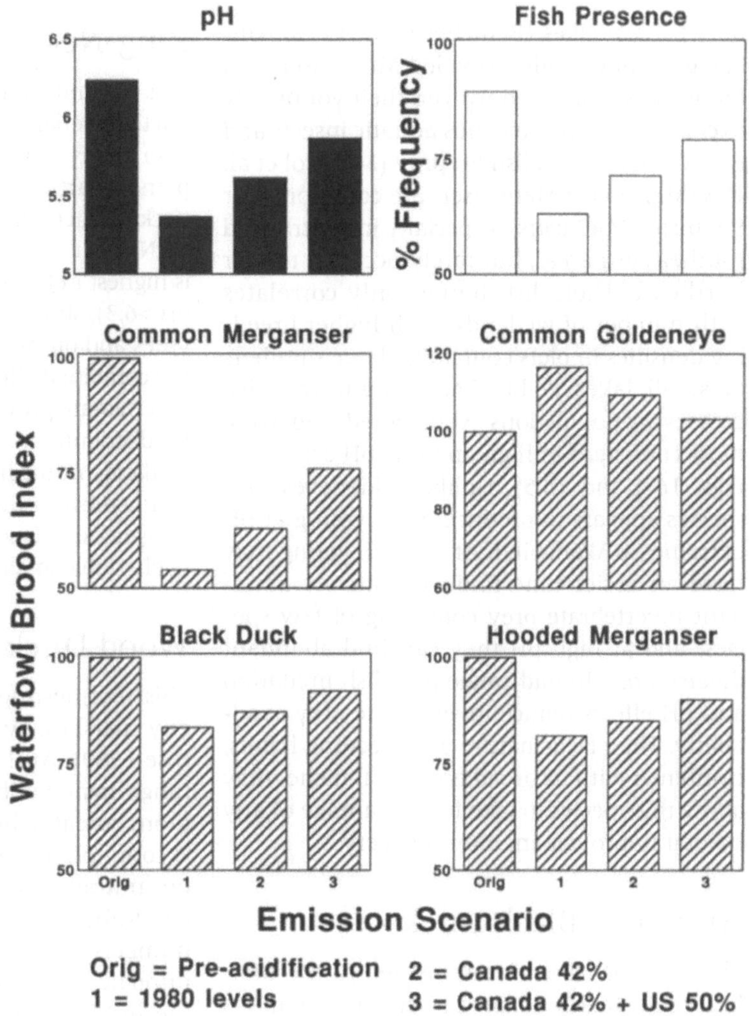

FIGURE 16.7. Predicted responses of pH, fish presence, and four waterfowl species in the Sudbury study area to several emission control scenarios. Waterfowl brood indices are scaled to original (pre-acidification) values.

served in the Sudbury area (Table 16.2). Here, goldeneyes are at the southern limit of their breeding range, but pockets of small, fishless, acid lakes support local concentrations of nesting pairs. The population fluctuated between 1985 and 1989, with no consistent trend. However, goldeneye breeding densities did increase on lakes where recovery should be rapid (pH 5.5–6.3), perhaps because invertebrate prey abundance and diversity increased, but fish have not yet invaded.

Research on goldeneyes at Wanapitei has confirmed that, given a variety of small oligotrophic lakes to choose from, fishless, often clustered lakes are preferred during all stages of breeding (Mallory et al. 1993, 1994; Wayland and McNicol, 1994). These associations are clearly linked to the rich supply of preferred invertebrate prey found in fishless lakes in the area, most of which are very acidic (pH <5.0) (McNicol et al. 1987b; McNicol and Wayland 1992). As pH improves and fish return to many of these lakes, the Common Goldeneye is the only species whose population is expected to decrease in the Sudbury area as a direct result of acidity shifts (Fig. 16.7).

Hooded Merganser

This merganser prefers small beaver-influenced lakes and wetlands in northeastern Ontario and is often associated with non-acidic, naturally fishless lakes (Blancher et al. 1992). Unlike goldeneyes, adult Hooded Mergansers can forage on small fish; however, their young rely exclusively on conspicuous aquatic insects and thus compete with fish for prey (McNicol et al. 1987b). Hooded Mergansers are common near Sudbury (3440 pairs), especially in central and southeastern areas, but rarely occur in the far northeast. Their distribution only correlates with number of wetlands, with higher breeding densities in plots containing large numbers of small lakes (Table 16.2). Significant increases in populations were noted and were most consistent in high and low pH areas (see Figs. 16.4 and 16.5). Unlike goldeneyes, this species appears to produce more young at pH extremes at Wanapitei. At low pH, young may have an easier time finding relatively abundant invertebrate prey consisting of few species, and at high pH they can find abundant diverse prey. In mid-range pHs, fish predation and pH effects reduce invertebrate prey availability. The plastic nature of its feeding habits, combined with its broad ecological niche, may allow this species to take full advantage of any habitat improvement after recovery.

American Black Duck

The Black Duck is closely related to the Mallard but is highly adapted to boreal conditions (Ross 1987). Although similarly abundant at Sudbury, its distribution is different from that of the Mallard. It is found throughout the area and shows a strong correlation to numbers of wetlands per plot (Table 16.2), a reflection of its preference for smaller breeding ponds (McNicol et al. 1987a). Its population has shown an upward tendency, consistent with recent patterns for northeastern Ontario. This trend probably reflects reductions of hunter bag limits instituted during the study in response to an overall population decline detected over the past 30 years (NAWMP 1986). Interestingly, the sharpest rise occurs in the lowest pH range. Although an explanation is presently being sought, this tendency suggests that additional birds are being forced into less desirable habitat.

Ring-Necked Duck

This ground-nesting species prefers small productive wetlands and marshes in the boreal region and is common in the Sudbury area (4300 pairs), particularly in the northeast. Ring-necked Ducks have a diverse diet (Bellrose 1980; McNicol et al. 1987b), and production of young is highest in lakes at each pH extreme (pH <5.0; pH >6.3), similar to the pattern for other insectivores and omnivores. Again, this may be due to the combined effects of fish predation and pH on invertebrate prey in lakes between pH 5 and 6.3. Its distribution, however, only correlates with wetlands numbers (Table 16.2). Given the lack of any relationship to pH, it is not surprising that Ring-necked Duck populations remained relatively stable in all pH classes during this study.

Wood Duck

This tree-nesting species prefers more southerly marshes, swamps, and small lakes (Bellrose 1980). At the northern limit of its present range near Sudbury, it is virtually absent in more boreal habitat to the north and west but is common in the southeast where it is keying on nutrient-rich clay substrates (Wickware and Rubec 1989). The Wood Duck has risen in numbers during the study, but this appears unrelated to pH changes. Instead, this reflects the sharp population increase this species has undergone in southern Ontario (Dennis et al. 1989), which has led to expansion at the edges of its range into appropriate habitat created by increasing amounts of beaver flowage in Ontario.

Mallard

The Mallard is the most widely distributed waterfowl species in the world and the most abundant duck in the Sudbury area. It is a marsh-dwelling species with a high tolerance of human activity and has a distribution very similar to the Wood Duck's, showing the same regional association with clay-dominated soils and a lack of relationship with other habitat variables. It selects the most fertile breeding habitat (Merendino 1993), which is the least likely to be affected by pH depression. The local population is stable, and trends in all three pH ranges appear similar.

Future Implications for Waterfowl

Most research on effects of acidification on waterfowl has focused on habitat and food chain relationships. However, waterfowl breeding in affected areas must be monitored over a long period to reliably establish whether a consistent trend of improvement in populations is occurring as lake chemistries recover. Nonetheless, we can use existing knowledge to make predictions on how species will respond to various emission control scenarios. The Waterfowl Acidification Response Modelling System (WARMS) is composed of an underlying acidification model (Jones et al. 1990) and fish and waterfowl models derived from data collected in the Sudbury and Algoma regions (Blancher et al. 1992). WARMS uses pH, lake area, dissolved organic carbon, total phosphorus, and fish presence to predict independent waterfowl species responses to changing lake chemistry. It provides estimates of pre-acidification, current and eventual (steady-state) values for pH, fish presence, and waterfowl breeding parameters under various emission scenarios.

WARMS predictions for pH, fish presence, and selected waterfowl species in the Sudbury area (based on 227 lakes) (Neary et al. 1990) for the following emission scenarios—(1) pre-acidification (background) levels of sulfur dioxide, (2) 1980 levels of sulfur dioxide, (3) Canadian emissions reduced by about 42% (by 1994), and (4) scenario 3 with about a 50% reduction in U.S. emissions (by year 2000)—are shown in Figure 16.7. Fish presence, pH, and waterfowl broods are all lower currently and under future scenarios than calculated values before lake acidification (except goldeneyes). However, as sulfur dioxide emissions decline, lake pH is predicted to improve dramatically. This is expected to result in the return of fish to many lakes where fish were lost due to acidification. The return of fish should provide more suitable foraging conditions for fish-eating species, such as the Common Merganser, when numbers of lakes supporting broods are expected to increase substantially.

Conditions for Black Ducks and Hooded Mergansers should also improve, although not of the magnitude of the Common Merganser. Increases in broods of these two species will be related to shifts in invertebrate assemblages (e.g., return of high-quality prey) resulting from chemical improvements and not strictly fish presence. Broods of the insectivorous Common Goldeneye are expected to decline toward pre-acidification levels, as fish return to many lakes and reduce the supply of invertebrate foods.

Even under the strongest emission reduction scenario (Canada 42% + United States 50%), WARMS predicts that waterfowl populations in the Sudbury area will not return to their pre-acidification levels. However, under any reduction in sulfur dioxide emissions, populations of all local waterfowl species should remain stable or move toward pre-acidification levels. Continued monitoring of Sudbury area lakes is required to refine these predictions, establish rates of recovery, and most important, verify that current abatement programs will indeed restore the capacity of sensitive aquatic habitats to sustain healthy populations of plants and animals, including waterfowl.

Acknowledgments. We thank Don Fillman for assistance with aerial surveys, Don Kurylo (GLFC lab) for chemical analyses, Chris Wedeles (ESSA Technologies Ltd.) for the WARMS analysis, and Bill Keller, John Cairns, Karen Laws, and John Gunn for their reviews of the manuscript. Bernie Neary (OMEE) kindly provided chemical data for the Sudbury area. This ongoing research is funded by the Long Range Transport of Airborne Pollutants program of Environment Canada.

References

Alvo, R., D.J.T. Hussell, and M. Berrill. 1988. The breeding success of common loons in relation to alkalinity and other lake characteristics in Ontario. Can. J. Zool. 66:746–752.

Bellrose, F.C. 1980. Ducks, Geese and Swans of North America. 3rd ed. Stackpole Books, Harrisburg, PA.

Bendell, B.E., and D.K. McNicol. 1987. Fish predation, lake acidity and the composition of aquatic insect assemblages. Hydrobiologia 150:193–202.

Bendell, B.E., and D.K. McNicol. In press. Lake acidity, fish predation and the distribution and abundance of some littoral insects. Hydrobiologia.

Blancher, P.J., D.K. McNicol, R.K. Ross, C.H.R. Wedeles, and P. Morrison. 1992. Towards a model of acidification effects on waterfowl in eastern Canada. Environ. Pollut. 78:57–63.

Dennis, D.G., G.B. McCullough, N.R. North, and B. Collins. 1989. Surveys of Breeding Waterfowl in Southern Ontario, 1971–87. Progress Note 180. Canadian Wildlife Service.

DesGranges, J.L., and M. Darveau. 1985. Effects of lake acidity and morphometry on the distribution of aquatic birds in southern Quebec. Holarct. Ecol. 8:181–190.

Eriksson, M.O.G. 1979. Competition between freshwater fish and goldeneyes *Bucephala clangula* (L.) for common prey. Oecologia 41:99–107.

Gilyazov, A.S. 1993. Air pollution impact on the bird communities of the Lapland Biosphere Reserve, pp. 383–390. *In* M.V. Kozlov, E. Haokioja, and V.T. Yarmishko (eds.). Aerial Pollution in Kola Peninsula. Proceedings of the International Workshop, April 14–16, 1992. Kola Scientific Center, Apatity, Russia.

Intera Tydac. 1991. SPANS 5.2 Reference Manual. Vol. 1. Intera Tydac Technologies, Ottawa, Ontario.

Jones, M.L., C.K. Minns, D.R. Marmorek, and F.C. Elder. 1990. Assessing the potential extent of damage to inland lakes in eastern Canada due to acidic deposition. II. Application of the regional model. Can. J. Fish. Aquat. Sci. 47:67–80.

Keller, W., J.M. Gunn, and N.D. Yan. 1992a. Evidence of biological recovery in acid stressed lakes near Sudbury, Canada. Environ. Pollut. 78:79–85.

Keller, W., J.R. Pitblado, and J. Carbone. 1992b. Chemical responses of acidic lakes in the Sudbury, Ontario, area to reduced smelter emissions, 1981–1989. Can. J. Fish. Aquat. Sci. 49(Suppl. 1):25–32.

Mallory, M.L., D.K. McNicol, and P.J. Weatherhead. 1994. Habitat quality and reproductive effort of Common Goldeneyes nesting near Sudbury, Canada. J. Wildl. Manage. 58:552–560.

Mallory, M.L., P.J. Weatherhead, D.K. McNicol, and M.E. Wayland. 1993. Nest site selection by Common Goldeneyes in response to habitat features influenced by acid precipitation. Ornis Scand. 24: 59–64.

McNicol, D.K., B.E. Bendell, and R.K. Ross. 1987a. Studies of the effects of acidification on aquatic wildlife in Canada: waterfowl and trophic relationships in small lakes in northern Ontario. Occ. Paper 62. Canadian Wildlife Service.

McNicol, D.K., P.J. Blancher, and B.E. Bendell. 1987b. Waterfowl as indicators of wetland acidification in Ontario. ICBP Tech. Publ. 6:149–166.

McNicol, D.K., P.J. Blancher, and R.K. Ross. 1990. Waterfowl as indicators of acidification in Ontario, Canada. Trans. Internat. Union Game Biol. 19:251–258.

McNicol, D.K., and M.L. Mallory. 1994. Trends in small lake water chemistry near Sudbury, Canada, 1983–1991. Water Air Soil Pollut. 73:105–120.

McNicol, D.K., and M. Wayland. 1992. Distribution of waterfowl broods in Sudbury area lakes in relation to fish, macroinvertebrates and water chemistry. Can. J. Fish. Aquat. Sci. 49(Suppl. 1):122–133.

Merendino, M.T. 1993. The relationship between wetland productivity and distribution of breeding Mallards, Black Ducks and their broods: historical and spatial analyses. Ph.D. thesis, University of Western Ontario, London.

Neary, B.P., P.J. Dillon, J.R. Munro, and B.J. Clark. 1990. The Acidification of Ontario Lakes: An Assessment of Their Sensitivity and Current Status with Respect to Biological Damage. Report. Ontario Ministy of the Environment, Toronto, Ontario.

North American Waterfowl Management Plan (NAWMP). 1986. North American Waterfowl Management Plan: A Strategy for Cooperation. Canadian Wildlife Service/U.S. Fish and Wildlife Service, Ottawa, Canada.

Reinecke, K., and R. Owen. 1980. Food use and nutrition of Black Ducks nesting in Maine. J. Wildl. Manage. 44:549–558.

Ross, R.K. 1987. Interim Report on Waterfowl Breeding Pair Surveys in Northern Ontario, 1980–1983. Progress Note Series 168. Canadian Wildlife Service.

Ross, R.K., D.G. Dennis, and G. Butler. 1984. Population trends of the five most common duck species breeding in southern Ontario, 1971–76, pp. 22–26. *In* S.G. Curtis, D.G. Dennis, and H. Boyd (eds.). Waterfowl Studies in

Ontario, 1973–81. Occ. Paper 54. Canadian Wildlife Service.

Scheuhammer, A.M. 1991. Effects of acidification on the availability of toxic metals and calcium to wild birds and mammals. Environ. Pollut. 71: 329–375.

Wayland, M., and D.K. McNicol. 1990. Status Report on the Effects of Acid Precipitation on Common Loon Reproduction in Ontario: The Ontario Lakes Loon Survey. Technical Report Series 92. Canadian Wildlife Service.

Wayland, M., and D.K. McNicol. 1994. Movements and survival of Common Goldeneye broods near Sudbury, Ontario, Canada. Can. J. Zool. 72:1252–1259.

Wickware, G.M., and C.D.A. Rubec. 1989. Ecoregions of Ontario. Ecological Land Classification Series 26. Sustainable Development Branch, Environment Canada, Ottawa, Ontario.

17

Acidification and Metal Contamination: Implications for the Soil Biota of Sudbury

Christine D. Maxwell

Considerable attention has been focused on the improvements in vascular plant communities in Sudbury that have occurred as a result of the land reclamation programs and reduced industrial emissions during the past two decades. Relatively little is known about the corresponding changes in soil microbial populations even though they are essential for the successful development of plant communities. In this chapter, the effects of industrial emissions on the diverse biological communities within the soil are discussed. The focus is on the effects of acidification and toxic metals on Sudbury soils, but because of the limited number of studies in this area, a more general but very brief review of other literature is included.

Soil Biota

The importance of the soil biota in soil formation, nutrient cycling, and the production of organic matter is a key factor in the functioning of ecosystems. Disruption of these activities by anthropogenic stresses such as smelter emissions can have a profound effect on the development of higher plant communities. Our general lack of knowledge of the effects of anthropogenic stress on soil microbial populations is attributable, in part, to the very nature of the soil environment, which has been described by Stotsky (1974) as undoubtedly the most complex microbial habitat.

Soil is composed of (1) mineral particles of various dimensions and chemical characteristics; (2) organic matter in many stages of decay; (3) a liquid phase in the form of soil water that contains soluble materials; (4) a gaseous phase, usually composed of oxygen, carbon dioxide, and nitrogen; and (5) an active living metabolizing soil community.

The components of the soil biota are diverse, including viruses, bacteria, cyanobacteria, fungi, algae, and the soil fauna. This biotic community contributes to the structural integrity of the soil through the process of aggregate formation and to the fertility of the soil through nitrogen fixation. Also, through the production of humus, the soil biota contribute to soil texture, water-holding capacity, and the complexation of minerals important in plant nutrition (Ehrlich 1981; Paul and Clark 1989). Soil microorganisms, in turn, are affected by many factors, including the chemical status of the soil (e.g., nutrient content and ionic composition), the physical conditions (e.g., temperature, moisture, gaseous composition, pH, redox potential, particulates), and the biological interactions between organisms (e.g., competition, parasitism, and predator–prey relationships).

Soil microorganisms are not randomly distributed in the soil profile. A general trend is for soil populations to show a decrease in size with increasing depth, and this is matched by a decrease in soil organic matter with depth. The A horizon is the most active zone from a biological point of view, because it harbors the roots of higher plants, the bacterial flora, fungi, algae, and the fauna. Bacteria frequently form microcolonies on the surface of aggregate particles, to which they are attached by means of fibrillae and mucigels. There is increased bacterial and fungal activity in the vicinity of plant roots, as a result of an increase in organic matter provided by live, senescent, and dead roots. This region has been termed the *rhizosphere*. The root surface itself harbors a distinct microbiota and is referred to as the *rhizoplane*. Most actively growing algae and cyanobacteria are restricted to the upper few millimeters of the soil surface where light permits photosynthesis.

Toxic Metals, Acid Conditions, and Soil Microorganisms–General Considerations

In unpolluted soils, the abundance of toxic metals (copper, nickel, etc.) is generally low (Table 17.1), and they are most often in forms unavailable to living organisms. By contrast, metals introduced as a result of anthropogenic activities are usually present as microscopically small particles, which may be sorbed, bound, chelated, or in salt form and are therefore more available to microorganisms (Doelman 1985). The availability, and hence potential toxicity, of metals is also governed by several physicochemical characteristics of the soil system, such as pH, oxidation-reduction potential, aeration, inorganic anions and cations, particulate and soluble organic matter, clay minerals, salinity, and temperature (Babich and Stotsky 1985).

Not all microorganisms are eliminated from industrially damaged areas. In fact, a wide range of relatively resistant or tolerant microorganisms can be found in soils contaminated by toxic metals. They possess physiological mechanisms that allow them to survive and reproduce (Gadd 1990). These mechanisms include the ability to produce polysaccharides, which act as biosorbents for metal cations, the ability to precipitate and crystallize metals on the cell surface, reduced permeability of the cell wall or cell membranes, or the ability to compartmentalize or bind metals to proteins intracellularly. In many bacterial strains, the genetic determinants of resistances to metals are located on plasmids, which are pieces of DNA that are separate from the chromosome (Gadd 1990).

Bacteria can carry out transformation of toxic metals through oxidation, reduction, methylation, and demethylation, and this may have

TABLE 17.1. Metal Concentrations (μg/g) in Sudbury Surface Soils from Surveys Conducted within 20 km of Smelters*

		Dudka et al. (submitted)		OMOE (1992)	
Metal	Background level	Mean	Range	Mean	Range
Arsenic	10	—	—	13	1–257
Cobalt	25	10	1–113	14	4–51
Copper	60	116	11–1891	155	14–2100
Lead	150	40	15–158	25	4–160
Nickel	60	104	5–2149	168	24–1500

*Surveys by Dudka et al. (submitted) and Ontario Ministry of the Environment (OMOE) (J. Negusanti, *unpublished data*). Maximum background levels for non-agricultural soils is also indicated (Compiled by Negusanti and McIlveen [1990]). Geometric means and ranges are presented.

—, no data.

some implications for the soil community in terms of mediating the effects of toxic metals (Shannon and Unterman 1993). The ability to transform metals as a detoxification mechanism could be potentially very useful in the remediation of metal-contaminated habitats; however, little is known about the importance of these organisms in the natural environment (Lovely 1993).

It is difficult to separate the effects of toxic metals from those of high soil acidity, because soil acidification increases the solubility of metals. Soil acidity in natural (unpolluted) soils is determined by several processes, some of which produce acids (e.g., decomposition of organic material through the production of organic and carbonic acids and the mineralization of plant nutrients) and some of which consume acids (weathering of soil minerals, and the anaerobic reduction of nitrogen and sulfur) (Abrahamsen 1987). Acid deposition has resulted in increased soil acidity in the vicinity of point sources such as smelters (Hutchinson and Whitby 1974); however, it is difficult to determine the relative contributions from atmospheric deposition and naturally occurring processes. There is some evidence to suggest the occurrence of seasonal changes in pH under some vegetational cover types as a result of the uptake and return of base cations to the soil (Johnson 1987). Whether small fluctuations in pH can significantly influence the soil biota remains a matter of conjecture.

Sudbury Soils: Implications for Soil Microorganisms

A diverse series of habitats exists for the soil microorganisms in the Sudbury area, ranging from heavily contaminated soils that are undergoing slow changes over time to less contaminated areas undergoing the same changes to sites in which amelioration has been rapid as a result of revegetation treatment.

The natural soils of the Sudbury area are podzols, which are acidic soils with a pH range of approximately 3.8–5.2. On hillsides close to the smelters where acidification and metal contamination are greatest, mineral soils may be found in small pockets, but the loss of vegetation has resulted in severe soil erosion, with the loss of typical soil profiles. In areas farther away, the typical soil profile is retained; however, high concentrations of toxic metals and low pH conditions often occur in the surface layers of the soil, the layers of greatest biological activity (Hutchinson and Whitby 1974). Among the elements found at elevated concentrations in surface soils near the smelters are copper, nickel, cadmium, cobalt, chromium, iron, manganese, sulfur, and zinc (Dudka et al., in press).

There is some evidence of reductions in the copper and nickel concentrations in surface soils over the past 20 years (Gundermann and Hutchinson 1993; Dudka et al., in press). The authors of the studies suggested that these changes in concentration of metals in surface soils are due partly to leaching and erosion of the physically unstable barren soils and possibly due to reduced atmospheric inputs. Despite these reductions, bioassay results using lettuce seedlings suggest that the concentrations of metals at sites close to the smelters are still inhibitory to root growth (Gundermann and Hutchinson 1993). These inhibitory effects may also be significant for the soil microbial community.

Diversity and Abundance of Soil Organisms

Soil Algae and Cyanobacteria

The photosynthetic components of the soil community, the soil algae and cyanobacteria, play several important roles. They are responsible for the addition of carbon and organic matter to the soil, and in addition, the growth of filamentous forms can result in the formation of crusts (sometimes with fungal hyphae and moss protonemata) that reduce water and wind erosion (Metting 1981).

The soil flora of three contaminated sites, near Sudbury, were characterized by a low

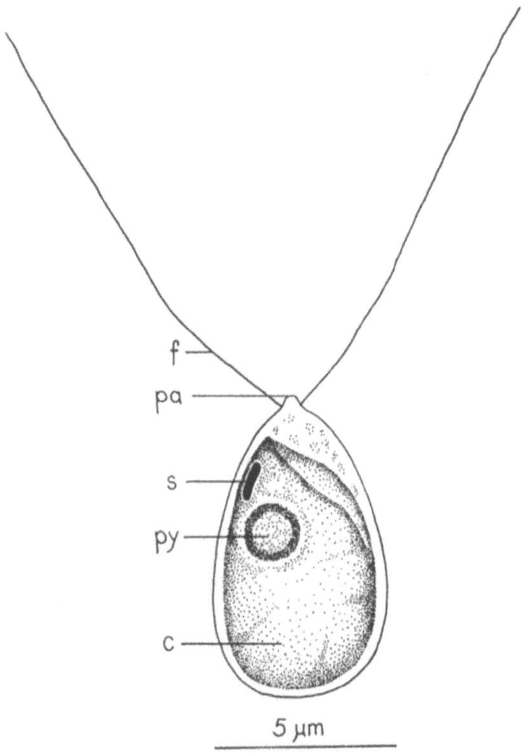

Figure 17.1. Acid-tolerant *Chlamydomonas acidophila. f,* flagellum; *pa,* papilla *s,* stigma; *py,* pyrenoid; *c,* chloroplast. (Drawing by M. Twiss.)

diversity of chlorophytes, one or two diatom species, and a notable absence of cyanobacteria (Maxwell 1991). Similar results have been reported for soils subject to emissions from various metallurgical plants in Russia (Shtina et al. 1985).

The chlorophyte-dominated flora of the Sudbury barrens is characteristic of acid soils. *Chlamydomonas acidophila* (Fig. 17.1), a single-celled motile alga, is ubiquitous in these soils. As the name suggests, this alga is acid-tolerant and is reported to occur in such widespread areas as Smoking Hills in the Canadian Arctic, where it is found in acidic ponds with a pH of 1.8 (Sheath et al. 1982), and strip-mine ponds in Ohio with a pH of 3.3 (Rhodes 1981). Not only is it acid-tolerant, but it also exhibits copper tolerance. Twiss (1990) compared the copper tolerance of isolates from three Sudbury soils, including a roast bed soil, and concluded that in comparison with a laboratory strain of *C. acidophila,* the isolates were indeed tolerant. A second unicellular green alga, *Chlorella saccharophila,* has also been recognized as a metal-tolerant organism in the Sudbury soils (Hutchinson et al. 1981).

The absence of cyanobacteria from soils in the Sudbury barren areas is not surprising. Several authors, including Brock (1973) and Dooley and Houghton (1974), have concluded that cyanobacteria are unable to live at pH levels less than 4.4, conditions that exist over much of the Sudbury area.

The response of the autotrophic soil microflora to liming and fertilizer treatments was surprisingly rapid. Maxwell (1991) followed the changes that occurred at a Wahnapitae site for 46 weeks. The first cyanobacteria were found just 2 weeks after the reclamation treatment, when a pH of 5.3 was recorded. The diversity of the cyanobacteria and chlorophytes increased with time. A nearby untreated control site showed no change over the 46 weeks (Table 17.2). Maxwell (1991) also examined soils that had been subjected to the revegetation treatment, 2, 4, and 5 years before her study. A progressive increase both in the diversity of green algae and cyanobacte-

TABLE 17.2. Changes in the Soil Microflora in 46 Weeks after Treatment with Lime and Fertilizer at a Contaminated Site at Wahnapitae

			Post-treatment (weeks)				
	Control site	Pretreatment	2	6	12	36	46
Cyanobacteria							
Anabaena sp.	—	—	—	x	x	x	x
Gloeocapsa sp.	—	—	—	—	—	—	—
Lyngbya sp. (2)	—	—	—	—	x	x	x
Nostoc muscorum	—	—	x	x	x	x	x
Oscillatoria sp. (1)	—	—	x	x	x	x	x
Chlorophyta							
Chlamydomonas acidophila	x	x	x	x	x	x	x
Chlorella sp.	—	—	—	—	—	x	x
Chlorococcum sp.	x	x	x	x	x	x	x
Chlorosarcina	x	—	—	—	—	—	—
Desmococcus sp.	x	x	x	x	x	x	x
Gloeocystis sp.	—	—	—	—	—	—	—
Oocystis sp.	—	—	—	—	—	—	—
Stichococcus subtilis	x	x	x	x	x	x	x
Chrysophyta							
Hantzchia	—	—	—	—	—	x	x
Navicula sp.	x	x	x	x	x	x	x
Euglenophyta							
Euglena sp.	x	x	x	x	x	—	x

x, Isolated in enrichment culture; —, not isolated in enrichment culture.

ria had occurred. Similar results have been documented in other studies of the photosynthetic components of the soil community. Shubert and Starks (1979) noted a distinct algal succession occurring with time on naturally revegetated mine spoil banks in North Dakota. Balezina (1975) also reported increases in the diversity and abundance of soil algae and cyanobacteria after the liming and the application of fertilizer to an agricultural soil in the former Soviet Union.

Soil Bacteria

Little information is available on the diversity and abundance of soil bacteria in Sudbury. That which is available relates to specific soil processes such as nitrogen fixation and nitrification and is covered in the next section. In a summary of the effects of various concentrations of metals on soil bacteria, Doelman (1985) reported significant decreases in populations both in field and laboratory experiments and noted that the bacteria are more sensitive to metals than fungi are. Metal contamination generally causes an increase in the fungal biomass to bacterial biomass ratio, as a result of the different sensitivities of prokaryotes and eukaryotes.

Hutchinson and Nakatsu (*personal communication*) found a vigorous acid and metal-tolerant bacterial flora in the Coniston Valley, with pseudomonads predominating. Tolerances to nickel, copper, cobalt, and aluminum were especially high. Nickel tolerance appeared to be controlled by a nickel-tolerant plasmid. Nordgren et al. (1983, 1985) reported decreases in bacterial numbers close to the brass mill at Gusum, where the soil is heavily contaminated with copper (10–15 mg/g) and zinc (15–20 mg/g), and close to a smelter in Northern Sweden, which emits a wide spectrum of metal particulates including lead, copper, zinc, and arsenic.

Fungi

The soil fungi are of particular importance in the decomposition of litter and the mineralization of organic matter, especially under acid conditions. Some basidiomycetes form symbiotic mycorrhizal associations, thereby directly affecting the nutrient status of vascular plants. Freedman and Hutchinson (1980) compared the abundance and diversity of soil fungi on contaminated sites close to the Copper Cliff smelter with uncontaminated sites, and although there were consistently fewer colonies isolated from contaminated sites, the numbers were not significantly different from those at uncontaminated sites.

Carter (1978) compared soil fungi in terms of species diversity and abundance at four sites in Sudbury. The soils at the sites differed in terms of copper and nickel concentrations, although care was taken to select sites that had similar conditions of moisture, pH, and organic matter and also similar leaf litter from red maple (*Acer rubrum*). At two highly contaminated sites, he found a lower species diversity index and reduced number of colony-forming units than at the two less-contaminated sites. Also, *Penicillium waksmanii* appeared to be restricted to contaminated sites. In laboratory tests, Carter demonstrated that the fungal flora of the contaminated sites were more tolerant of high nickel concentrations than those of the less-contaminated sites (approximately 0.4 mg/g of copper, nickel). Similar effects on soil fungi have been reported for metal-contaminated areas in Sweden (Ruhling et al. 1984).

Soil Fauna

The soil fauna is composed of protozoans and metazoans, living both in the soil and in the surface litter. The fauna play an important part in the early breakdown of litter, greatly facilitating the activities of the soil microflora. Those that fragment litter include the millipedes, mites, isopods, and collembolans.

There is limited information on the soil fauna in Sudbury. Behan-Pelletier and Winterhalder (*unpublished manuscript*) recorded 35 species of oribatid mites, representing 22 families from sites around Sudbury. In the barren areas, they found a very limited oribatid fauna with only three species. The semibarren fauna was more diverse with 33 species, although this number is very low in comparison with uncontaminated forest sites in Ontario and Quebec. Revegetated sites had 13 species, indicating a slow recovery.

Reports from other contaminated areas, such as Palmerton, Pennsylvania (Strojan 1978), show a reduced number and diversity of arthropods in forest litter compared with control sites. Springtail species (collembolans) that live in the litter or surface layer of the soil decrease in the vicinity of the brass mill at Gusum, Sweden (Bengtsson and Rundgren 1984), but species that live deeper in the soil are not affected.

Carbon Dioxide Fixation and Litter Decomposition

The fixation of carbon dioxide in terrestrial ecosystems is mainly carried out by higher plants, soil algae, and cyanobacteria. Several groups of bacteria, the green bacteria, the methanogenic bacteria, and acetogenic bacteria, also fix carbon dioxide, but little is known about their contribution of carbon fixation to the ecosystem. Organic carbon from plants enters the soil through the process of litter decomposition or, more directly, through the activity of the soil algae and cyanobacteria. Mycorrhizal fungi can also provide substantial quantities of organic matter, derived from the photosynthetic activities of the host plants, and also, there are recent reports of direct carbon dioxide fixation by both ectomycorrhizal fungi (Lapeyrie 1988) and vesicular-arbuscular fungi (Bechard and Piche 1989).

The amount of organic matter entering the soil depends on the amount of litter input and the rate of decomposition. On the barren slopes close to the smelters where there is little or no vegetation, organic matter in the soil is either in the form of residual humus or derived from the reduced flora of soil algae or perhaps from mosses such as *Pohlia.* Such organisms are often

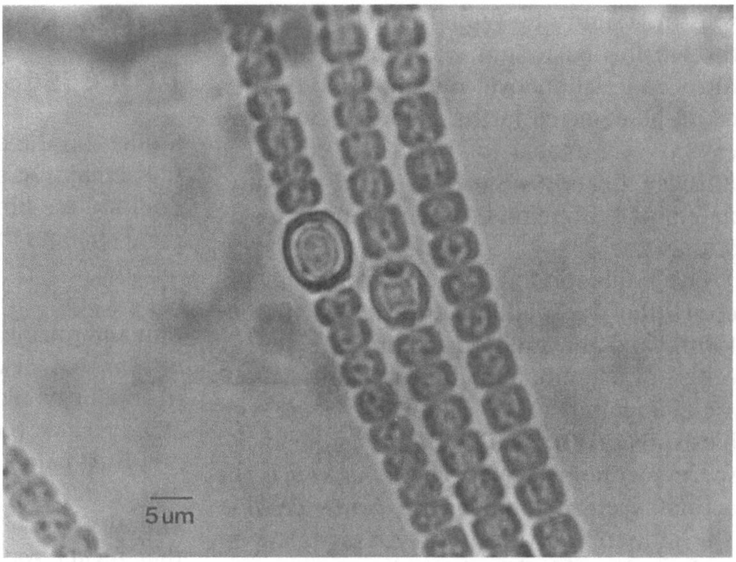

FIGURE 17.2. Photomicrograph of a filament of *Nostoc* showing a heterocyst. (Photo by D. Woodfine.)

the primary colonizers of bare soil. Land reclamation programs should permit a progressive increase in this limited flora, although the changes may be slow (see Chapter 13).

Metals from smelter emissions tend to accumulate in litter layers (Hutchinson and Whitby 1974; Dumontet et al. 1992), and their toxicity can result in the decreased activity of the decomposers. Measurements of soil respiration and soil metabolic activity have both shown reduced microbial activity in litter at contaminated sites near the Copper Cliff smelter in Sudbury (Freedman and Hutchinson 1980), near to the smelter at Gusum, Sweden (Nordgren et al. 1983), and also downwind from the smelter at Rouyn-Noranda, Quebec (Dumontet et al. 1992). Studies using litter bags have also shown reduced decomposition rates at contaminated sites compared with uncontaminated sites (Freedman and Hutchinson 1980; de Catanzaro 1983). In a laboratory study, copper was shown to be more inhibitory to the decomposition process than nickel.

The Nitrogen Cycle—Nitrogen Fixation

The biological fixation of nitrogen from the atmosphere is carried out by prokaryotes that possess the enzyme nitrogenase. The organisms fall into two main categories, the asymbiotic forms, which may be autotrophic or heterotrophic, and the symbiotic forms, which form associations with higher plants.

Many of the free-living photosynthetic cyanobacteria are capable of nitrogen fixation. The ability to fix nitrogen is found most often in filamentous forms such as *Nostoc* and *Anabaena;* these possess heterocysts, although some non-heterocystous forms may also have the ability (Fig. 17.2).

Maxwell (1991) found that cyanobacteria were absent on contaminated sites in Sudbury and concluded that pH was a limiting factor. Rapid colonization of amended sites occurred after the revegetation treatment, and among the colonists were two heterocystous forms, *Nostoc muscorum* and *N. paludosum.* Their presence suggests that nitrogen fixation could be occurring; however, experimental evidence for *Nostoc punctiforme* (Granhall 1970) shows the fixation process occurring in a range from pH 5.0 to 10.0, with an optimum of pH 7.6. If this is the case with species found on revegetated sites, the pH may still be too low to permit significant nitrogen fixation in Sudbury soils.

A free-living heterotrophic bacterium, similar to *Azotobacter* was isolated from contaminated soils in Sudbury by Winterhalder (1987). Although this bacterium was capable of fixing nitrogen in laboratory conditions, no fixation could be detected in the field. De Catanzaro (1983) was unable to detect any biological nitrogen fixation occurring in the contaminated soils of remnant jack pine stands in Sudbury.

The symbiotic fixation of nitrogen far exceeds the contributions of the asymbiotic forms, and the rates of fixation are several orders of magnitude higher. Two associations are of importance, the legume-*Rhizobium* symbiosis, (Fig. 17.3) and the non-legume angiosperms-actinomycete symbiosis, in which the actinomycete is often of the genus *Frankia*. The efficiency of the two systems is apparently comparable, but the actinomycete association has the advantage of being able to function in more acidic conditions (pH 4.2) than the rhizobial association.

There are no native legumes occurring on the barren and semibarren soils in the Sudbury area; however, two exotic acid-tolerant legumes, Birdsfoot trefoil (*Lotus corniculatus*) and Alsike clover (*Trifolium hybridum*), have been introduced successfully as part of the revegetation program. Nodulation by *Rhizobium* and nitrogen fixation have been demonstrated for both species, despite the fact that the pH of the soils was considered to be low for the survival of rhizobia (Winterhalder 1987). Another leguminous exotic, the woody black locust (*Robinia pseudo-acacia*) has also been introduced to the Sudbury soils, where it is hardy and vigorous (see Chapter 8).

Frankia is a non-motile filamentous actinomycete that forms root nodules with several plants native to the Sudbury area, including sweet fern (*Comptonia peregrina*) and the alders (*Alnus crispa* and *Alnus rugosa*). Whereas *Frankia* itself is non-motile, there is a recent report that spores of *Frankia*, which are resistant to desiccation, may be transported by the activities of birds (Paschke and Dawson 1993) and that the nitrogen-fixing ability of this actinomycete allows actinorhizal vascular plants to become very invasive.

Other Nitrogen Cycling Processes

Studies on the effects of acidity and metals on the remaining stages of the nitrogen cycle in Sudbury are limited, and none relates to temporal changes that have occurred over the past decade.

De Catanzaro (1983) compared mineralization and nitrification in the contaminated soils of remnant jack pine (*Pinus banksiana*) stands in Sudbury with uncontaminated sites at Windy Lake (60 km northwest of Sudbury) and Burt Lake (50 km west of Kirkland Lake) and found higher levels of ammonium in the Sudbury soils. Laboratory studies confirmed that nickel was causing a stimulation of the mineralization process. Similar results have been shown for forest soils in Germany, contaminated with various metals (Necker and Kunze 1986). Stimulation of mineralization is not necessarily beneficial, because ammonium, if not taken up and used by higher plants or immobilized by microorganisms, may be lost in leaching.

The study of nitrification in soils is important from the standpoint of soil fertility, because higher plants readily take up nitrate ions into their roots for assimilation into organic compounds. Nitrate not used by higher plants may be leached from the soil or used by microorganisms capable of assimilatory nitrogen reduction (denitrification). Nitrification is a two-stage process involving two chemoautotrophic bacteria, *Nitrosomonas* and *Nitrobacter*. The former oxidizes NH_4^+ to NO_2^-, the latter NO_2^- to NO_3^-. It is also considered to be a process that is highly pH-sensitive. Optimum pH values range from pH 6.6 to 8.0. Below pH 6.0, nitrification rates decrease, and less than 4.5, they cease. Other environmental factors such as soil aeration, moisture, temperature, and organic matter are also important in determining nitrification rates.

De Catanzaro (1983) found both *Nitrosomonas* and *Nitrobacter* in soils at contaminated sites in Sudbury in higher numbers than in uncontaminated forest soils. The contaminated sites had soil pH values of 3.2–3.4, whereas

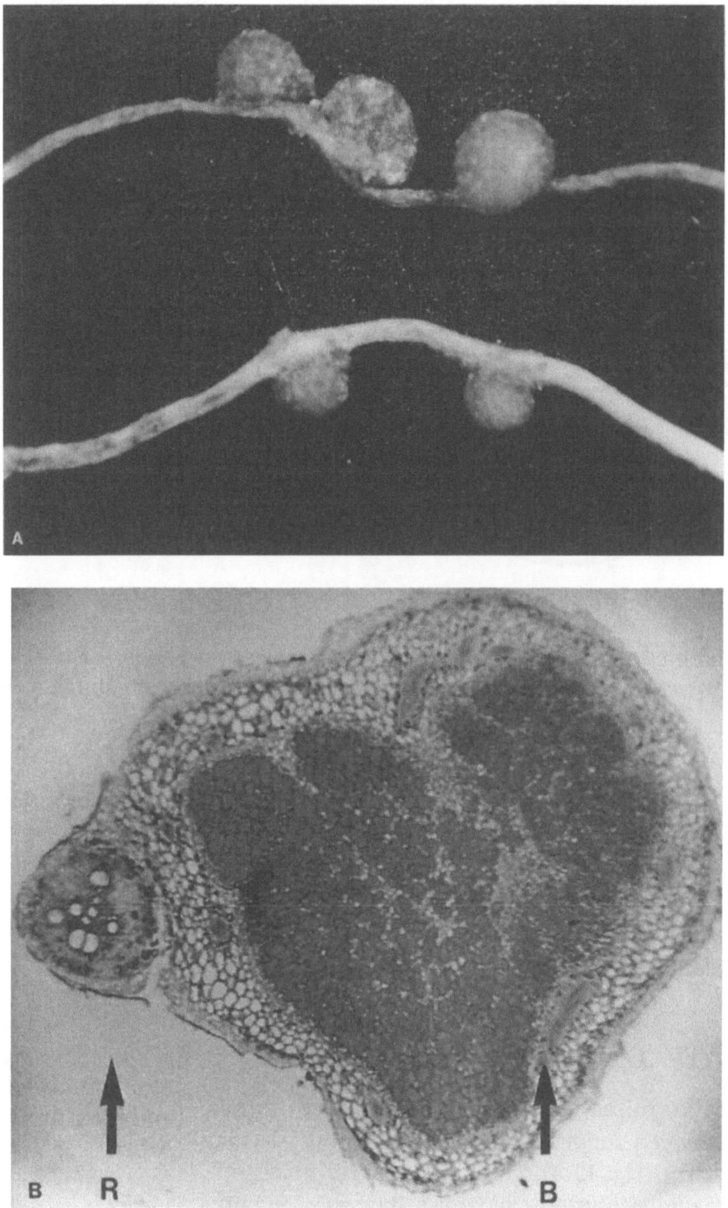

FIGURE 17.3. (**A**) Root nodule on *Lotus corniculatus;* (**B**) photomicrograph of a cross section of a root (*R*) and root nodule showing the location of the rhizobial bacteria (B). The diameter of the nodule is approximately 1 mm. (Photo by D. Lietz.)

the uncontaminated soils were also acidic (pH 3.4–3.6). Whereas pronounced seasonal trends were seen, nitrate levels in organic Sudbury soils were generally higher than those of uncontaminated sites, perhaps as a result of the higher numbers of nitrifiers. Incubation studies in which 100 $\mu g\ g^{-1}$ and 500 $\mu g\ g^{-1}$ nickel additions were made to uncontaminated soils showed an increase in nitrification at the lower concentration and a decrease at the higher concentration. This suggests that low concentrations of nickel may stimulate nitrification, whereas high concentrations will inhibit the process.

There are two possible explanations for the occurrence of nitrification in such acidic soils. In the first, there may be microsites in the soil in which the pH is higher than that of the soil solution. These could result from the decomposition of nitrogen-rich material with the production of ammonia in localized areas. In

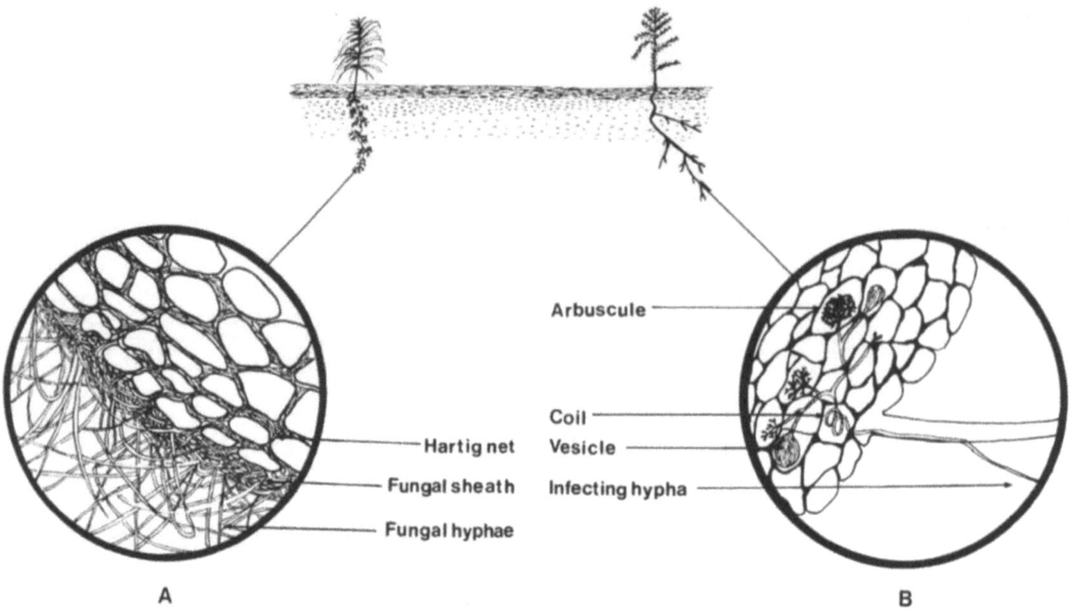

FIGURE 17.4. Two common types of mycorrhizae: (**A**) Ectomycorrhiza on the root of a pine seedling; and (**B**) vesicular-arbuscular mycorrhiza on the root of juniper. (Drawing by D. Woodfine.)

the second case, several heterotrophic bacteria, actinomycetes, and soil fungi are known to produce NO_3^- from NH_4^+ in laboratory conditions; however, the significance of these findings in field conditions is not known (Paul and Clark 1989).

Mycorrhizae

Mycorrhizae are symbiotic associations formed between fungi and the roots of most vascular plants (Fig. 17.4). The importance of mycorrhizae in ecosystem dynamics has only recently been recognized. Potentially, they can affect all aspects of a functioning ecosystem, including carbon allocation and nitrogen and phosphorus cycling. The association can confer greater abilities to withstand drought, nutrient stress, and environmental perturbation on the higher plants, allowing them to colonize areas previously unavailable to them (Allen 1991).

The activity and diversity of mycorrhizal fungi may be severely reduced by pollutants in smelter emissions. In profoundly disturbed sites, mycorrhizae may be completely absent. Ruhling et al. (1984) reported a decrease in most mycorrhizal forms, with an increase in copper concentrations in the soil at Gusum; however, some species, such as *Laccaria laccaria,* were apparently tolerant of fairly high levels of contamination.

Ectomycorrhizae may increase the tolerance of their hosts to metals. Jones and Hutchinson (1986) investigated the role of several species of ectomycorrhizal fungi in conferring copper and nickel tolerance on the paper birch (*Betula papyrifera*), which is widespread on contaminated sites in Sudbury. They reported that the ectomycorrhizal fungi conferred a degree of metal tolerance on the birch seedlings; however, the four mycorrhizal isolates used in this experiment differed in the degree of protection afforded to the seedlings. There were also differences in tolerance to nickel and copper. At lower nickel levels (34 μM nickel) one of the fungal isolates, *Scleroderma flavidum,* stimulated growth, whereas at higher levels (85 μM nickel), growth was 86% of control seedlings grown without nickel. Higher concentrations of nickel in seedling roots, compared with stems, suggested that nickel was retained in the fungal tissues. In similar experiments with copper, the infected seedlings showed a reduction of growth in higher concentrations (63 μM copper) and no difference from controls under low copper concentrations (32 μM copper). The results suggested that copper and nickel may cause toxicity by different mechanisms.

In further experiments, Jones and Hutchinson (1986) examined some of the physiological aspects of this mycorrhizal relationship and concluded that the effectiveness of *Scleroderma flavidum* in protecting seedlings from nickel toxicity could be related in part to the larger fungal biomass produced by this mycobiont in comparison with *Lactarius rufus* and also to significant differences in the mechanisms by which nickel is translocated from root to shoot (Jones et al. 1988).

Wilkins (1991) investigated the growth of several tree species infected with a variety of ectomycorrhizal fungi in the presence of several metals and found that the presence of mycorrhizae decreases the metal accumulation in the shoots and concentrates the metals in the extramatrical hyphae. Vesicular-arbuscular mycorrhizae have also been shown to be effective in decreasing the toxic effects of high soil metal concentrations (Angle et al. 1988). As with ectomycorrhizae, the accumulation of metals in shoots is generally reduced, but in this case, the concentration in roots may increase significantly. Metal retention in the root may be attributed to the complexation of metals to proteins in the walls of the fungal hyphae (Dehn and Schuepp 1989). Vesicular-arbuscular mycorrhizae may be involved in the metal tolerance demonstrated by some of the grasses invading the barren areas near to the smelters.

Mycorrhizal fungi are frequently reported to be absent from highly disturbed sites, and most of the vascular plants present on such sites are described as either non-mycorrhizal species or facultatively mycorrhizal species. Information on the mycorrhizal status of many colonizing species is limited for Sudbury, but this would be a potentially interesting area for further research.

Conclusion

The recovery of vegetation after environmental destruction is dependent on several key factors. First, there must be a reduction or cessation of the conditions that led to the degradation. Second, an active, integrated soil community must be re-established; and third, important biogeochemical cycles such as the carbon and nitrogen cycles must become functional.

Over the past 20 years, there have been significant reductions in emissions in Sudbury and improvements in some vascular plant communities, with and without the help of soil amelioration. To many people, the "greening" of Sudbury provides concrete evidence of recovery; however, as this chapter has shown, we know little of the activities of the soil microbial communities despite the fact that the success of the higher plant communities is dependent on the activities of the belowground biota. The biotic communities of Sudbury soils remain the least investigated of all communities in the area. The main reasons for this may be the inherent complexity of the soil biota, the spatial heterogeneity of soil conditions, and the temporal aspects of colonization and species interactions.

Acknowledgments. I wish to thank B. Mountney, G.W. Brown, D.G. Woodfine, and F. Wilhelm for technical assistance and R. Jones, G. Ferroni, J. Gunn, J. Cairns, and R. Benoit for helpful comments on the manuscript.

References

Abrahamsen, G. 1987. Air pollution and soil acidification, pp. 321–331. *In* T.C. Hutchinson and K.M. Meema (eds.). Effects of Atmospheric Pollutants on Forests, Wetlands and Agricultural Ecosystems. Springer-Verlag, Berlin.

Allen, M.F. 1991. The Ecology of Mycorrhizae. Cambridge University Press, Cambridge.

Angle, J.S., M.A. Spiro, A.M. Heggo, M. El-Kherbawy, and R.I. Chaney. 1988. Soil microbial–legume interactions in heavy metal contaminated soil at Palmerton, PA, pp. 321–336. *In* Proceedings of the University of Missouri 22nd Annual Conference on Trace Substances in Environmental Health, St. Louis, MO.

Babich, H., and G. Stotsky. 1985. Heavy metal toxicity to microbe-mediated ecological processes: a review and potential application to regulatory policies. Environ. Res. 36:111–137.

Balezina, L.S. 1975. Effect of mineral and organic fertilizers on the development of algae in a sod-podzolic soil. Mikrobiologiya 44:347–350.

Bechard, G., and Y. Piche. 1989. Fungal growth stimulation by CO_2 and root exudates in vesicular-arbuscular mycorrhizal symbiosis. Appl. Environ. Microbiol 55:2320–2325.

Bengtsson, G., and S. Rundgren. 1984. Ground living invertebrates in metal-polluted forest soils. Ambio 13:29–33.

Brock, T. D. 1973. Lower pH limit for existence of blue green algae: evolutionary and ecological implications. Science 179:480–483.

Carter, A. 1978. Some aspects of the fungal flora in nickel-contaminated and non-contaminated soils near Sudbury, Ontario, Canada. M.Sc. thesis, University of Toronto, Ontario.

De Catanzaro, J.B. 1983. Effects of nickel contamination on nitrogen cycling in boreal forests in northern Ontario. Ph.D. thesis, University of Toronto, Ontario.

Dehn, B., and H. Schuepp. 1989. Influence of VA mycorrhizae on the uptake and distribution of heavy metals in plants. Agric. Ecosyst. Environ. 29:79–83.

Doelman, P. 1985. Resistance of soil microbial communities to heavy metals. *In* V. Jènsen, A. Kjøller, and L.H. Sørensen (eds.). Microbial Communities in Soil. Elsevier Applied Science Publishers, New York.

Dooley, F., and J.A. Houghton. 1974. The nitrogen-fixing capabilities and the occurrence of blue green algae in peat soils. Br. Phycol. J. 8:289–293.

Dudka, S., R. Ponce-Hernandez, and T.C. Hutchinson. In press. Current level of total element concentrations in the surface layer of Sudbury's soils. Sc. Total Environ.

Dumontet, S., H. Dinel, and P.E.N. Lévesque. 1992. The distribution of pollutant heavy metals and their effect on soil respiration and acid phosphatase activity in mineral soils of Rouyn-Noranda region, Québec. Sc. Total Environ. 121:231–245.

Ehrlich, H.L. 1981. Geomicrobiology. Dekker, New York.

Freedman, B., and T.C. Hutchinson. 1980. Effects of smelter pollutants on forest leaf litter decomposition near a nickel-copper smelter at Sudbury, Ontario. Can. J. Bot. 58:1722–1736.

Gadd, G.M. 1990. Metal tolerance, pp. 178–210. *In* C. Edwards. (ed.). Microbiology of Extreme Environments. Open University Press. Milton Keynes, England.

Granhall, U. 1970. Acetylene reduction by blue green algae isolated from Swedish soils. Oikos 21:330–332.

Gundermann, D.G., and T.C. Hutchinson. 1993. Changes in soil chemistry 20 years after the closure of a nickel-copper smelter near Sudbury, Ontario pp. 559–562. *In* R.J. Allen and J.O. Nriagu (eds.). Proceedings of the International Conference on Heavy Metals in the Environment, Toronto, Ontario. Vol. 2. CEP Consultants, Edinburgh.

Hutchinson, T.C., C. Nakatsu, and D. Tam. 1981. Multiple metal tolerance and co-tolerance in algae, pp. 300–304. *In* W.H.O. Ernst (ed.). Proceedings of the Third International Conference on Heavy Metals in the Environment, Amsterdam. CEP Consultants, Edinburgh.

Hutchinson, T.M., and L.M. Whitby. 1974. Heavy-metal pollution in the Sudbury mining and smelting region of Canada. Environ. Conserv. 1:123–132.

Johnson, D.W. 1987. A discussion of changes in soil acidity due to natural processes and acid deposition, pp. 333–345. *In* T.C. Hutchinson and K.M. Meema (eds.). Effects of Atmospheric Pollutants on Forest Wetlands and Agricultural Ecosystems. Springer-Verlag, Berlin.

Jones, M., J. Dainty, and T.C. Hutchinson. 1988. The effect of infection by *Lactarius rufus* or *Scleroderma flavidum*. Can. J. Bot. 66:934–940.

Jones, M.D., and T.C. Hutchinson. 1986. The effect of mycorrhizal infection on the response of *Betula papyrifera* to nickel and copper. New Phytol. 102: 429–442.

Lapeyrie, F.F. 1988. Oxalate synthesis from soil bicarbonate by the mycorrhizal fungus *Paxillus involutus.* Plant Soil 110:3–8.

Lovely, D.R. 1993. Dissimilatory metal reduction. Annu. Rev. Microbiol. 47:263–290.

Maxwell, C.D. 1991. Floristic changes in soil algae and cyanobacteria in reclaimed metal-contaminated land at Sudbury, Canada. Water Air Soil Pollut. 60:381–393.

Metting, B. 1981. The systematics and ecology of soil algae. Bot. Rev. 47:195–312.

Necker, U., and C. Kunze. 1986. Incubation experiments on nitrogen mineralization by fungi and bacteria in metal amended soil. Angnew Bot. 60:81–94.

Negusanti, J.J., and W.D. McIlveen. 1990. Studies of the Terrestrial Environment in the Sudbury

Area, 1978–1987. Ontario Ministry of the Environment, Ontario.
Nordgren, A., E. Bååth, and B. Söderström. 1983. Microfungi and microbial activity along a heavy metal gradient. Appl. Environ. Microbiol. 45: 1829–1837.
Nordgren, A., E. Bååth, and B. Söderström. 1985. Soil microfungi in an area polluted by heavy metals. Can. J. Bot. 63:448–455.
Paschke, M.W., and J.O. Dawson. 1993. Avian dispersal of *Frankia*. Can. J. Bot. 71:1128–1131.
Paul, E.A., and F.E. Clark. 1989. Soil Microbiology and Biochemistry. Academic Press, San Diego, CA.
Rhodes, R.G. 1981. Heterothallism in *Chlamydomonas acidophila negoro* isolated from acidic strip mine ponds. Phycologia 20:81–82.
Ruhling, A., E. Bååth, A. Nordgren, and B. Söderström. 1984. Fungi in metal contaminated soil near the Gusum brass mill, Sweden. Ambio 13:32–34.
Shannon, M.J.R., and R. Unterman. 1993. Evaluating bioremediation: distinguishing fact from fiction. Annu. Rev. Microbiol. 47:715–738.
Sheath, R.G., M. Havas, J.A. Hellebust, and T.C. Hutchinson. 1982. Effects of long-term natural acidification on the algal communities of tundra ponds at the Smoking Hills, N.W.T., Canada. Can. J. Bot. 60:58–72.
Shtina, E.A., L.B. Neganova, T.A. Yel'shina, I. Shilova, and M.F. Andonova. 1985. Soil algae in polluted soils. Sov. Soil Sci. 17:18–27.
Shubert, L.E., and T.L. Starks. 1979. Algal succession on orphaned coal mine spoils, pp. 661–669. *In* M.K. Wali (ed.). Ecology and Coal Resource Development. Pergamon Press, New York.
Stotsky, G. 1974. Activity, ecology and population dynamics of microorganisms in soil, pp. 57–65. *In* A. Laskin and H. Lechavalier (eds.). Microbial Ecology. CRC Press, Cleveland, OH.
Strojan, C.L. 1978. The impact of zinc smelter emissions on litter arthropods. Oikos 31:41–46.
Twiss, M.R. 1990. Copper tolerance of *Chlamydomonas acidophila* (Chlorophyceae) isolated from acidic, copper-contaminated soils. J. Phycol. 26:655–659.
Wilkins, D.A. 1991. The influence of sheathing (ecto) mycorrhizas of trees on the uptake and toxicity of metals. Agric. Ecosyst. Environ. 35:245–260.
Winterhalder, K. 1987. The Sudbury Regional Land Reclamation Programme—an ecologist's perspective, pp. 81–92. *In* P.J. Beckett (ed.). Proceedings of the 12th Annual Meeting of the Canadian Land Reclamation Association. CLRA, Guelph, Ontario.

18

Birch Coppice Woodlands near the Sudbury Smelters: Dynamics of a Forest Monoculture

Gerard M. Courtin

The combination of environmental stresses to which the Sudbury region has been subjected has led to a series of distinct zones of vegetation surrounding the smelters. Closest to the smelters is a 170-km^2 zone of barren land that is nearly completely devoid of plant life. Adjoining the barrens is a 720-km^2 semibarren area, a zone of transition between the barrens and the natural plant community of the region (see Chapter 2). The two zones, the barren zone and transition zone (Amiro and Courtin 1981), are the direct result of human activity, and neither is found naturally as a successional stage of the eastern hemlock-white pine-northern hardwood forest (Braun 1950) that was once typical of the area.

The transition zone is divided into two largely monocultural communities, one dominated by white birch (*Betula papyrifera*) and the other by red maple (*Acer rubrum*), although each species may be found in the community dominated by the other (Amiro and Courtin 1981). The birch transition community is the more widespread of the two communities and occurs across a variety of slope and soil conditions, whereas the maple transition community is mainly restricted to glacio-fluvial deposits. Both communities are composed of trees with unique growth characteristics that become more prominent as one approaches the edge of the barrens. The multiple stems and slow growth of the trees in the birch transition forest is one of the more conspicuous of these altered features. Such trees are described as "coppiced."

Coppicing is a term usually used to describe a form of biomass rejuvenation that has been a management practice in Europe for hundreds and sometimes even thousands of years (Buckley 1992). In the preface to his recent book, Buckley (1992) described the process as follows:

> All coppice woodlands have one thing in common—they are repeatedly cut down, in the very reasonable expectation that the trees will regrow by themselves. Whether this regrowth is the result of new shoots sprouting from cut stumps or tree roots is unimportant: the point is that the new canopy forms rapidly and, for the main part, vegetatively from the old, without the need for any great management effort. The process is almost infinitely repeatable.

In North America, coppicing has rarely if ever been used as a forestry management practice; however, the birch transition forest of Sudbury does fulfil the above definition, except that the agency of biomass removal is not direct as would be the case with cutting, but rather, it is the indirect result of human impact. Nevertheless, the impact is cyclic, and the effects that result are both biotic and abiotic. In this chapter, I describe some of these processes by which this unique plant community is maintained in a stressed ecosystem.

FIGURE 18.1. (**A**) Aerial view from 10 m of a typical birch transition community to indicate the general morphology of the woodland and the absence of surface vegetation. (**B**) Example of a large birch coppice (stem diameter >5 cm at a height of 20 cm).

Site Characteristics

The birch transition forest (Amiro and Courtin 1981) is characterized by small aggregates of highly coppiced birch that are usually widely spaced such that the gaps in the canopy are often larger than the coppice aggregates themselves (Fig. 18.1). Relatively few other vascular and non-vascular species exist among the birch coppices (Table 18.1). Oke (1987) spoke of such strong vegetational contrasts as having a micro "oasis effect," and therefore, the present assemblage of coppices and the gaps between them, because of their relatively small size, is considered to form a mosaic of micro-oases and microdeserts. Each birch coppice rarely achieves a height of more than 6 m and a diameter of 12 cm, and few individual stems exceed 30 years of age (James and Courtin 1985; Trépanier 1985). It is important to note that stools from which the individual coppices arise are considerably older than the maximum age of the coppices, as is the case in traditional European coppice woodlands. Attempts to age the Sudbury stools have proved impossible, owing to the chaotic ring pattern that results from the cyclic regrowth of the aerial portion (Courtin, *unpublished data*). Although no records exist other than photographs (see, for example, Wallace and Thomson 1993), I estimate that the forest in its present form probably arose during the time of open-bed roasting that occurred during the first quarter of the century (see Chapter 2).

Birch transition sites typically have very acid soils (pH < 4.2) that result in high ex-

TABLE 18.1. Importance Indices of the Major Species of Trees, Shrubs, Herbs, and Bryoids Found in the Birch Transition Forest[a]

	Mean importance index	Standard error
Trees		
Acer rubrum	1.5	(0.6)
Betula papyrifera	27.7	(1.9)
Populus tremuloides	4.4	(1.4)
Shrubs		
Acer rubrum	5.2	(1.5)
Betula papyrifera	21.8	(3.4)
Populus tremuloides	4.2	(1.6)
Salix bebbiana	6.2	(1.4)
Salix discolor	0.9	(0.6)
Herbs		
Agrostis scabra	13.3	(2.4)
Betula papyrifera	9.8	(3.1)
Calamagrostis canadensis	5.9	(1.7)
Comptonia peregrina	6.7	(2.6)
Cornus canadensis	3.7	(1.6)
Deschampsia flexuosa	12.5	(3.8)
Diervilla lonicera	4.3	(2.2)
Vaccinium angustifolium	23.9	(4.9)
Bryophytes and lichens		
Pohlia nutans	37.8	(4.5)
Polytrichum commune	5.9	(1.6)
Stereocaulon spp.	5.0[b]	
Other		
Bare rock	29.0	(4.6)
Bare soil	28.6	(4.3)
Dead material	67.5	(3.0)

[a]From Amiro 1979. The importance index is based on the density, frequency, and cover of a given species relative to all other species in a given layer.
[b]Estimated.

tractable levels of copper, nickel, aluminum, and manganese (James 1982). Sudbury's transition and barren zone soils are prone to heave, owing to segregated ice and needle ice formation. The reason is that the so-called heavable fraction, the silt and very fine sand textural classes, compose approximately 50% of the mineral material in the upper 50 cm of the soil column. The remainder is made up of coarse and fine sand (approximately 35%) and the balance is clay (Sahi 1983). Schramm (1958) stated that soils are susceptible to heaving when they contain substantial amounts (about 30%) of particles smaller than, and including, the very fine sands (i.e., < 70 µm in diameter).

Effects of Microclimate

The microclimate within the birch transition forest is governed almost entirely by the structure of the stand (Fig. 18.2). In summer, the canopy openings are areas where solar irradiance penetrates to the soil surface during the day, which results in high soil surface and tree canopy temperatures (Table 18.2), whereas the micro-oases are cooler by virtue of surface shading, evaporative cooling from the soil, and transpiration. The soil surface with its temperatures as high as 70°C provide an environment that causes most of the birch seedlings recruited the same spring to die by desiccation. In autumn, it is the nighttime reradiation regime that strongly influences the micro-deserts. On clear calm nights, any solar radiation stored as heat during the day is rapidly lost to the night sky, with the result that freezing of the surface soil layers gives rise to the formation of either segregated ice lenses or needle ice (Fig. 18.3). This vertical displacement of the soil by ice uproots the majority of seedlings that either managed to survive the summer drought and heat or were recruited to the site in early autumn, a period characterized by frequent rains. It is also the presence of wet or even saturated soils that provides the moisture source for the frost effects observed in autumn. The processes just described are summarized schematically in Figure 18.4.

Just as there is a mosaic pattern to the entire woodland, so too is there a mosaic pattern within the microdeserts. The ice formation and heaving of the soil occurs in very localized areas. At the edge of these areas of extreme surface instability (Fig. 18.5A), the occasional white birch seedling manages to survive past the first, critical growing season but grows extremely slowly for several years because of soil drought in summer and severe pruning of the roots through frost action in the autumn. These trees can be referred to as "bonsai" birch

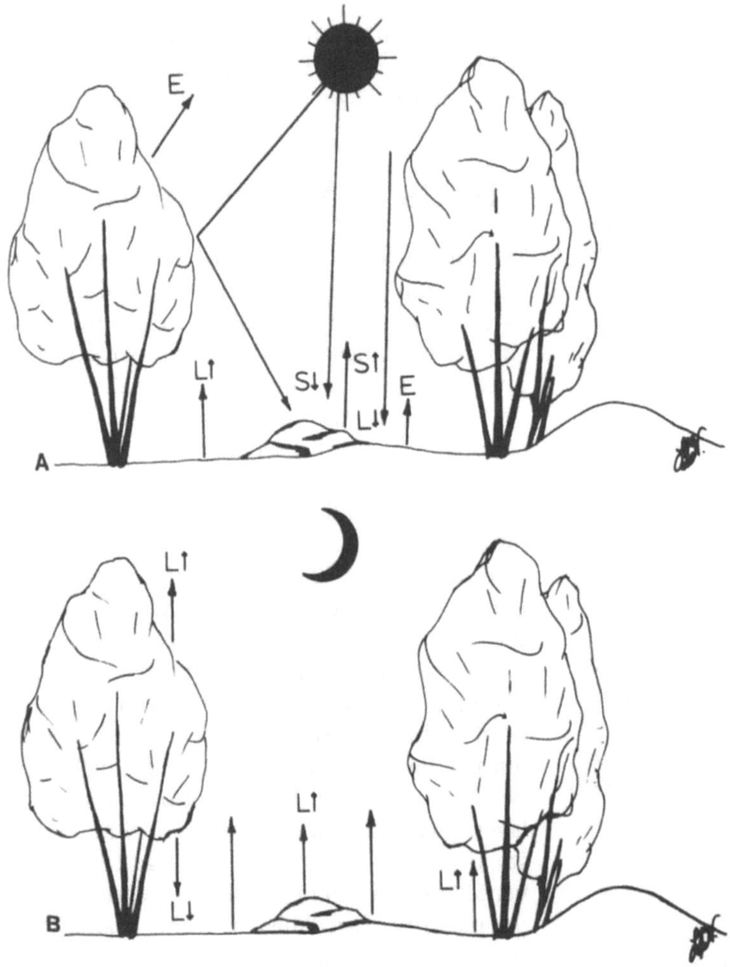

Figure 18.2. Schematic representation of the major shortwave (*S*), longwave (*L*), and evapotranspiration (*E*) fluxes operating in the birch coppice woodland during the day (**A**) and during the night (**B**).

Table 18.2. Typical Values of White Birch Physiological Response and Environmental Conditions on Days with Fine Weather between June 11 and July 31, 1981*

Parameter	Range of values
Temperature (°C)	
Air	21–26
Leaf	27–28
Soil surface	40–70
Stomatal conductance (s/m)	12–125
Shoot water potential (MPa)	−1.25 – −1.3
Gravimetric water content (g/gDW)	
0–5 cm	0.12–0.40
10–15 cm	1.12–0.40

*From James 1982; Courtin, *unpublished data.*

(Fig. 18.5B). Their stunted growth mimics, in part, the Japanese horticultural practice for producing dwarf trees.

Effects of Insects

The dominance of white birch across much of the Sudbury landscape has led to an insect problem analogous to that found in agriculture. A monoculture provides an abundance of the same food source, and insect outbreaks rapidly reach epidemic proportions. The chronic stress due to contaminated soils that plants are subjected to in industrial areas may also increase the trees' susceptibility to insect attack (Louda 1988; Riemer and Whittaker 1989). Also, the ability of pathogens or other natural

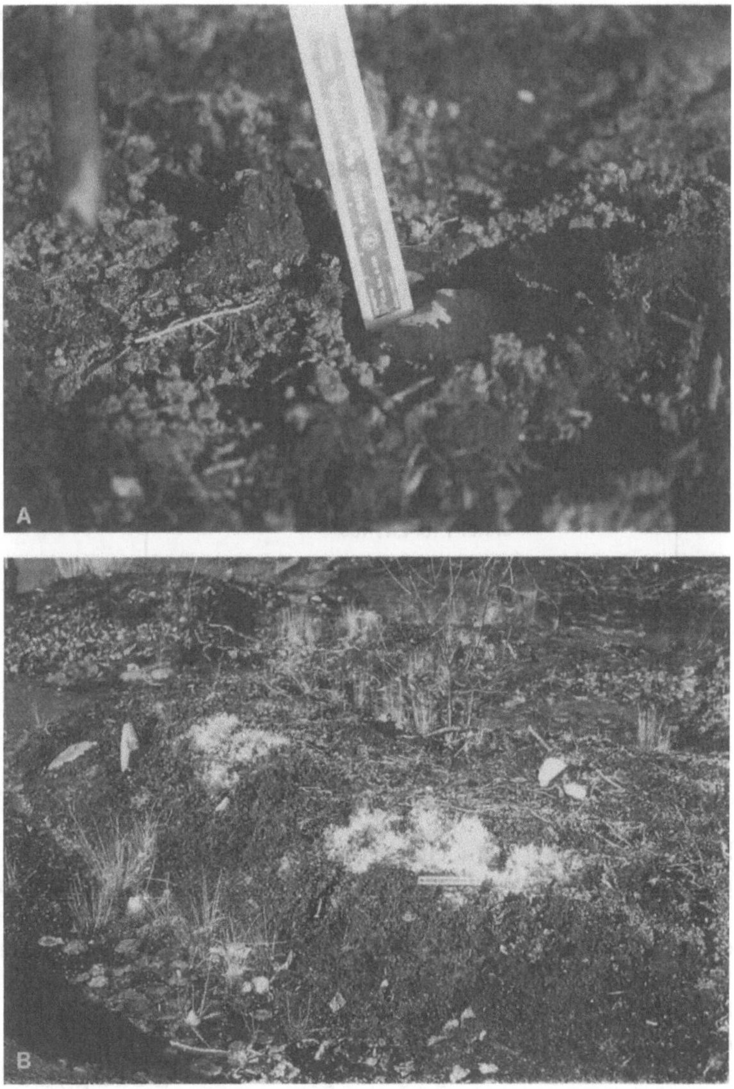

FIGURE 18.3. Examples of frost disturbance within the birch coppice woodland owing to heaving of the entire surface soil through the formation of segregated ice lenses (**A**) and efflorescence of needle ice covering an area of about 2 m^2 (**B**). Ice lenses typically displace the soil vertically up to 3 cm, whereas needle ice often exceeds 10 cm in height.

enemies to control herbivorous insects may be reduced at industrially affected sites (Haukioja 1992).

Several insects representing different orders and each having a cycle of abundance that is different from the others have subjected birch in the Sudbury area to an almost unremitting onslaught of either damage or defoliation (Fig. 18.6). The leaf miners (*Messa nana, Fenusa pusilla*) were numerous in the late 1970s through 1975, but their impact on the Sudbury birch is unknown. In 1986, a single severe infestation of birch sawfly (*Arge pectoralis*) completely defoliated many trees. At least some of these trees have died subsequently, indicating that the loss of leaves (even though late in the season after the majority of photosynthetic storage for the following spring was completed) caused a shock which they were unable to overcome. At least three major infestations of forest tent caterpillar (*Malacasoma distria*) have occurred over the past two decades. The primary target has been trembling aspen (*Populus tremuloides*), but the insect tended to move to birch once the aspen food source was exhausted. Each infestation has

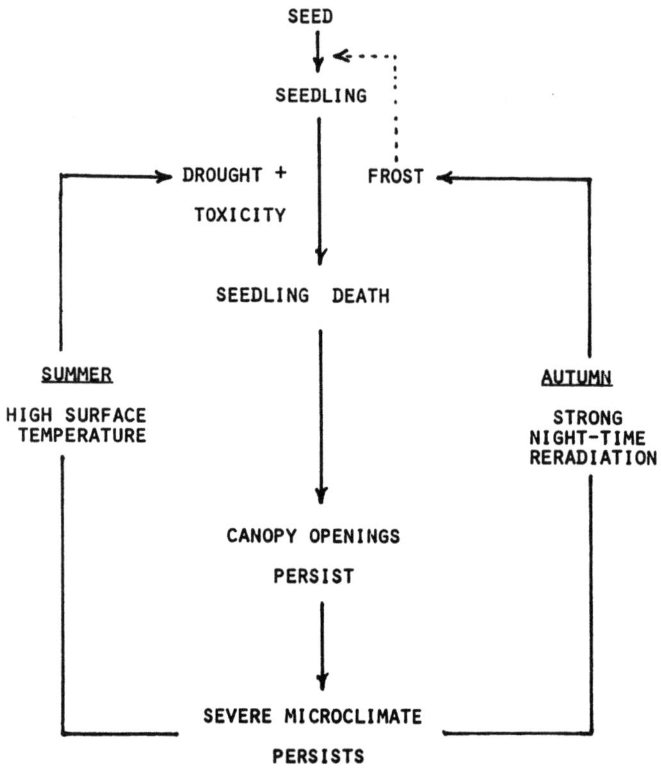

Figure 18.4. Model of the barriers to seedling establishment in the birch coppice woodland.

lasted 2–3 years, until checked by natural predators.

The most serious problems have been caused by bronze birch borer (*Agrilus anxius*). Its method of attacking the cambium layer beneath the bark causes girdling and subsequent death of the stem. In the spring after the first attack, the emerging canopy is comprised of very small leaves, and there is an abundance of female catkins. This proliferation of reproductive tissue in birch after insect attack is perhaps analogous to the production of "agony cones" in stressed conifer species.

Three years ago, the gypsy moth (*Lymantria dispar*) reached Sudbury from eastern Ontario, and although the heaviest damage has been on red oak (*Quercus borealis*), the larvae also vigorously attack trembling aspen and white birch. A generalization that can be made with respect to all the above insects is that attack is always more severe at the edges of stands than within stands. The reason is presumed to be the ease of accessibility on the one hand and a warmer microclimate on the other. In the coppice woodland, with its large canopy openings, most trees are therefore vulnerable to insect attack.

Comparison with Other Simplified Ecosystems: High Arctic Tundra

One result of the combination of anthropogenic stressors that exist in the Sudbury area is the simplification of the entire food web. The dominant trophic level is the primary producer but with low species diversity. The spatial distribution of the vegetation leads to a very harsh microclimate that has a tremendous impact on the remainder of the food web. The dominant herbivores are the insects and the dominant carnivores are the insects that feed on the herbivorous insects. The cop-

FIGURE 18.5. Examples of effects of frost disturbance in the birch coppice woodland. (**A**) Frost hummock showing patches of bare soil, patches of fragmented moss, and tree roots that have been heaved to the soil surface. (**B**) A *Bonsai* birch "seedling" that is 7 years old but only 15 cm high. Note the distorted trunk (just above the thumb) owing to repeated heaving, the very shallow root system with few feeder roots, and the branching just above the root collar that is an indication that coppicing has already begun.

pice woodland with its large canopy openings offers little shelter for vertebrates so that the only representatives are a few insect-eating birds in summer and a few seed-eating birds in winter. No sign has been found of either voles or mice, and the occasional larger mammals such as red fox (*Vulpes vulpes*) and red squirrel (*Tamiasciurus hudsonicus*) uses such sites simply as a travel route. Again, because of microclimate extremes, the litter accumulates only in sheltered depressions, and decomposition is very slow because of drought and heat in summer and surface reradiative cooling in spring and autumn. In essence, the ecosystem simplicity that one observes near Sudbury not only rivals that of the High Arctic tundra but possibly exceeds it (Table 18.3).

Although the driving forces are different in most cases, the results are very similar. Chamaephytes dominate in the arctic and phanaerophytes dominate in Sudbury, but both are dwarfed. Reproductive success through seeds

Figure 18.6. Insects that attack white birch in Sudbury area. (**A**) leaf miner (*Fenusa pusilla*) larvae within birch leaf; (**B**) leaf miner (*Messa nana*) larvae within birch leaf; (**C**) birch sawfly (*Arge pectoralis*) larvae; (**D**) forest tent caterpillar (*Malacasoma distria*) larvae; (**E**) gypsy moth (*Lymantria dispar*) larvae; (**F**) bronze birch borer (*Agrilus anxius*) adult. (Photos by J.D. Shorthouse.)

in both systems is low. Decomposition is slow, in the arctic because of cold and in Sudbury because of drought, metal toxicity, and low pH. Even the dominant characteristic of patterned ground through frost action in the arctic (Washburn 1956) is found in a modified form in Sudbury.

Discussion

Long-term changes in the birch transition forest toward greater environmental stability and species diversity is predicated on two possible mechanisms operating either separately or in concert. Amelioration of the severe surface

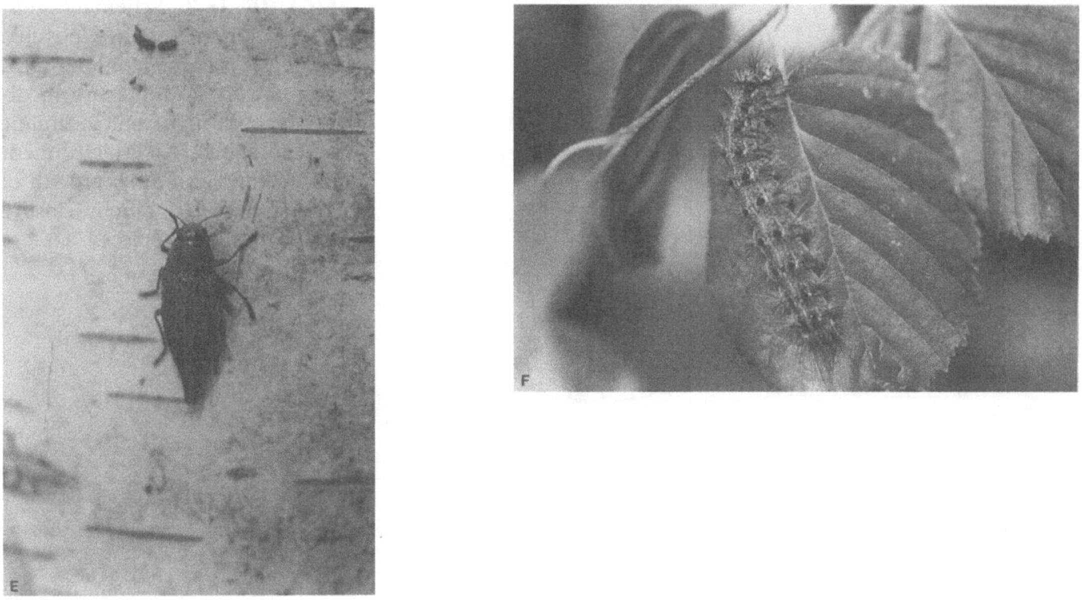

Figure 18.6. (*Continued*).

microclimatic conditions could take place if the ground vegetation of herbs, shrubs, tree seedlings, and mosses develops sufficiently to provide (1) a closed canopy to reduce intense surface radiation and reradiation, (2) a litter layer that does not become redistributed by wind, and (3) a network of roots that bind the soil. Alternatively, the same ends would be achieved either by complete closure of the tree canopy or a combination of both increased tree cover and ground vegetation cover (Fig. 18.7). With the progressive curbing of air pollution in recent years, the way seemed to be paved for major improvements in health, vigor, and diversity of Sudbury's vegetation. In fact, in the early 1970s, birch did appear to be growing better than in the past. At that time, however, there were no quantitative data on

Table 18.3. Characteristics of Industrially Disturbed Sudbury and High Arctic Tundra Ecosystems

Characteristic	*Industrially disturbed*	*Tundra*[a]
Temperature extremes	Microclimatic owing to site characteristics	Macroclimatic owing to latitude
Low species diversity	Historical	Climatic
Low productivity	Acid, metal-toxic soils	Climatic
Slow decomposition	High surface temperatures and drought	Low summer temperatures; short snow-free season
Low fertility	Low cation exchange capacity owing to low pH	Slow decomposition owing to cold and anaerobic conditions
Drought	Physically induced owing to topography	Physiologically induced owing to frost action and anaerobic soils
Frost disturbance	Diel freeze-thaw owing to clear sky reradiation; formation of needle ice and ice lenses	Seasonal freeze-thaw owing to macroclimate; formation of patterned ground

[a]Studies from Devon Island (Bliss 1977; Courtin and Labine 1977) are used to provide the overviews for the tundra.

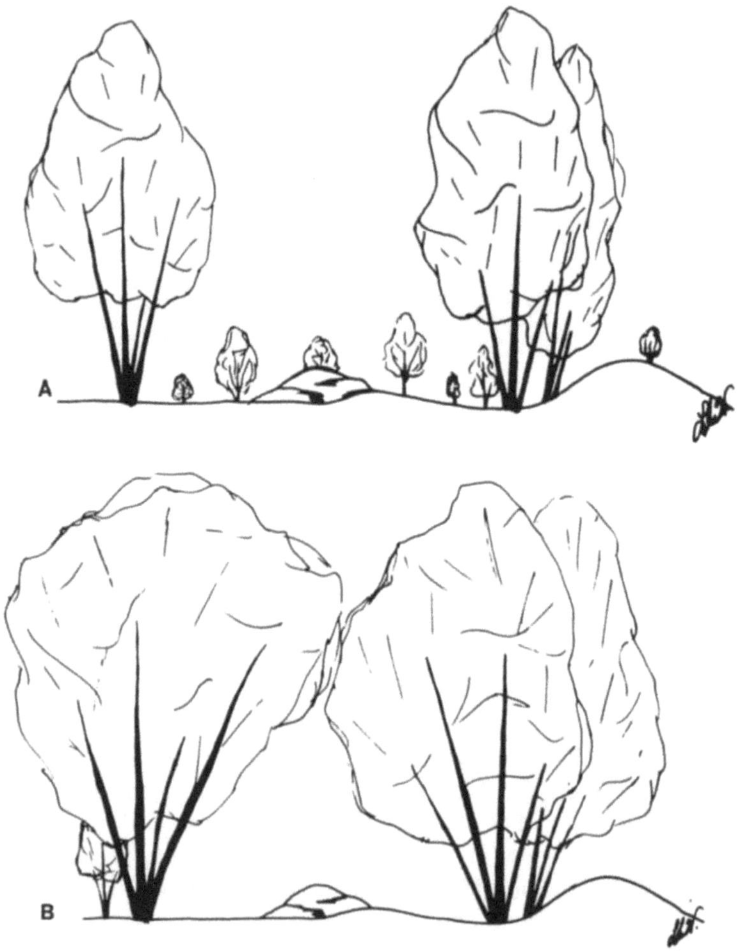

FIGURE 18.7. Schematic representation of the two possible strategies proposed for long-term site stability and increase in diversity through microclimatic amelioration. (**A**) Establishment of seedlings and other growth of low stature; (**B**) canopy closure of the existing coppices.

productivity, and it appears that if there was an improvement, it was short-lived. My observations of the birch coppice woodland over a period of 25 years and research into its dynamics since 1977 indicate that little if any unidirectional change toward a stable and more diverse community has taken place (Fig. 18.8).

The most plausible explanation for the long-term lack of change may be found in the nature of the coppice birch woodland itself, first because of its structure and second because it is a monoculture. Drought and frost, combined with acid, metal-toxic soils, operate to minimize the recruitment of birches from seed and totally eliminate all seedlings of red maple, the other dominant tree species. Hence the first mechanism proposed for microclimatic amelioration (i.e., the revegetation of canopy openings) is unlikely to occur rapidly. However, there has been an increase in recent years in the cover of the acid-loving mosses *Pohlia nutans* and *P. cruda,* and Sahi (1983) observed that the most frost-stable areas were those that were moss-covered. Cracks develop in the moss through desiccation, and birch seeds lodge in these cracks and germinate. The mulching effect of the moss seems to provide a moist and stable seed bed that, in time, may increase the numbers of birch that establish from seed.

The second mechanism, that of canopy closure of the existing coppices, is unlikely to occur, because in such a monoculture, the forest appears particularly prone to the cyclic and often devastating effect of defoliating insects. Entire coppices have been observed to die

FIGURE 18.8. Birch transition coppice woodland. (**A**) Photograph taken in 1980; (**B**) photograph taken in 1993. Few changes have taken place in the intervening years.

after complete defoliation by birch sawfly in late summer, and bronze birch borer kills any stem that it attacks because its larvae totally girdle the tree in 2 years. In recent years, bronze birch borer has increased in numbers to the point where as many as 90% of stems in some stands have been killed. However, attack by insects stimulates growth of basal sprouts to re-initiate the coppice cycle in the same way that cutting does in conventional coppice woodlands.

Without the intervention of soil amelioration (liming and fertilizing) and the re-introduction of native conifers (white and red pine, white spruce) it is unlikely that any rapid unidirectional change will take place in the upland birch woodlands. The planting of the adjacent barrenlands to conifers (see Chapter 8) will provide a future seed source, but seedlings will probably suffer the same fate as deciduous seedlings that attempt to become established. With soil amelioration, however, changes could be much more rapid. Winterhalder (1983) reported that planting trials on the barrens that received lime demonstrated better growth than unlimed trial plots. He suggested that survival in summer was drought-dependent, whereas winter survival was a function of the degree of frost-heaving. Liming alone, therefore, might well reduce the degree of frost damage sufficiently and increase the rate of seedling growth to the point at which revegetation of canopy opening is great enough to reduce significantly the extremes of surface microclimate. There certainly is evidence that liming improves growing conditions, but it is unlikely that improved growth will necessarily mean greater plant diversity unless other tree species, especially conifers, are planted, as has been the case for the reclaimed barrens (see Chapter 8). According to Winterhalder (*personal communication*), the liming of organic soils stimulated the growth of 10 species that had been lying dormant as part of the seed bank. Winterhalder (1983) found that the birch transition seed bank only contained white birch and tickle grass (*Agrostis scabra*). The paucity of a natural seed source is a further indication that recovery of the birch ecosystem, without human intervention, will probably be extremely slow.

The observed effects of smelter emissions on vegetation are not unique to Sudbury. Other climatically similar areas such as northern Russia, where birch is also a dominant species, have suffered serious damage from nickel and copper smelting (Kryuchkov 1993). It is hoped that the work on barriers to vegetation recovery performed in Sudbury may aid other workers attempting restoration of mine-damaged lands elsewhere.

Acknowledgments. The research described, which has spanned a period of 16 years, could not have been done without the help of many generations of students. The following undergraduates were invaluable either as summer assistants to me or to my graduate students or through their undergraduate theses, or both: P.D. Thibodeau, M. Kalliomaki, L.I. Wallenius, M. Trépanier, and J. Fyfe. Their work, their insight, and especially their friendship and dedication are gratefully acknowledged. The winter fieldwork and computer analysis performed by secondary school co-op students E. Morris and D. Pauzé were most appreciated. Photographer M. Roche and artist L.I. Wallenius assisted in the preparation of the figures. Review comments were provided by John Cairns, Jr., John Gunn, and Mikhail Kozlov.

References

Amiro, B.D. 1979. Plant community patterns in an industrially disturbed ecosystem. M.Sc. thesis, Laurentian University, Sudbury, Ontario.

Amiro, B.D., and G.M. Courtin. 1981. Patterns of vegetation in the vicinity of an industrially disturbed ecosystem, Sudbury, Ontario. Can. J. Bot. 59:1623–1639.

Bliss, L.C. (ed.). 1977. Truelove Lowland, Devon Island, Canada: A High Arctic Ecosystem. The University of Alberta Press, Edmonton, Alberta.

Braun, E.L. 1950. The Eastern Deciduous Forest. Hafner Publishing Co., New York.

Buckley, G.P. 1992. Ecology and Management of Coppice Woodlands. Chapman and Hall, London.

Courtin, G.M., and C.L. Labine. 1977. Microclimatological studies on Truelove Lowland, pp. 73–106. *In* L.C. Bliss (ed.). Truelove Lowland, Devon Island, Canada: A High Arctic Ecosystem. The University of Alberta Press, Edmonton, Alberta.

Haukioja, E. 1992. Research on ecological effects of aerial pollution: a Finnish perspective, pp. 67–69. *In* M.V. Kozlov, E. Haukioja, and V.T. Yarmishko (eds.). Aerial Pollution in Kola Peninsula. Proceedings of the International Workshop; April 14–16, 1992, St. Petersburg, Russia. Kola Scientific Center, Apatity, Russia.

James, G.I. 1982. Factors influencing the birch transition community of the industrial barrens, Sudbury, Ontario. M.Sc. thesis, Laurentian University, Sudbury, Ontario.

James, G.I., and G.M. Courtin. 1985. Stand structure and growth form of the birch transition community in an industrially damaged ecosystem, Sudbury, Ontario. Can. J. For. Res. 15:809–817.

Kryuchkov, V.V. 1993. Degradation of ecosystems around the "Severonikel" smelter complex, pp. 35–46. *In* M.V. Kozlov, E. Haukioja, and V.T. Yarmishko (eds.). Aerial Pollution in Kola Peninsula. Proceedings of the International Workshop, April 14–16, 1992, St. Petersburg, Russia. Kola Scientific Center, Apatity, Russia.

Louda, S.M. 1988. Insect pests and plant stresses as considerations in revegetation of disturbed ecosystems, pp. 51–67. *In* J. Cairns (ed.). Rehabilitation of Damaged Ecosystems. CRC, Boca Raton, FL.

Oke, T.R. 1987. Boundary Layer Climates. 2nd Ed. Methuen, London.

Riemer, J., and J.B. Whittaker. 1989. Air pollution and insect herbivores: observed interactions and possible mechanisms, pp. 73–105. *In* E.A. Bernays (ed.). Insect Plant Interactions. CRC Press, Boca Raton, FL.

Sahi, S.V. 1983. Frost heaving and needle ice formation and their effect upon seedling survival at selected sites in Sudbury, Ontario. M.Sc. thesis, Laurentian University, Sudbury, Ontario.

Schramm, J.R. 1958. The mechanism of frost heaving of tree seedlings. Proc. Am. Philos. Soc. 102:333–350.

Trépanier, M. 1985. Stem analysis of white birch in the birch transition community, Sudbury, Ontario. B.Sc. (Hon) thesis, Laurentian University, Sudbury, Ontario.

Wallace, C. M., and A. Thomson (eds.). 1993. Sudbury: From Rail Town to Regional Capital. Dundurn Press, Toronto, Ontario.

Washburn, A.L. 1956. Classification of patterned ground and review of suggested origins. Bull. Geol. Soc. Am. 67:823–866.

Winterhalder, K. 1983. Limestone application as a trigger factor in the revegetation of acid, metal-contaminated soils of the Sudbury area, pp. 201–212. *In* Proceedings of the 8th Annual Meeting, Canadian Land Reclamation Association, August 1983. University of Waterloo, Waterloo, Ontario.

19

Potential Role of Lowbush Blueberry (*Vaccinium angustifolium*) in Colonizing Metal-Contaminated Ecosystems

Joseph D. Shorthouse and Giuseppe Bagatto

The industrially damaged lands around the smelters of Sudbury are inhabited by several species of plants that can tolerate high levels of toxic metals (Hogan and Rauser 1978; Cox and Hutchinson 1980). One of these plants, the sweet lowbush blueberry (*Vaccinium angustifolium*), has colonized large areas of the smelter-affected area. It is particularly abundant within the birch transition forest described in the previous chapter (Chapter 18). This chapter reviews the attributes of *V. angustifolium* that made it successful in the Sudbury industrially damaged lands. There are actually two species of blueberry near Sudbury, *V. angustifolium* and *V. myrtilloides*, of which *V. angustifolium* is the more common of the two. Although they are similar (see descriptions of each in Vander Kloet [1988]) and grow in the same habitats, care was taken to restrict our studies to *V. angustifolium*.

Accumulation of Metals and Mechanisms of Tolerance

Extremely high levels of metals occur naturally in some parts of the world, and a surprising array of plants are able to thrive at these sites. For example, about 220 taxa are found on copper-rich soils in Upper Shaba in Zaïre, and one tree in nickel-rich sites in New Caledonia thrives while accumulating up to 25% nickel in its sap on a dry weight basis (Baker 1987). However, many studies of metal accumulation and adaptation have been conducted, not at naturally contaminated sites but in areas damaged by industrial activities, particularly areas in which smelting and refining of metals occur. From these studies, it has been determined that tolerance to metals can be achieved by mechanisms that allow plants to accumulate metals in their tissues without causing damage or, alternatively, by mechanisms that confer tolerance by excluding metal uptake. Tissue accumulation is often mediated by cellular processes that sequester metals and render them biologically inactive. Metal-sequestering proteins and molecules such as metallothioneins and phytochelatins have been implicated in this process (Rauser and Winterhalder 1984; Tomsett and Thurman 1988). Alternatively, metal exclusion by plants is often mediated by specific energy-dependent transport processes operating at the root–soil interface or in association with mycorrhizal symbionts, which aid in preventing movement of metals into the root (Jones and Hutchinson 1986; Baker 1987).

Ecosystems influenced by mining and smelting are useful for studying metal tolerance because they provide a natural laboratory where selection for metal tolerance and accumulation in plants can be examined in detail. Elevated levels of metals in soils at such sites

Figure 19.1. Lowbush blueberry growing on a barren site in Sudbury's industrially damaged ecosystem. (Photo by J.D. Shorthouse.)

may act as powerful agents for rapidly selecting genetically adapted, metal-tolerant ecotypes of various species of plants (Freedman and Hutchinson 1981; Baker 1987). Research in this area has consequently been driven by a need to understand the impact of metal pollution on plants and has a potential application in development of metal-tolerant ecotypes for habitat restoration.

Biology of Sweet Lowbush Blueberry

Vaccinium angustifolium is a broad-leaved, deciduous, low shrub endemic to North America, whose range in Canada extends from the east coast of Newfoundland to central Manitoba (Vander Kloet 1988). It tolerates a wide range of climatic conditions, is particularly adapted to a temperate climate, and grows best in acidic soils with pH 4.0–5.5 (Hall et al. 1964) and moderate-to-low fertility. It has woody stems and averages 20 cm in height. It is an effective colonizer of disturbed sites, including those modified by fire, smelters, and mine tailings (Sheppard 1991). Because of the concern about metal contamination in the berries, there have been several studies dealing with contaminant accumulation in this plant (see Sheppard [1991] for references).

The ability of lowbush blueberry to both colonize and survive (i.e., relict species) smelter-damaged lands near Sudbury has proved beneficial to both revegetation programs within the region and to the residents of the area who enjoy picking and consuming the berries. It is commonly one of the first plants to become established on sites devoid of almost all other plant species (Fig. 19.1).

Lowbush blueberry is one of many plants involved in natural succession from cleared land to forest. Most of the new shoots of mature plants develop from dormant buds on underground rhizomes. The tips of growing shoots die in the early part or middle of the summer, and the buds develop into either vegetative or flowering types. The extensive rhizome system of mature blueberry can play an important role in preventing slope erosion.

In Sudbury, lowbush blueberries have colonized part of the smelter damaged area and are commonly associated with white birch (*Betula papyrifera*) (Fig. 19.2). This association may benefit both species. Leaves of young birch growing in blueberry patches become entangled in the stems of blueberries, which helps retain moisture the following season. Thick mats of blueberry also accumulate snow, which similarly benefits both species. However, as birch increase in size, their shade becomes harmful to blueberry shrubs, and they disappear (Hall et al. 1979); but as described in

FIGURE 19.2. Mat of lowbush blueberry growing in association with white birch in Sudbury's birch transition zone. (Photo by J.D. Shorthouse.)

the previous chapter (Chapter 18), much of the coppiced birch forest in Sudbury is maintained as an open canopy, and as a result, shading by birch here is less of a problem for blueberries.

Lowbush blueberry reproduces by both seeds and rhizomes. Each berry contains an average of 13 seeds (Hall et al. 1979), and the seeds are spread in the droppings of birds such as the American robin (*Turdus migratorius*) and mammals such as black bears (*Ursus americanus*) and red fox (*Vulpes vulpes*). Plants usually flower and produce rhizomes 4 years after germination. The berries ripen in early July in Sudbury. Blueberry plants proliferate most rapidly once an organic layer is formed (Trevett 1956), and this presumably explains why mats near Sudbury are thickest near birch trees, where leaves and other organic debris collect. Asexual reproduction occurs when the rhizomes are cut or killed by fire, shading, burrowing, or frost action. The high incidence of frost-heaving within Sudbury's denuded areas (see Chapter 18) therefore likely contributes to asexual reproduction in this plant (Fig. 19.3). Rhizomes are the key to a continuing high level of productivity because it is primarily from rhizomes that new fruiting wood is initiated. This extensive rhizome sys-

FIGURE 19.3. Blueberry plantlets separated by frost-heaving. (Photo by J.D. Shorthouse.)

Figure 19.4. Larva of gypsy moth defoliating lowbush blueberry on one of Sudbury's barren sites. (Photo by J.D. Shorthouse.)

tem can also play an important role in preventing soil erosion.

Few insect herbivores attack lowbush blueberry in the Sudbury region; however, forest tent caterpillars (*Malacosoma disstria*) and caterpillars of gypsy moths (*Lymantria dispar*) defoliate plants during their peak years (Fig. 19.4). Important defoliators in the maritimes are the black army cutworm (*Actebia fennica*), the chain-spotted geometer (*Cingilia catenaria*), and a pest of the fruit, the blueberry maggot (*Rhagoletis mandax*). A stem gall found on vegetative shoots and induced by the chalcid wasp *Hemadas nubilipennis* is especially common on Sudbury plants (West and Shorthouse 1989). Shorthouse et al. (1986) also suggested that this insect may benefit the blueberry plant because the pruning effect produces a bushier plant.

Accumulation of Metals within Lowbush Blueberry near Sudbury, Ontario

Lowbush blueberry appears tolerant of elevated concentrations of certain metals. This species is therefore useful for studying the effects of metal pollution (Sheppard 1991). We examined the pattern of copper and nickel accumulation in tissues of lowbush blueberry collected in mid-July, when the plants were fruiting, at six sampling sites ranging from 7 to 74 km northwest from the Copper Cliff smelter (Bagatto and Shorthouse 1991). Data from this study and comparable information on a variety of other plants in the area are provided in Table 19.1.

The patterns of tissue accumulation of the metals (Fig. 19.5) is likely related to the physiological fates of the metals. High levels of copper (69 μg/g) and nickel (40 μg/g) in the roots confirm the observation that roots are sites of preferential copper and nickel accumulation when external supplies of these metals are excessive (Marschner 1986). These levels of copper and nickel are considerably elevated compared with root concentrations at 74 km (6 μg/g dry wt. [Cu] and 6 μg/g dry wt. [Ni]) and copper concentrations of 2–11 μg/g dry wt. reported in the literature (Ingestad 1973; Peterson et al. 1988). The considerably lower concentration of copper in leaves from the site closest to the smelter (22 μg/g) is below the 30 μg/g concentration considered toxic for most crop species (Robson and Reuter 1981).

It is interesting that the concentration of copper did not differ significantly between tissue types 23 km from the smelter and beyond. This is the expected pattern in habitats with typical background levels of copper where it functions as a plant nutrient participating in enzymatically bound copper redox reactions (Marschner 1986). Furthermore, organic mat-

TABLE 19.1. Copper and Nickel Concentrations (ppm-dry wt. or μg/g dry wt.) in Plants Collected within 15 km of the Sudbury Smelters

Species	Tissue	Nickel	Copper	Reference
Vaccinium angustifolium	Leaf	92	75	Hutchinson and Whitby 1974
Vaccinium angustifolium	Leaf	70	50	Freedman and Hutchinson 1980a
Vaccinium angustifolium	Leaf	33	21	Bagatto and Shorthouse 1991
Vaccinium angustifolium	Stem	49	54	Bagatto and Shorthouse 1991
Vaccinium angustifolium	Root	40	69	Bagatto and Shorthouse 1991
Vaccinium angustifolium	Berries	7	4	Bagatto and Shorthouse 1991
Acer rubrum	Leaf	98	37	Hutchinson and Whitby 1974
Acer rubrum	Leaf	60	15	Lozano and Morrison 1981
Acer rubrum	Leaf	100	140	Freedman and Hutchinson 1980a
Deschampsia flexuosa	Leaf	902	726	Hutchinson and Whitby 1974
Comptonia peregrina	Leaf	113	57	Hutchinson and Whitby 1977
Betula papyrifera	Leaf	95	148	Hutchinson and Whitby 1977
Betula papyrifera	Leaf	100	25	Lozano and Morrison 1981
Betula papyrifera	Leaf	170	100	Freedman and Hutchinson 1980a
Quercus rubra	Leaf	100	16	Lozano and Morrison 1981
Populus tremuloides	Leaf	150	15	Lozano and Morrison 1981
Populus tremuloides	Leaf	370	90	Freedman and Hutchinson 1980a
Osmunda claytoniana	Fronds	11	61	Burns and Parker 1988
Matteuccia struthiopteris	Fronds	111	47	Burns and Parker 1988
Typha latifolia	Shoot	60	13	Taylor and Crowder 1983
Typha latifolia	Fruit	19	17	Taylor and Crowder 1983
Typha latifolia	Rhizome	40	30	Taylor and Crowder 1983
Typha latifolia	Root	52	38	Taylor and Crowder 1983
Quercus borealis	Leaf	70	20	Freedman and Hutchinson 1980a
Pinus resinosa	Leaf	40	40	Freedman and Hutchinson 1980a
Salix humilis	Leaf	220	260	Freedman and Hutchinson 1980a
Myrica asplenifolia	Leaf	110	90	Freedman and Hutchinson 1980a
Diervilla lonicera	Leaf	60	90	Freedman and Hutchinson 1980a
Polygonum cilinode	Leaf	90	110	Freedman and Hutchinson 1980a
Solidago canadensis	Leaf	200	180	Freedman and Hutchinson 1980a
Epilobium angustifolium	Leaf	110	160	Freedman and Hutchinson 1980a
Deschampsia caespitosa	Leaf	240	370	Freedman and Hutchinson 1980a
Agrostis stolonifera	Leaf	130	70	Freedman and Hutchinson 1980a
Equisetum sylvaticum	Leaf	450	250	Freedman and Hutchinson 1980a
Polytrichum commune	Leaf	620	910	Freedman and Hutchinson 1980a

ter and pH increase with distance from the smelter (Freedman and Hutchinson 1980a,b), thereby decreasing the availability of copper ions in the soil. Of interest, Sheppard (1991) found that copper was at significantly higher concentrations in berries than in leaves and stems at many of the sites he examined across central Canada; however, Sheppard did not examine the levels of copper and nickel in the fruits of lowbush blueberry from near the Sudbury smelters. Concentrations of copper and nickel in the berries of *V. angustifolium* far from the North American sources of pollution are similar to those of berries growing near the Sudbury smelters (Bagatto and Shorthouse 1991).

The high concentration of nickel in blueberry roots likely represents cell wall adhesion in this tissue. The behavior of nickel in plants may be the result of it being a nonfunctional analog of copper and zinc as shown in studies of root transport (Cataldo et al. 1978). Although root concentrations of nickel were lower than that of copper, leaves and stems from all sites had proportionally higher nickel concentrations than copper. Also, the pattern

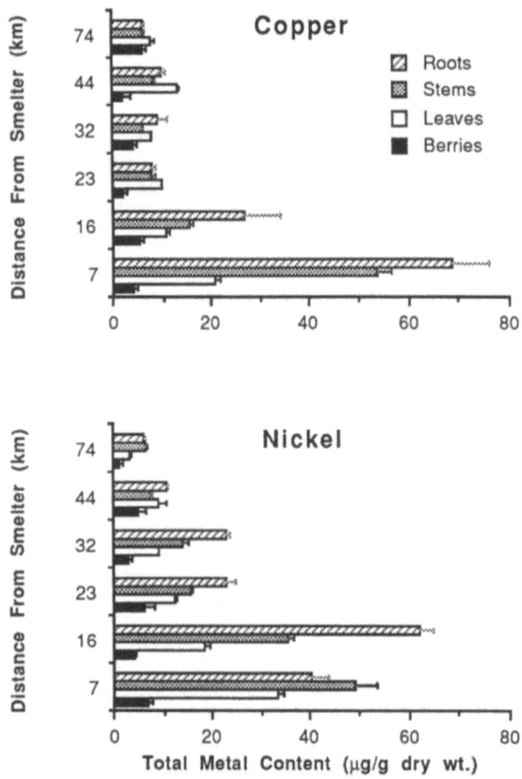

FIGURE 19.5. Concentrations of copper and nickel in tissues of lowbush blueberry at various distances from the Inco smelter.

of nickel accumulation at all sites was the same (roots>stems>leaves>berries), whereas with copper, this pattern changed at sites beyond 23 km from the smelter. The general nonessentiality of nickel as a plant nutrient and its greater availability within acidic soils of the Sudbury area would account for this finding.

Nickel is presumably taken up as a nonfunctioning analog of some other plant nutrient (copper or zinc), and its concentration in the plant would depend on the concentration of nickel in the environment. The levels of nickel in lowbush blueberry closest to the smelter far exceed requirements for nickel; however, they fall within the range of 3–300 μg/g considered non-toxic, depending on the species of plant, organ, developmental stage, and nutritional supply (Hutchinson and Whitby 1974).

We conclude that lowbush blueberry accumulate substantial amounts of copper and nickel in structural and vegetative tissues when growing in contaminated soils. However, the high levels of these metals, particularly in the roots and stems, appear non-toxic to the plants. It was particularly reassuring that the levels of copper and nickel were low in berries collected near the smelters, compared with plants from other parts of Canada, as plants in this area produce heavy crops of berries that are popular with local residents.

The only other site known to us where copper and nickel in the tissues of berries growing near smelters have been examined is that of the "Severonikel" complex in the Kola Peninsula of Russia (Barkan et al. 1993). Here, extremely high levels of copper and nickel were found in the berries of *Vaccinium vitis-idaea*, *V. myrtillus*, *Rubus chamaemorus*, and *Empetrum hermaphroditum*. Levels of copper and nickel were highest in the upper organic soils closest to the smelters, and the levels in berries were dependent on metal contents of the soil. Levels of copper as high as 53 μg/g dry wt. and nickel 96 μg/g dry wt. were found in berries of *V. vitis-idaea* within 6 km of the smelter, with copper at levels of 115 μg/g dry wt. and nickel at 83 μg/g dry wt. in the berries of *E. hermaphroditum* at the same site, all of which are about 10 times above the safe levels recommended by the Russian Health Protection Ministry and are unsuitable for food.

Summary

The historical atmospheric deposition of metal particulates from smelters has created a serious environmental stress on plant communities, and most species of plants in the Sudbury area appear to suffer accordingly. Unlike other forms of abiotic stress such as drought or anomalous weather, which are usually short-term, metal pollution represents a continuous source of stress for plants and animals. The lasting character of metal-induced stress implies that its effects on plants and animals may differ from those of other stresses and may result in certain adaptations by exposed organisms. Consequently, studies of the effects of metals on different components of ecosystems are not only of scientific interest but also are economically important and represent an important area of research. There is a shortage of knowledge about the ways in which plants such as lowbush blueberry cope biochemically, physiologically, ecologically, and genetically with elevated levels of pollutants.

Sulfur dioxide and heavy metals have potentially different effects on plants and insects. Moderate doses of sulfur dioxide can lead to enhanced performance of phytophagous insects feeding on plants fumigated by sulfur dioxide, whereas the effects of heavy metals on phytophagous insects are assumed to be mainly negative (Alstad et al. 1982). Little attention has been paid to possible interactions between sulfur dioxide and metals and the consequences for plants and insects (Riemer and Whittaker 1989). It has been suggested that sulfur dioxide emissions cause soil acidification, which, in turn, increases solubility of metals in the soil and enhances uptake by plants (Lobersli and Steinnes 1988). At certain levels, soil acidification can benefit some species such as blueberry, but the added metal load in stressed lands such as Sudbury can undoubtably create a complex set of effects, many of which may be detrimental to the plant.

Although there is little information about the adaptation to metal stress, it is known that populations of plants will evolve through natural selection in response to specific ecological conditions of its local environment and that adaptive genetic differentiation may occur (Bradshaw 1972). We suspect that populations of Sudbury blueberries have developed tolerance to copper and nickel and represent distinct metal-tolerant ecotypes, as has been shown for some grass species in the Sudbury area (Cox and Hutchinson 1980; Rauser and Winterhalder 1984; Archambault 1991). More important, the absence of elevated levels of metals in the berries of Sudbury plants may mean that the same will occur at other smelter sites. If so, the introduction of Sudbury lowbush blueberry to polluted sites in other parts of the world with similar climates should be considered. Not only would the Sudbury plants help colonize denuded sites, but local residents would be provided with edible berries as well.

In conclusion, we suggest that lowbush blueberry plays an important role in areas such as the industrially damaged lands near Sudbury because

- **they are adapted to high levels of copper and nickel and perhaps other metals**
- **they thrive on acidic soils**
- **high levels of metals accumulated within plant tissues are not passed to the fruit**
- **they have few herbivores**
- **once established, they produce a thick humus, which retains soil moisture and reduces soil erosion**
- **they can improve growing conditions for trees such as birch**
- **they exhibit good dispersal of seeds**
- **they can reproduce asexually as a result of heavy frost-heaving**

We also suggest that future research aimed at enhancing the colonizing ability of lowbush blueberry be undertaken in the following areas:

testing for metal- and acid-tolerant ecotypes
examining interactive relationships that affect the establishment and succession of other plants
identification of mechanisms for metal tolerance (e.g., role of mycorrhizae)

Acknowledgments. Research on lowbush blueberry in the Sudbury area was supported by grants from the Natural Sciences and Engineering Research Council of Canada and the Laurentian University Research Fund awarded to JDS. We thank E.K. Winterhalder and J.M. Gunn for suggestions for improving this chapter.

References

Alstad, D.N., G.F.J. Edmunds, and L.H. Weinstein. 1982. Effects of air pollutants on insect populations. Annu. Rev. Entomol. 27:369–384.

Archambault, D.J.-P. 1991. Metal tolerance studies on populations of *Agrostis scabra* Willd. (tickle grass) from the Sudbury area. M.Sc. thesis, Laurentian University, Sudbury, Ontario.

Bagatto, G., and J.D. Shorthouse. 1991. Accumulation of copper and nickel in plant tissues and an insect gall of lowbush blueberry, *Vaccinium angustifolium*, near an ore smelter at Sudbury, Ontario, Canada. Can. J. Bot. 69:1483–1490.

Baker, A.J.M. 1987. Metal tolerance. New Phytol. 106 (Suppl.):93–111.

Barkan, V.S., M.S. Smetannikova, R.P. Pankratova, and A.V. Silina. 1993. Nickel and copper accumulation by edible forest berries in surroundings of "Severonikel" smelter complex, pp. 189–196. *In* Aerial Pollution in Kola Peninsula. Proceedings of the International Workshop, April 14–16, 1992, St. Petersburg, Russia. Kola Scientific Center, Apatity, Russia.

Bradshaw, A.D. 1972. Some evolutionary consequences of being a plant. Evol. Biol. 5:25–47.

Burns, L.V., and G.H. Parker. 1988. Metal burdens in two species of fiddleheads growing near the ore smelters at Sudbury, Ontario, Canada. Bull. Environ. Contam. Toxicol. 40:717–723.

Cataldo, D.A., T.R. Garland, and R.E. Wildung. 1978. Nickel in plants. II. Distribution and chemical form in soybean plants. Plant Physiol. 62: 566–570.

Cox, R.M., and T.C. Hutchinson. 1980. Multiple metal tolerances in the grass *Deschampsia cespitosa* (L.) Beauv. from the Sudbury smelting area. New Phytol. 84:631–647.

Freedman, B., and T.C. Hutchinson. 1980a. Pollutant inputs from the atmosphere and accumulations in soils and vegetation near a nickel-copper smelter at Sudbury, Ontario, Canada. Can. J. Bot. 58:108–132.

Freedman, B., and T.C. Hutchinson. 1980b. Long-term effects of smelter pollution at Sudbury, Ontario, on forest community composition. Can. J. Bot. 58:2123–2140.

Freedman, B., and T.C. Hutchinson. 1981. Sources of metal and contamination of terrestrial environments, pp. 35–94. *In* N.W. Lepp (ed.). Effect of Heavy Metal Pollution on Plants. Applied Science Publishers, London.

Hall, I.V., L.E. Aalders, N.L. Nickerson, and S.P. Vander Kloet. 1979. The biological flora of Canada. I. *Vaccinium angustifolium* Ait., sweet lowbush blueberry. Can. Field. Naturalist 93:414–430.

Hall, I.V., L.E. Aalders, and L.R. Townsend. 1964. The effects of soil pH on the mineral composition and growth of the lowbush blueberry. Can. J. Plant Sci. 44:433–438.

Hogan, G.D., and W.E. Rauser. 1978. Tolerance and toxicity of cobalt, copper, nickel and in zinc clones of *Agrostis gigantea*. New Phytol. 83:665–670.

Hutchinson, T.C., and L.D. Whitby. 1974. Heavy metal pollution in the Sudbury mining and smelting region of Canada. 1. Soil and vegetation contamination by nickel, copper and other metals. Environ. Conserv. 1:123–132.

Hutchinson, T.C., and L.M. Whitby. 1977. The effects of rainfall and heavy metal particulates on a boreal forest ecosystem near the Sudbury smelting region of Canada. Water Air Soil Pollut. 7: 421–438.

Ingestad, T. 1973. Mineral nutrient requirements of *Vaccinium vitis idaea* and *V. myrtillus*. Physiol. Plant. 29:239–246.

Jones, M.D., and T.C. Hutchinson. 1986. The effect of mycorrhizal infection on the response of *Betula papyrifera* to nickel and copper. New Phytol. 102: 429–442.

Lobersli, E.M., and E. Steinnes. 1988. Metal uptake in plants from a birch forest area near a copper smelter in Norway. Water Air Soil Pollut. 37:25–39.

Lozano, F.C., and I.K. Morrison. 1981. Disruption of hardwood nutrition by sulfur dioxide, nickel, and copper air pollution near Sudbury, Canada. J. Environ. Qual. 10:198–204.

Marschner, H. 1986. Mineral Nutrition in Higher Plants. Academic Press, London.

Peterson, L.A., E.J. Stang, and M.N. Dana. 1988. Blueberry response to NH_4-N and NO_3-N. J. Am. Soc. Hort. Sci. 113:9–12.

Rauser, W.E., and E.K. Winterhalder. 1984. Evaluation of copper, nickel, and zinc tolerances in four grass species. Can. J. Bot. 63:58–63.

Riemer, J., and J.B. Whittaker. 1989. Air pollution and insect herbivores: observed interactions and possible mechanisms, pp. 73–105 *In* E.A Bernays (ed.). Insect Plant Interactions. CRC Press, Boca Raton, FL.

Robson, A.D., and D.J. Reuter. 1981. Diagnosis of copper deficiency and toxicity, pp. 287–312. *In* J.F. Loneragan, A.D. Robson, and R.D. Graham (eds.). Copper in Soils and Plants. Academic Press, London.

Sheppard, S.C. 1991. A field and literature survey, with interpretation, of elemental concentrations in blueberry (*Vaccinium angustifolium*). Can. J. Bot. 69: 63–77.

Shorthouse, J.D., A. West, R.W. Landry, and P.D. Thibodeau. 1986. Structural damage by female *Hemadas nubilipennis* (Hymenoptera: Pteromalidae) as a factor in gall induction on lowbush blueberry. Can. Entomol. 118:249–254.

Taylor, G.J., and A.A. Crowder. 1983. Uptake and accumulation of heavy metals by *Typha latifolia* in wetlands of the Sudbury, Ontario region. Can. J. Bot. 61:63–73.

Tomsett, A.B., and D.A. Thurman. 1988. Molecular biology of metal tolerance of plants. Plant Cell. Environ. 11:383–394.

Trevett, M.F. 1956. Observations on the Decline and Rehabilitation of Lowbush Blueberry Fields. Miscellaneous Publication 626. Maine Agricultural Experiment Station, Orono, ME.

Vander Kloet, S. P. 1988. The Genus *Vaccinium* in North America. Agriculture Canada Publication 1828. Ministry of Supply and Services, Ottawa.

West, A., and J.D. Shorthouse. 1989. Initiation and development of the stem gall induced by *Hemadas nubilipennis* (Hymenoptera: Pteromalidae) on lowbush blueberry, *Vaccinium angustifolium* (Ericaceae). Can. J. Bot. 67:2187–2198.

20

Urban Lakes: Integrators of Environmental Damage and Recovery

John M. Gunn and W. (Bill) Keller

Most of the world's human populations now live in rapidly expanding urban areas (Richardson 1991). These cities and towns vary widely in size and appearance, but they share a common feature: They are near water. Urban waters, in the form of lakes, streams, groundwater aquifers, and the nearshore areas of oceans, satisfy a wide variety of human needs (drinking water, transportation, industrial use, agricultural use, etc.). However, with few exceptions, urban waters have been and are being badly degraded by human activities (NRC 1992). To many, it may therefore be surprising that urban waters, particularly lakes, have received very little study by ecologists in North America (Gilbert 1989; McDonnell and Pickett 1990).

Like other aspects of the "massive, unplanned experiment" (McDonnell and Pickett 1990) that urbanization represents, the study of the ecology of urban waters is complicated by the wide variety of factors affecting urban aquatic ecosystems. These confounding variables make it difficult for ecologists to construct quantitative models of the functioning of urban systems or even to establish clearly the cause of specific problems, such as fish kills, algal blooms, or the accumulation of contaminants. These problems degrade the value of these important aquatic resources and deserve greater attention.

One potentially useful way of viewing urban lakes is to consider them as measures of the conditions within the associated catchment areas (see Chapter 24). Water, the "universal solvent," samples the available nutrients, soluble contaminants, and suspendible particles from the terrestrial system as it flows to the lake. Additional inputs of solutes, gases, and particles come through the atmosphere or are added directly by human activities. The resulting chemical solution, and the life that it supports, forms a rather large integrated sample, useful for assessing the "health" of the associated urban area.

Recent studies of the urban lakes within the city of Sudbury (Fig. 20.1) provide examples of the damaging effects of industrial and urban activities. They also illustrate the resilience of natural systems and the benefits to be achieved by pollution abatement programs.

Sudbury Urban Lakes

Within the city of Sudbury, there are 33 lakes of more than 10 ha in size, as well as a wide variety of smaller lakes and ponds. The larger lakes vary in surface area from 14 to 1331 ha and in maximum depth from 3 to 36 m. They make up more than 10% of the surface area of this glacially scoured landscape, giving Sudbury one of the highest concentrations of urban lakes of any city in Canada (Fig. 20.2). The lakes are clustered into two main watersheds that drain to the French or Spanish rivers, tributaries of Lake

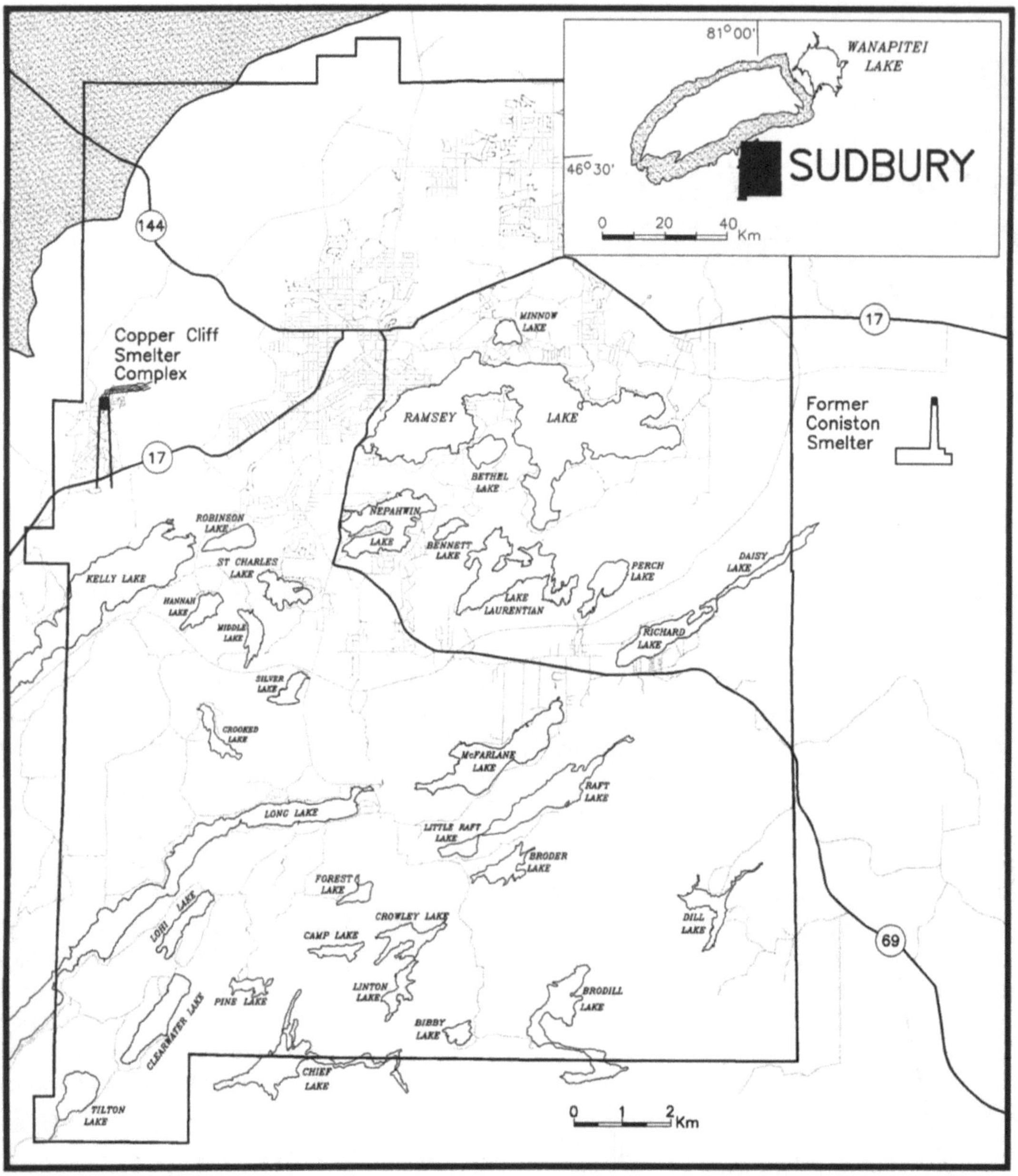

FIGURE 20.1. Municipal boundaries of Sudbury, showing urban lakes and the locations of nearby metal smelters.

FIGURE 20.2. Sudbury—the city of lakes. Copper Cliff smelter is in the background. (Photo by E. Snucins.)

Huron. A string of lakes, called the Long Lake chain, cuts diagonally across the city, following an earthquake fault line (Fig. 20.1), and drains to the Whitefish River.

Stress Factors

Six main stresses affect the chemistry and biology of urban lakes within this industrial city: (1) acidification, (2) metal contamination, (3) eutrophication, (4) shoreline or other watershed alterations, (5) introductions and invasions of exotic species, and (6) altered hydrology.

Acidification

The pH of the lakes ranges widely, from 4.3 to 8.9. Because of elevated sulfur deposition, due to the local smelting industry, 11 of the lakes are acidified to the point (pH <6.0) that damage to sensitive aquatic organisms is expected (see Chapter 5). Although there is some evidence of an association between acidic lakes and proximity to smelters (Fig. 20.3), a more surprising observation is that not all lakes are acidic even though they are close to such enormous sources of sulfur dioxide. This variability demonstrates the key role that the geochemistry of the bedrock and surficial soils within lake catchments plays in determining lake water quality (Jeffries et al. 1984). Catchments buffer the effects of incoming acid precipitation. In fact, the alkalinity of a few Sudbury lakes has actually increased under the enhanced weathering rates produced by acidic precipitation (Dixit et al. 1992). However, the overwhelming effect of the smelters has been a decline in lake water alkalinity and pH throughout the greater Sudbury area (see Chapter 5).

Metal Contamination

All the lakes in the city show the effects of industrial emissions of metals to the atmosphere (Fig. 20.3). Copper (4–320 μg/L) and

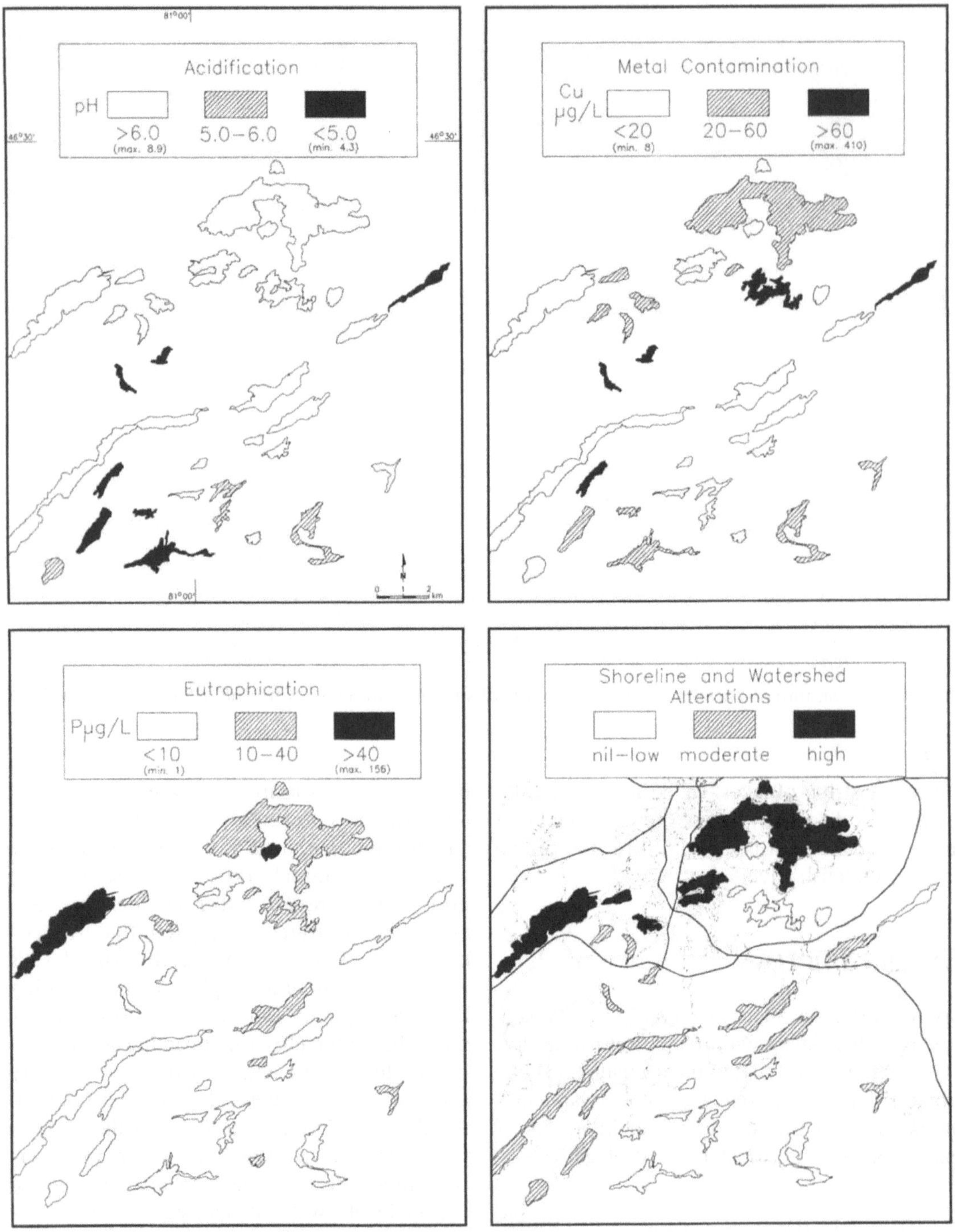

FIGURE 20.3. Effect of various environmental stresses on Sudbury urban lakes.

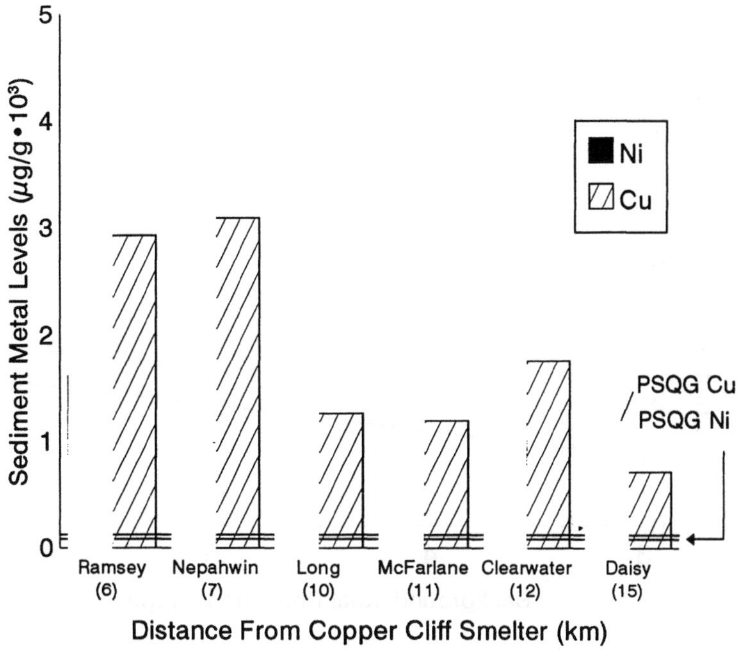

FIGURE 20.4. Concentrations of copper and nickel in the surface (2 cm) sediments of several Sudbury urban lakes in 1993. While concentrations are clearly highest in lakes nearest the Copper Cliff smelter, all lakes have very high concentrations, well above Ontario provincial government sediment quality guidelines (PSQG: copper 110 μg/g, nickel 75 μg/g; MOEE 1993).

nickel (25–590 μg/L) concentrations approach or exceed Ontario government water quality protection objectives (copper 5 μg/L, nickel 25 μg/L; MOE 1984) in all lakes. Silver Lake, a highly oligotrophic lake located about 6 km from the Copper Cliff smelter, is probably one of the world's most heavily contaminated lakes, considering lakes affected only by atmospheric deposition. During water quality surveys since 1981, concentrations of copper, nickel, zinc, and aluminum in Silver Lake of up to 460, 900, 120, and 1400 μg/L, respectively, have been recorded.

Most urban lakes in close proximity to the smelter have high accumulation of toxic metals in bottom sediments (Fig. 20.4). One particularly contaminated lake, Kelly Lake, receives treated liquid effluent from the Copper Cliff smelter site. The lake also receives inputs of metals from several other tributaries, as well as through atmospheric inputs and from a municipal sewage treatment plant. Sediments in Kelly Lake are not only heavily contaminated with copper and nickel but also have high concentrations of rare metals such as palladium, iridium, and platinum (Crocket and Teruta 1976).

Cultural Eutrophication

Two lakes (Bethel, Kelly) show the classical symptoms of advanced cultural eutrophication (Vollenweider 1968; Vallentyne 1974; Schindler 1977), as a result of nutrient inputs from private and municipal sewage plants (see Fig. 20.3). These symptoms include high levels of phosphorus and nitrogen, high standing crops of algae, particularly nuisance blue-green algae, and low dissolved oxygen levels with associated fish kills. The sewage treatment plant inputs to Bethel Lake were stopped in 1986, and although lake phosphorus concentrations declined from more than 200 μg/L in the late 1970s and early 1980s to less than 100 μg/L by the early 1990s, the lake is still highly eutrophic. Similarly, despite construction of a municipal sewage treatment plant in 1972 and implementation of phosphorus removal at the plant in 1987, Kelly Lake is still eutrophic.

Eutrophication also occurs by inputs of nutrients from a variety of diffuse sources (lawn and garden fertilizer, septic fields, eroding soil, etc.). This problem is not as severe in the nutrient-poor landscape of Sudbury as it is in

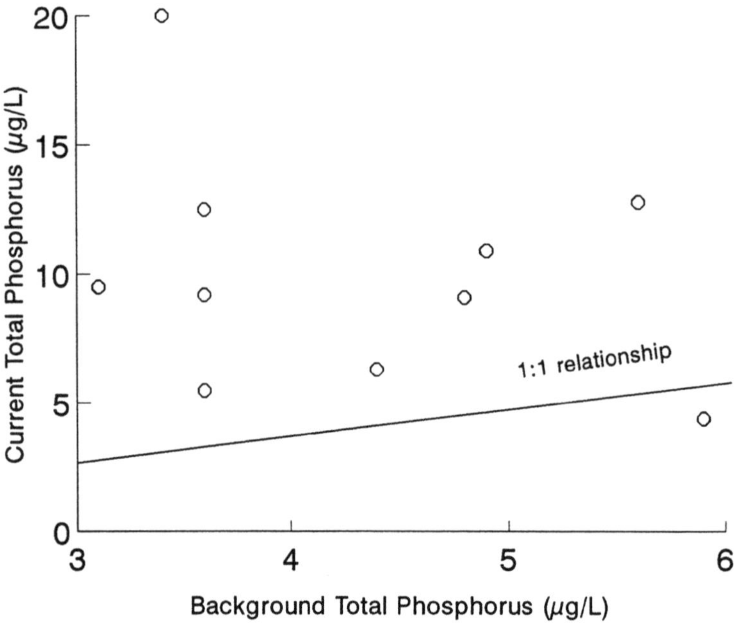

FIGURE 20.5. Comparison between recent measured concentrations of total phosphorus in several Sudbury urban lakes (Ministry of Environment and Energy, *unpublished data*) and pristine (predevelopment) concentrations estimated from a watershed-based total phosphorus model (Dillon et al. 1986).

many agricultural areas, but most lakes with houses and cottages within their catchments show nutrient enrichment (Fig. 20.5).

Shoreline and Watershed Alterations

Sudbury's city lakes vary widely in the extent and type of residential and industrial development around their shorelines—from nearly complete circling of the lakes by houses in the center of the city, to lakes with only a few seasonal cottages, to parkland lakes with no current development in the lake catchment areas (see Fig. 20.3). Typically, housing development has occurred in the low-lying areas along a web of stream valleys and lake shorelines at the base of the rocky hills and knobs that characterize Sudbury. Loss of wetlands has been a common feature of much of the development.

In a few cases, lakes have been eliminated through infilling by industrial waste (Fig. 20.6). However, most shoreline alterations are of a less-extreme type (construction of beaches, docks, breakwalls, etc.). These incremental changes also eventually lead to dramatic deleterious effects, particularly on the biology of urban lakes, through the loss of reproductive or other critical habitats for fish (Bryan and Scarnecchia 1992).

Exotic Species

Like the more familiar terrestrial examples, such as the House Sparrow (*Passer domesticus*) and the many introduced plants, urban lakes are affected by exotic species, intentionally or unintentionally introduced by humans. The introduction of the eurasian water milfoil (*Myriophyllum spicatum*), a nuisance aquatic plant that has proliferated in lakes within the Long Lake chain, and the establishment of a marine fish species, the highly competitive rainbow smelt (*Osmerus mordax*) in Nepahwin Lake, are two examples of exotic species in Sudbury urban lakes. In neither case is the source of the colonizers known.

Altered Hydrology

Urbanization creates dramatic effects on the flow path of water and the extent and timing of release and storage of water in the catchment. Usually, urban development encourages the rapid discharge of precipitation and meltwater

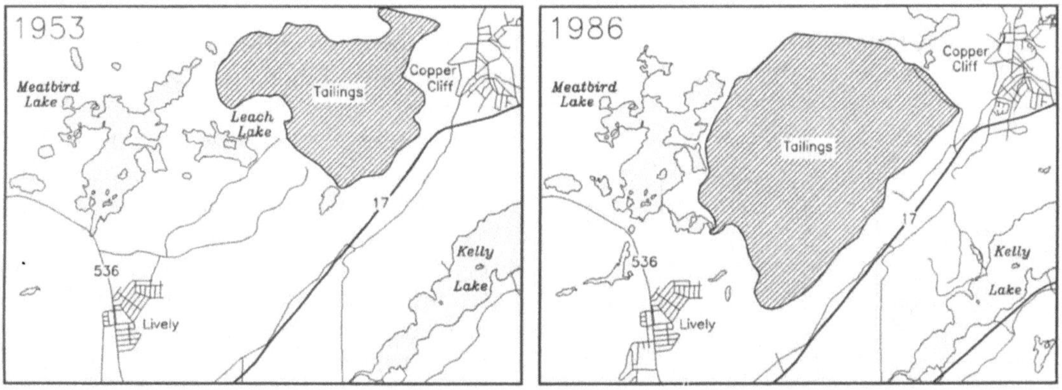

FIGURE 20.6. Infilling of lakes near the Copper Cliff smelter by expanding tailings and slag piles, 1953–1986.

through channelization, elimination of wetlands, and reduced infiltration of water by increasing the amount of impermeable surfaces (pavement, roofs, hard-packed fill). In Sudbury, the historical loss of soil and vegetation cover from the hillsides further decreases water retention and reduces evapotranspiration.

Flooding in the spring and lowered groundwater or lake levels in the summer are the well-known consequences of the alterations described above. Sudbury urban lakes have all these usual problems, necessitating engineered flood control structures and water-level control dams on many of the lakes. However, these solutions are far from perfect. Dams affect fish and other animal movement and the supply of water for downstream lakes. Storm water drainage (Fig. 20.7) brings high levels of road salt, metals, and other contaminants to the lakes (Pye et al. 1983). For example, chloride concentrations increased from about 40 to 50 and 75 to 100 mg/L in lakes Ramsey and Nepahwin, respectively, between the late 1970s and the late 1980s (Ministry of Environment and Energy, *unpublished data*). It is estimated that the background concentration of chloride in lakes before urbanization is usually under 2 mg/L (Neary et al. 1990).

FIGURE 20.7. Stormwater drainage. (Photo by J.M. Gunn.)

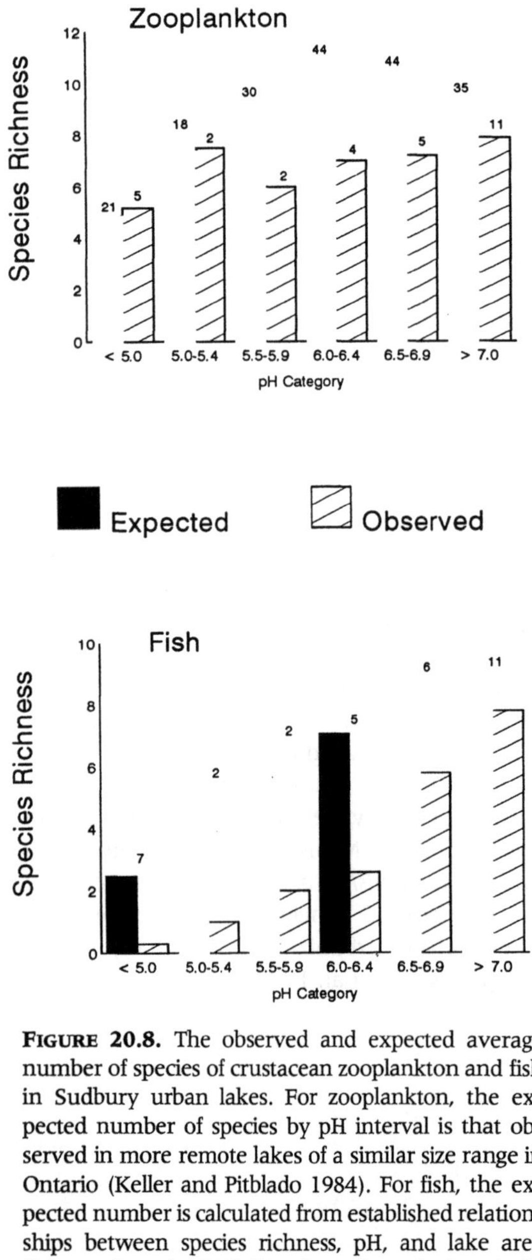

FIGURE 20.8. The observed and expected average number of species of crustacean zooplankton and fish in Sudbury urban lakes. For zooplankton, the expected number of species by pH interval is that observed in more remote lakes of a similar size range in Ontario (Keller and Pitblado 1984). For fish, the expected number is calculated from established relationships between species richness, pH, and lake area (Matuszek et al. 1990). Numbers above the bars indicate the number of lakes in the interval.

Integrators of Ecosystem Function and Health

Distinguishing the magnitude or relative importance of the multiple stresses on urban lakes is obviously difficult. However, many management questions deal with the overall condition of lakes rather than the details of specific causes (e.g., are the lakes improving? deteriorating? What recreational opportunities do they provide?). For these types of questions, biological monitoring provides the most direct measure of the effects of stresses operating in urban ecosystems. Some of the advantages of biological monitoring over inferring, or modeling, ecosystem responses from chemical or physical variables include (Schindler 1987; Fausch et al. 1990; Brinkhurst 1993; Johnson et al. 1993)

1. biological communities integrate the effects of temporal and spatial habitat degradation
2. monitoring of carefully chosen sensitive "indicator species" can be useful for identifying the type of stress(es) operating
3. biological attributes (e.g., presence of fish species) often have high societal values, aiding in the interpretation of the findings to the public
4. biological monitoring by experts can be more cost-effective than reliance on intensive sampling of organic and inorganic contaminants

In contrast with the limited information available on the effects of urbanization, there has been a great deal of work on the effects of acidification on the biology of freshwater lakes in north temperate areas (Baker et al. 1990) (see Chapter 5). This literature provides a basis for beginning to examine the effects of multiple stresses on Sudbury lakes by first attempting to separate the effects of low pH from other stresses. For example, an assessment of the species richness (the number of species present) of fish and crustacean zooplankton suggests that there are far fewer species present in Sudbury urban lakes than would be predicted from the measured acidity of these waters (Fig. 20.8). This result is surprising, because human activities often introduce new species (e.g., through release of bait fish). In this case, two different hypotheses can be proposed for the observed scarcity of species:

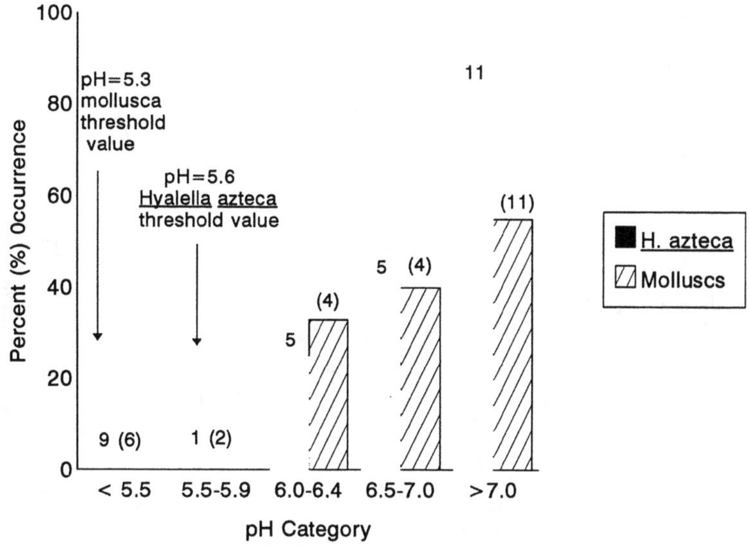

FIGURE 20.9. The distribution of amphipods (*Hyalella azteca*) and molluscs from the 1990–1993 biological survey of the Sudbury urban lakes. Occurrence was determined from a standard sampling effort on each lake. The *vertical lines* are the lower limit of pH tolerance of *Hyalella azteca* (Stephenson and Mackie 1986) and the minimum pH of occurrence for most common mollusc species (Eilers et al. 1984). Numbers above bars indicate the number of lakes sampled in the pH intervals for *Hyalella azteca* and common mollusc species (*in brackets*).

1. **species are missing because of the presence of additional stresses (e.g., metals, habitat loss)**
2. **chemical changes have occurred, but biological responses lag behind**

There is some support for both of these hypotheses.

Additional Stress Factors

The high concentrations of metals in the water column and sediments (see Fig. 20.4) in Sudbury urban lakes probably create toxic conditions for many aquatic species (Campbell and Stokes 1985; Spry and Weiner 1991; Wren and Stephenson 1991). This may be the reason that certain sensitive benthic macroinvertebrates, such as molluscs and the amphipod *Hyalella azteca*, are absent from many relatively high pH lakes (Fig. 20.9). The common sediment burrowing mayfly (*Hexagenia* sp.) is absent from Ramsey, Nepahwin, McFarlane, and Long lakes, the only Sudbury urban lakes surveyed for this species to date (W. Keller, *unpublished data*). Although metal contamination may be particularly severe in an industrial city, elevated metal concentrations are a common problem in many urban areas (Purves and MacKenzie 1969; Culbard et al. 1988).

Loss of habitat diversity through input of eroded soils from hillsides, or construction of human structures such as beaches and breakwalls, may also contribute to the low species richness in urban lakes.

Improvements in Lake Chemistry and the Effects of Biological Lag Time

There is very encouraging evidence from long-term monitoring studies on Clearwater Lake (Fig. 20.10) that water quality has begun to improve with reductions in industrial emissions (see Chapter 5). Although such intensive monitoring data do not exist for many of the urban lakes, it can be shown through paleolimnological reconstructions, using diatom and chrysophycean fossils preserved in lake sediments (Dixit et al. 1989) (see Chapters 3 and 5) that water quality has recently improved in several Sudbury lakes. Because of the limited dispersal ability of some organisms (see Chapter 15), biological communities may be delayed in recovering in urban lakes where water quality has improved. The discrepancy between observed and expected community and population status in Figures 20.8 and 20.9 may therefore partially reflect the lag time needed for the biological components

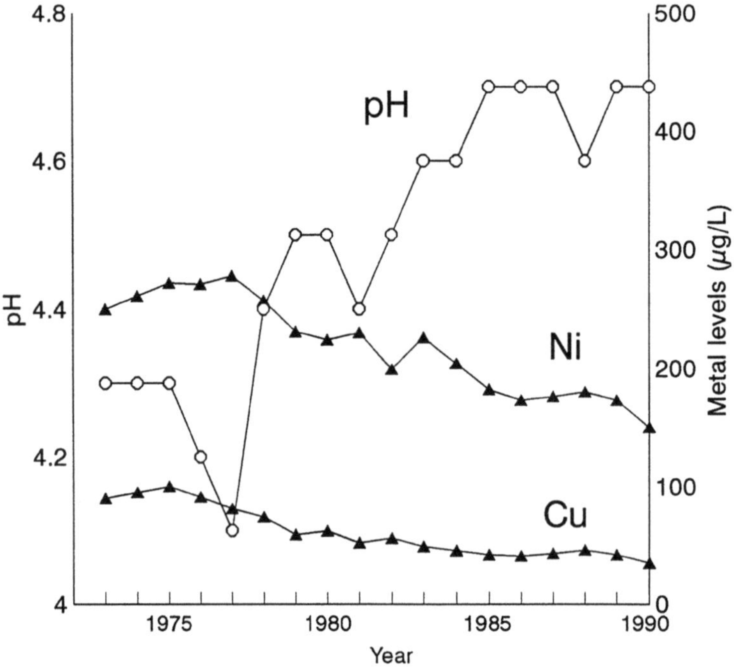

FIGURE 20.10. Changes in lake water pH and dissolved metals in Clearwater Lake, 1973–1990 (from Bodo and Dillon, in press).

of the system to catch up with the chemical changes.

The sediment accumulation of metals remains a potential encumbrance to recolonization by sensitive benthic invertebrates. But, as shown in Chapter 14, sediments can become an effective sink for contaminants, where over a very long time metals become buried under layers of less-contaminated sediments. However, with the abundance of metal particulates in most urban environments (Purves and MacKenzie 1969), lake sediments may always remain as biologically impoverished habitats.

Summary

Given the stresses that urban lakes are subjected to, it may be surprising to some readers that there is no such thing as a "dead lake". Even the most severely stressed lakes contain a variety of aquatic life forms. Generally, highly stressed lakes have simplified aquatic communities, represented mainly by rather tolerant species (Stokes et al. 1973; Baker et al. 1990); but the overall productivity of the urban lakes in industrial cities may be very similar to more pristine systems with the same levels of nutrients (Dillon et al. 1979). There may be some unique problems in Sudbury lakes because of extreme metal contamination of sediments, but it is probably safe to conclude that general functional characteristics of the lakes, such as respiration rates, primary productivity, and nutrient cycling, are not markedly affected. Such characteristics are therefore not particularly useful early indicators of ecosystem damage (Schindler 1987) or recovery.

The structure and composition of biological populations and communities that inhabit urban lakes can provide more sensitive and more useful indicators of ecosystem health than the functional characteristics descibed above (Karr 1991). The presences of naturally reproducing sensitive species (Marshall et al. 1987; Baker et al. 1990), without skin or spinal abnormalities (Hinton and Lauren 1990) and without elevated burdens of toxic contaminants (Campbell and Stokes 1985; Wren and Stephenson 1991), are some of the encouraging signs that human activities are not adversely affecting urban lakes. These are some of the types of measures that should be

FIGURE 20.11. Winter angling on Ramsey Lake in downtown Sudbury. (Photo by W. Keller.)

used in long-term biological monitoring programs to examine changing conditions in urban lakes.

Although there are still many severe problems in Sudbury urban lakes, there are also convincing signs of improving ecosystem health. One of the most tangible indicators is the variety of fish species that occupy the lakes. Among the 30 species present are sportfish species such as smallmouth bass (*Micropterus dolomieui*), northern pike (*Esox lucius*), largemouth bass (*M. salmoides*), lake trout (*Salvelinus namaycush*), and walleye (*Stizostedion vitreum*). Most of these species have been recently introduced or re-introduced and have established reproducing populations that are the focus of an intensive recreational fishery (Fig. 20.11). A positive aspect of this fishery, and a still intriguing scientific issue, is the fact that mercury and other metal contaminants are found in low concentration in fish and other biota in these lakes (Bowlby et al. 1988; Wren and Stokes 1988).

It is hoped that with further industrial cleanup and more "environmentally friendly" planning (see Chapter 24), the many still-degraded systems can some day be restored as valuable natural assets in this urban environment.

Acknowledgments. We thank Jim Carbone, Nancy Dolson, Rob Kirk, Rod Sein, Ed Snucins, and the many summer students who helped with the urban lakes survey. Bill Lautenbach and Marty Puro provided review comments. Michael Conlon and Léo Larivière assisted with the graphics.

References

Baker, J.P., D.P. Bernard, S.W. Christensen, and M.J. Sale. 1990. Biological effects of changes in surface water acid-base chemistry. State of Science Technical Report 13. National Acid Precipitation Assessment Program. Report for the U.S. Environmental Protection Agency, Corvallis, OR.

Bodo, B.A., and P.J. Dillon. In press. De-acidification trends in Clearwater Lake near Sudbury, Ontario 1973–1992. *In* Proceedings—Stochastic and Statistical Methods in Hydrology and Environmental Engineering. Kluwer Academic.

Bowlby, J.N., J.M. Gunn, and V.A. Liimatainen. 1988. Metals in stocked lake trout (*Salvelinus namaycush*) in lakes near Sudbury, Ontario. Water Air Soil Pollut. 39:217–230.

Brinkhurst, R.O. 1993. Future directions in freshwater biomonitoring using benthic macroinvertebrates, pp. 442–460. *In* D.M. Rosenburg and V.H. Resh (eds.). Freshwater Biomonitoring and Benthic Macroinvertebrates. Chapman & Hall, New York.

Bryan, M.D., and D.L. Scarnecchia. 1992. Species richness, composition, and abundance of larvae and juveniles inhabiting natural and developed shorelines of a glacial Iowa lake. Environ. Biol. Fish. 35:329–341.

Campbell, P.G.C., and P.M. Stokes. 1985. Acidification and toxicity of metals to aquatic biota. Can. J. Fish. Aquat. Sci. 42:2034–2049.

Crocket, J.H., and Y. Teruta. 1976. Pt, Pd, Au and Ir content of Kelley Lake bottom sediments. Can. Miner. 14:58–61.

Culbard, E.B., I. Thornton, M. Watt, S. Wheatley, S. Moorcroft, and M. Thompson. 1988. Metal contamination in British urban dusts and soils. J. Environ. Quality 17:226–234.

Dillon, P.J., K.H. Nicholls, W.A. Scheider, N.D. Yan, and D.S. Jeffries. 1986. Lakeshore Capacity Study: Trophic Status. Technical Report. Ontario Ministry of Municipal Affairs, Ontario.

Dillon, P.J., N.D. Yan, W.A. Scheider, and N. Conroy. 1979. Acidic lakes in Ontario: characterization, extent and responses to base and nutrient additions. Arch. Hydrobiol. Beih. 13:317–336.

Dixit, S.S., A.S. Dixit, and J.P. Smol. 1989. Lake acidification recovery can be monitored using chrysophycean microfossils. Can. J. Fish. Aquat. Sci. 46:1309–1312.

Dixit, S.S., A.S. Dixit, and J.P. Smol. 1992. Assessment of changes in lake water chemistry in Sudbury area lakes since pre-industrial times. Can. J. Fish. Aquat. Sci. 49 (Suppl. 1):8–16.

Eilers, J.M., G.L. Lien, and R.G. Berg. 1984. Aquatic Organisms in Acidic Environments: A Literature Review. Technical Bulletin 150. Department of Natural Resources, Madison, WI.

Fausch, K.D., J. Lyons, J.R. Karr, and P.L. Angermeier. 1990. Fish communities as indicators of environmental degradation. Am. Fish. Soc. Sympos. 8:123–144.

Gilbert, O.L. 1989. The Ecology of Urban Habitats. Chapman and Hall Ltd., New York.

Hinton, D.E., and D.L. Lauren. 1990. Integrative histopathological approaches to detecting effects of environmental stressors on fishes. Am. Fish. Soc. Sympos. 8:51–66.

Jeffries, D.S., W.A. Scheider, and W.R. Snyder. 1984. Geochemical interactions of watersheds with precipitation in areas affected by smelter emissions near Sudbury, Ontario. pp. 196–241. *In* J. Nriagu (ed.). Environmental Impacts of Smelters. John Wiley and Sons, New York.

Johnson, R.K., T. Wiederholm, and D.M. Rosenberg. 1993. Freshwater biomonitoring using individual organisms, populations, and species assemblages of benthic macroinvertebrates, pp. 40–158. *In* D.M. Rosenberg and V.H. Resh (eds.). Freshwater Biomonitoring and Benthic Macroinvertebrates. Chapman and Hall, New York.

Karr, J.R. 1991. Biological integrity: a long neglected aspect of water resource management. Ecol. Applic. 1:66–84.

Keller, W., and J.R. Pitblado. 1984. Crustacean plankton in northeastern Ontario lakes subjected to acidic deposition. Water Air Soil Pollut. 23: 271–291.

Marshall, T.R., R.A. Ryder, C.J. Edwards, and G.R. Spangler. 1987. Using the Lake Trout as an Indicator of Ecosystem Health—Application of the Dichotomus Key. Technical Report 49. Great Lakes Fishery Commission, Ann Arbor, MI.

Matuszek, J.E., J. Goodier, and D.L. Wales. 1990. The occurrence of Cyprinidae and other small fish species in relation to pH of Ontario lakes. Trans. Am. Fish. Soc. 119:850–861.

McDonnell, M.J., and S.T.A. Pickett. 1990. Ecosystem structure and function along urban-rural gradients: an unexploited opportunity for ecology. Ecology 71(4):1232–1237.

Ministry of Environment (MOE). 1984. Water management—goals, policies, objectives and implementation procedures of the Ministry of the Environment. Technical Report. Toronto, Ontario.

Ministry of Environment and Energy (MOEE). 1993. Guidelines for the Protection and Management of Aquatic Sediment Quality in Ontario. Technical Report. Toronto, Ontario.

National Research Council (NRC). 1992. Restoration of Aquatic Ecosystems: Science, Technology, and Public Policy. National Academy Press, Washington, DC.

Neary, B.P., P.J. Dillon, J.R. Munro, and B.J. Clark. 1990. The Acidification of Ontario Lakes: An Assessment of Their Sensitivity and Current Status with Respect to Biological Damage. Technical Report. Ontario Ministry of Environment, Dorset, Ontario.

Purves, D., and E.J. MacKenzie. 1969. Trace element contamination of parklands in urban areas. J. Soil Sci. 20:288–290.

Pye, V.I., R. Patrick, and J. Quarles. 1983. Groundwater Contamination in the United States. University of Pennsylvania Press, Philadelphia.

Richardson, N. 1991. Urbanization: building human habitats. pp. 13-1–13-31. *In* S. Burns, M. Sheffer, and R. Lanthier (eds.). The State of Canada's Environment. Government of Canada, Ottawa.

Schindler, D.W. 1977. Evolution of phosphorus limitation in lakes. Science 195:260–262.

Schindler, D.W. 1987. Detecting ecosystem response to anthropogenic stress. Can. J. Fish. Aquat. Sci. 44(Suppl. 1):6–25.

Spry, D.J., and J.G. Weiner. 1991. Metal bioavailability and toxicity to fish in low alkalinity lakes: a critical review. Environ. Pollut. 71: 243–304.

Stephenson, M., and G.L. Mackie. 1986. Lake acidification as a limiting factor for the distribution of the freshwater amphipod *Hyalella azteca*. Can. J. Fish. Aquat. Sci. 43:288–292.

Stokes, P.M., T.C. Hutchinson, and K. Krauter. 1973. Heavy metal tolerance in algae isolated from contaminated lakes near Sudbury, Ontario. Can. J. Bot. 51:2155–2168.

Vallentyne, J.R. 1974. The Algal Bowl—Lakes and Man. Misc. Special Publication 22. Department of the Environment, Ottawa.

Vollenweider, R.A. 1968. Scientific fundamentals of the eutrophication of lakes and flowing waters, with particular reference to nitrogen and phosphorus as factors in eutrophication. Report DAS/CSI/68.27. Organization of Economic Cooperation and Development, Paris.

Wren, C.D., and G.L. Stephenson. 1991. The effect of acidification on the accumulation and toxicity of metals to freshwater invertebrates. Environ. Pollut. 71:205–241.

Wren, C.D., and P.M. Stokes. 1988. Depressed mercury levels in biota from acid and metal stressed lakes near Sudbury, Ontario. Ambio 17:28–30.

Section E

Planning for the Future

Maurice F. Strong

From an environmentalist's point of view, the story of the Sudbury area is one of both bitter and sweet irony. It is certainly embittering to reflect that this southerly outcropping of the ancient Canadian Shield, an ecosystem created and preserved by nature over hundreds of millions of years, was converted to a barren wasteland in less than a century by the human hand. But at the risk of making a virtue of a necessity, it is sweet to contemplate that this devastated pocket of planet earth may yet become a global model of determined, if belated, environmental enlightenment and a case history of ecological rejuvenation.

At the conclusion of the United Nations Conference on Environment and Development at Rio de Janeiro in June 1992, I said that our task was now to "move down from the Summit and into the trenches" to the level of practical action and meaningful decisions that will be needed to fulfil the vision of Rio and to implement the agreements reached there. As the Earth Summit concluded, the people of the Sudbury community had already been toiling in the ecological trenches for nearly two decades, working to restore their blighted patch of earth to a condition that would again resemble a functioning ecosystem.

One of the happiest and most proud moments of the entire Earth Summit, for me as a Canadian, came when the Sudbury community was given the 1992 United Nations Local Government Honours Award for its work to reverse the process of environmental degradation.

Sudbury was one of the earliest regions to feel the baneful brunt of unsustainable industrial practices. It was also one of the first to recognize that, no matter how afflicted their environment was, it was not beyond repair. And so the people of Sudbury, through their local and provincial governments, and with guidance and assistance from the academic and industrial sectors, set out to show the world how it is done.

Unfortunately, such enlightenment is not yet universal, despite the heightened awareness of global environmental issues created by the Rio Earth Summit and its predecessors. In the 20 years between the first World Conference on the Human Environment in Stockholm in 1972 and the Rio Summit, much progress was made in some areas, including our understanding of the complex system of interaction through which human activities have an effect on the environment and resources of the planet. A host of new institutions were established at various levels to deal with the policy, regulatory, scientific, economic, and other dimensions of these issues. Virtually every nation, including developing countries, established national ministries or agencies with responsibility for environmental policies and regulations.

The initiatives were accompanied by some significant progress in addressing several substantive environmental concerns, notably the "close-in" problems of air, water, and land pollution in industrialized countries.

But despite this progress, it became evident by the mid-1980s that, overall, the conditions of the earth's environment and some of its most vital ecosystems had continued to deteriorate and some of the primary risks such as global warming and ozone depletion had become more acute and menacing than they appeared at Stockholm. At the same time, developing countries were experiencing problems of pollution and environmental degradation rapidly approaching the levels of the more industrialized countries while lacking the resources to cope with them. It was also becoming increasingly evident that there was a direct and inextricable link between economic development and its environmental impacts.

Against this background, the United Nations General Assembly decided in December 1983 to establish the World Commission on Environment and Development to examine the condition of and prospects for the economy and the environment in the perspective of the year 2000 and beyond. The commission, under the leadership of Norwegian Prime Minister Gro Harlem Brundtland, in its landmark report released in 1987, "Our Common Future," made a compelling case for sustainable development as the only viable pathway to a secure and promising future for the human community and produced a set of recommendations for achieving it.

This report provided the basis for the UN General Assembly's decision to convene, on the twentieth anniversary of the Stockholm conference, the United Nations Conference on Environment and Development. Studies made in preparation for Rio made it starkly clear that fundamental changes in economic behavior offer the only real prospect of effecting a transition to a secure and sustainable future as we move toward the twenty-first century. And the impressive case history of Sudbury shows that it can be done.

The desolation of the Sudbury area is not an isolated case. With the demise of the communist regimes of eastern Europe and the

Soviet Union, we are only beginning to see the full extent of the massive environmental devastation they produced. Closer to home, in the Atlantic provinces and New England states, an entire industry and a way of life are threatened by profligate and exploitative fishing practices.

But Sudbury's immense achievement is a beacon of encouragement. It holds out the prospect that even the most resolute ravaging by humankind can be reversed if action is taken early enough. Beyond that broad object lesson, the Sudbury case also demonstrates that economic benefits can flow from good environmental practices. As is pointed out elsewhere in this book, the reclamation project has provided short-term jobs for more than 3000 persons over the past 15 years. The improved landscape is also serving to assist in attracting prospective businesses, as well as tourists.

This is a novel and welcome twist on one of the main themes of the Earth Summit—that good environmental performance is fully compatible with positive economic performance. Japan and other countries have demonstrated this in terms of converting a large and dynamic economy to sustainable principles. Now, Sudbury is proving the point with an impressive remedial project.

21

Developments in Emission Control Technologies/Strategies: A Case Study

Dan F. Bouillon

Initiated by government legislation and encouraged by economic factors (Box 21.1), developments in emission control technologies and strategies are taking place at an accelerating pace within industrialized nations. The sophistication of these developments is increasing at the same time. However, global application of the technology is lacking, particularly in developing countries.

To meet more stringent environmental protection regulations, many industries are developing new and usually very broad-scale emission control technologies and strategies. As illustrated in the following case study, such development can achieve environmental protection goals while at the same time produce economic benefits for industry.

Strategies

Early control strategies largely adopted the approach that "dilution was the solution to pollution." One example of this approach was the installation of tall stacks, such as the study company's 381-m stack. These stacks were designed to improve the local air quality by increased dispersion of waste gases.

To date, most pollution control technologies and strategies have been directed at intercepting pollutants before they leave the plant, a so-called end of pipe solution to environmental problems (AWMA 1992). The standard air quality control technologies of baghouses, cyclones, electrostatic precipitators, and scrubbers are examples of such efforts. However, these approaches can be considered "react and cure" modes of developing environmental control technology and strategies (Watson 1992). In the future, a proactive approach must be taken to prevent or at least reduce environmental pollutants at the source. The case study illustrates this concept in its infancy.

Pollution abatement programs need not be limited to dealing only with environmental problems at a particular industrial site. In fact, once they are developed, they can be profitably expanded to solve other companies' problems as well. The case study includes an example.

Public Awareness and Development of Legislation

In the 1960s and 1970s, the deteriorating state of the global environment led to a public outcry for action. The period saw the birth of vocal non-government environmental organizations such as Greenpeace. The study of environmental sciences was also added to the educational system during this period. Along with increased media coverage of environmental issues, these factors combined to create

Box 21.1. Trading in Pollution Permits

As part of the sulfur dioxide emission reduction program established through the 1990 revisions of the U.S. Clean Air Act, an innovative market-driven process of pollution control was established in the United States. To begin the process of initiating trading in pollution permits, the government establishes a cap on the total emissions of a pollutant in a given airshed. This cap is divided into multiple equal units of the pollutant. The release of each unit of a pollutant within the given airshed requires a permit for that amount.

All emitters of the pollutant in the airshed would need to acquire enough permits to match their output. Permits would be transferable at prices determined in a free market. New industries would need to purchase existing permits. Supply-and-demand economics would drive emission reductions. These reductions would continue until the cost of additional reduction equals the price of a pollution permit.

Interested groups could purchase permits and remove them from the free market. This would drive further pollution reduction in the airshed. The government could also lower emissions in the airshed by reducing the unit size of each permit without increasing the number of permits (Lee 1993). Not all companies agree with this approach, considering it buying the right to pollute.

pressure on the various levels of governments to act on environmental problems.

Governments around the globe began to develop, modify, and enforce new and broader environmental legislation (IUAPPA 1991). In 1967, the Ontario government introduced the Environmental Protection Act, which provided the framework for further legislation dealing with specific environmental problems. An example of this type of legislation was the Countdown Acid Rain program, discussed in Chapter 4, which mandated the reduction of acid gas emission from selected companies and facilities in Ontario. The 1990 revisions of the U.S. Clean Air Act contain similar provisions for emissions controls but also introduced innovative economic incentives to achieve them (Box 21.1). The case study presented here demonstrates how one company, Inco Limited, responded to these pressures to reduce sulfur dioxide emissions at its Copper Cliff smelter.

Inco Limited's Sulfur Dioxide Emission Reduction Program

Inco's Sudbury operations are one of the world's largest facilities for the integrated mining, milling, smelting, and refining of sulfide ore, with current production rates of 115,000 tonnes nickel and 120,000 tonnes copper per year. The company is also a significant producer of cobalt, selenium, tellurium, gold, silver, platinum group metals, liquid sulfur dioxide, and sulfuric acid as byproducts of nickel and copper processing (Figure 21.1). The company mines approximately 12 million tonnes of ore yearly to achieve these production levels.

Ore Chemistry and Composition

The Sudbury ore is approximately 25% mixed iron, copper, and nickel sulfide minerals in a 6:1:1 ratio, respectively. The remainder, which is waste rock or "gangue," is discarded as "tailings." Most of the ore's sulfur content is associated with the iron minerals. The copper (chalcopyrite) and nickel (pentlandite) minerals contain approximately equal amounts of metal and sulfur. The copper mineral is easily separated from the ore, but the nickel mineral is finely spread throughout the iron minerals. A small fraction of the nickel is also tied up in the crystal structure of the iron minerals. Therefore, large quantities of waste iron minerals must be processed to extract nickel, and for each tonne of nickel produced, 8 tonnes of sulfur will be processed (Fig. 21.1). These facts affect both the milling and smelting processes.

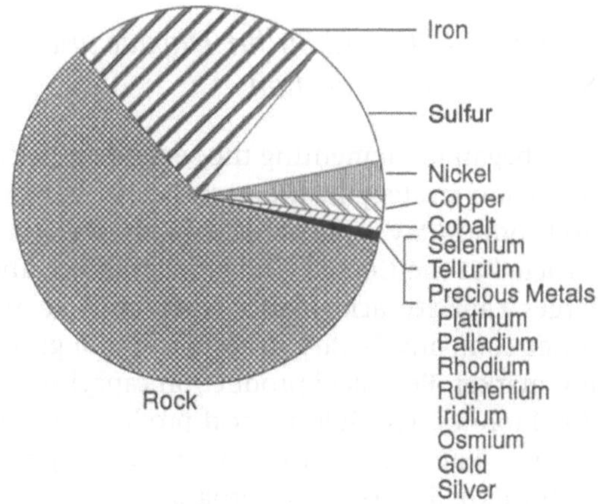

FIGURE 21.1. Typical Sudbury district ore analysis.

Sulfur, Smelting, and Sulfur Dioxide

As described in Chapter 2, the early mining companies in the Sudbury area used "roast yards" to make a crude matte out of bulk sulfide ore. Large volumes of high-concentration sulfur dioxide gas were released at ground level from the roast yards. The consumption of wood for the roast yards and low-level release of sulfur dioxide compounded existing vegetation damage caused by clear-cut forestry (Chapter 2).

In 1929, the first Inco smelter was built at Copper Cliff, and the use of "roast yards" stopped. As the size of the smelter increased and the demand for products grew, the sulfur dioxide emissions increased correspondingly. Inco's sulfur dioxide emissions peaked at more than 5600 tonnes/day or about 2 million tonnes/year in the early 1960s. At this time, almost all the smelter fumes were released into the atmosphere through three approximately 150-m stacks. Lesser amounts also escaped through vents in the roof of the smelter as fugitive emissions.

The resulting ground-level sulfur dioxide fumigations created serious local air quality conditions and contributed to widespread vegetation damage. Growing public awareness and concern regarding the sulfur dioxide pollution problem led to pressure on all levels of government toward control and abatement legislation aimed specifically at Inco's sulfur dioxide emissions.

Sulfur Dioxide Emissions Control Strategies

Three strategies are available to reduce the sulfur dioxide emissions that are the result of smelting of sulfide ore:

1. **remove sulfur before processing in the smelting furnaces**
2. **recover the sulfur dioxide as marketable products such as elemental sulfur, liquid sulfur dioxide, and sulfuric acid**
3. **use a process such as a hydrometallurgical process that does not convert sulfur to sulfur dioxide**

Strategy 1—Iron Mineral Rejection

Much of Inco's current effort at sulfur dioxide emissions reduction is directed at sulfur removal before smelting. This has largely been achieved through more efficient separation of the waste high sulfur iron minerals from the valuable copper and nickel minerals at the milling stage of ore processing. Improved milling techniques, developed after many years of research, have allowed for increased rejection of the iron minerals while minimizing the loss of the valuable copper and nickel minerals. Adoption of this separation approach reduces sulfur dioxide emissions by removing approximately 67% of the original sulfur in the ore before the ore enters the smelting furnaces.

Strategy 2—Liquid Sulfur Dioxide and Sulfuric Acid Production

Inco began implementing the second emission reduction strategy in the early 1930s when a 400-tonnes/day sulfuric acid plant was constructed at the Copper Cliff smelter. More and larger capacity acid plants were built at another company facility in response to a growing market. Peak acid production capacity was 3000 tonnes/day. Sulfuric acid production now plays a key role in the company's ongoing sulfur dioxide abatement program.

In 1952, liquid sulfur dioxide production began at the Copper Cliff smelter using standard compression and refrigeration technology. The plant has a production capacity of about 400 tonnes/day, with an average annual production of about 80,000 tonnes. The product is in high demand by many different users such as specialty chemical manufacturers, the largest single market being the pulp and paper industry.

To produce elemental sulfur from the sulfur dioxide is not economically feasible with current technology at the Copper Cliff smelter. Moreover, the process requires extensive use of fossil fuels. Combustion of fossil fuels generates large amounts of carbon dioxide, a greenhouse gas, and nitrogen oxide, an acid-forming gas. The increased emissions of nitrogen oxide would therefore defeat a prime objective of the sulfur dioxide abatement program—to reduce acid rain.

Strategy 3—Alternative Processes

The third strategy, to use a chemical leaching or extraction method (hydrometallurgical processes), that does not produce sulfur dioxide is not technically or economically feasible for the Copper Cliff operations. The capital costs and technological problems associated with the hydrometallurgical processes are too great. The current processes used in these methods would also result in the production of vast quantities of unwanted waste materials such as iron hydroxide and calcium sulfate (gypsum).

The gypsum produced during chemical extraction of metals would not meet market requirements in density and chemistry. Similarly, waste iron hydroxide would be of poor quality (i.e., particle size and contamination with other material). Given the tonnage of ore used in the Copper Cliff operation, these unusable waste products would then require a large dedicated disposal site, which simply transfers the pollution liability from air to land. Such transfers of environmental problems from one site to another, called "media transfers," cannot be considered true environmental solutions (see Chapter 22).

History of Sulfur Dioxide Abatement Efforts at Copper Cliff

Smelting Process Changes

About 1950, Inco and Outkumpu Oy, the Finnish mining company, developed separate flash furnace technologies for use in copper sulfide smelting. This new technology benefited the mining companies in two ways.

1. **Once the furnace was brought up to operating temperature by fossil fuel burners, flash smelting became an "autogenous" process. Heat released by the combustion of sulfur in the feed into sulfur dioxide replaced that provided by fossil fuels in the earlier process. The incoming feed provided the flow of sulfur fuel to maintain the smelting reactions.**
2. **Pure oxygen was used for combustion. As a result, the offgas from the flash furnace contains about 70–75% sulfur dioxide. No carbon dioxide and minimal nitrogen oxides were coproduced. This high sulfur dioxide strength offgas allowed the use of existing fixation technology to capture, as marketable products, sulfur dioxide that would otherwise have been released to atmosphere. Liquid sulfur dioxide was produced by compression and refrigeration.**

3. **Process efficiency was increased while minimizing the production of nitrogen oxides. Fossil fuel consumption use was kept to a minimum, with the main use being heating the furnace to operating temperature. The minimal need to use fossil fuel enhanced the operations' economic competitiveness.**

Iron Processing

To maximize the recovery of copper and nickel from the ore, the iron minerals were processed through the smelter until 1954. This increased the environmental liability of the operation through increased production of sulfur dioxide and created a waste storage problem. The unwanted iron also had to be removed from the furnaces as a waste iron silicate or "slag." Because the furnaces have a finite capacity, treating and handling unwanted iron limits the production rates of nickel and copper. Increasing demand for the company's metal products pushed the development of a separate treatment process for the iron minerals to remove this load from the smelter. Improved techniques and technology allowed more of the iron minerals to be rejected at the milling stage.

In 1954, a nickel bearing (about 0.8–1.0%), high sulfur (about 32%) fraction of the rejected iron minerals was routed to the new iron ore recovery plant. The first stage in the new process was the controlled roasting of the iron minerals down to about 0.5% sulfur, using fluid bed roaster technology. These roasters generated offgases at about 10–12% sulfur dioxide, which is ideal for conversion into sulfuric acid. The 400-tonnes/day sulfuric acid plant was moved over to this new facility from the Copper Cliff smelter in 1958. Two new acid plants were constructed. Final total production capacity was 3000 tonnes/day of 100% sulfuric acid equivalent. The various strengths of sulfuric acid and grades of sulfuric acid concentrate (oleum) produced by these plants found ready and varied markets, such as the Elliot Lake uranium mines (located 150 km west of Sudbury), lead acid battery makers, and detergent manufacturers. The second stage was removal of the nickel, cobalt, and a residual copper via an ammonia-carbon dioxide leach. The remaining iron oxide was pelletized for sale to the steel industry.

Recent Control Technologies and Strategies

Early company developments in environmental control technologies and strategies were driven by economics. The early 1970s saw the establishment of government regulations enforcing environmental control initiatives. To meet these initiatives, the pace of technological developments began to accelerate.

To meet these more stringent environmental requirements, Inco designed a comprehensive sulfur dioxide abatement program in the mid-1980s. At a cost of implementation of more than $Can 600 million, it became one of the largest industrial pollution abatement programs in the world. The following is a brief description of this new program.

Rather than attempt to use entirely new technology, the company adapted the proven copper smelting flash furnace to treat a combined nickel-copper concentrate (Figs. 21.2 and 21.3). From the new bulk flash furnaces, a portion of the high-strength sulfur dioxide gas is then directly compressed into liquid sulfur dioxide while the remainder is diluted with air to meet the requirements of a new 2900 tonnes/day (100% sulfuric acid basis) sulfuric acid plant (Fig. 21.4). The sulfuric acid plant is crucial to achieving legislated sulfur dioxide emission limits. By the mid-1980s, the economics changed and the company was losing money by making sulfuric acid at the iron ore recovery plant. However, the facility remained in operation to maintain the company's share of the sulfuric acid market. The company needed this market share for the sulfuric acid that would be produced to meet the government's 1994 emission limit. The old facility was closed in May 1991 when the new acid plant at the smelter began operation.

The finished flash furnace product (Bessemer matte) is separated into a nickel sulfide

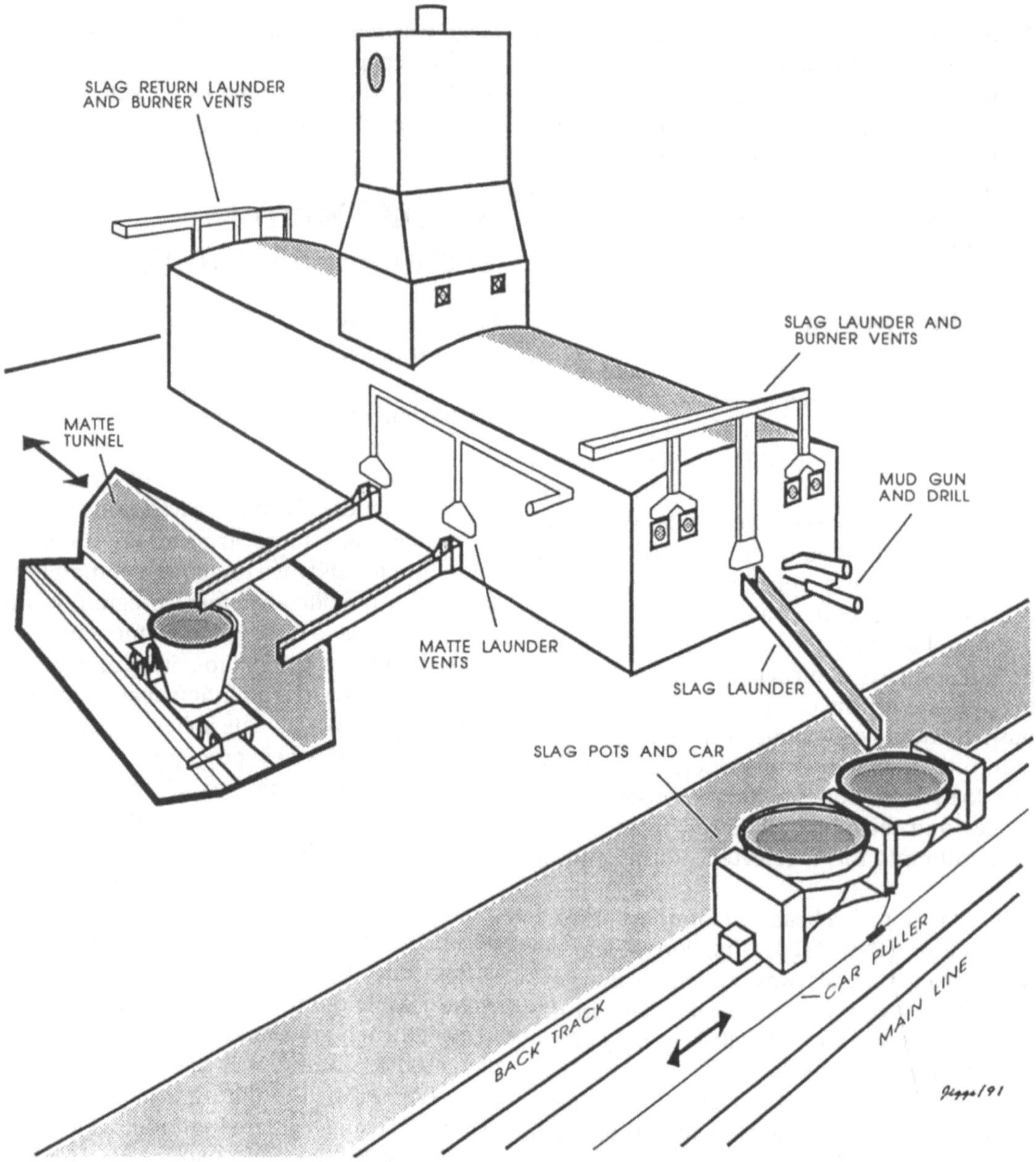

FIGURE 21.2. Inco Copper Cliff smelter flash furnace—external view of northeast side.

fraction and a pure copper sulfide. The copper sulfide is then treated in a proprietary melter, which also produces a high sulfur dioxide strength offgas (Fig. 21.5). The melter offgas is combined with those of the bulk flash furnaces for fixation as either liquid sulfur dioxide or sulfuric acid.

In addition to the environmental benefits of the new smelter, there are considerable economic benefits from the new facility. Economic benefits consist of lowered energy costs, increased worker productivity, and the conversion of former emitted sulfur dioxide gases into salable products. The payback time for the

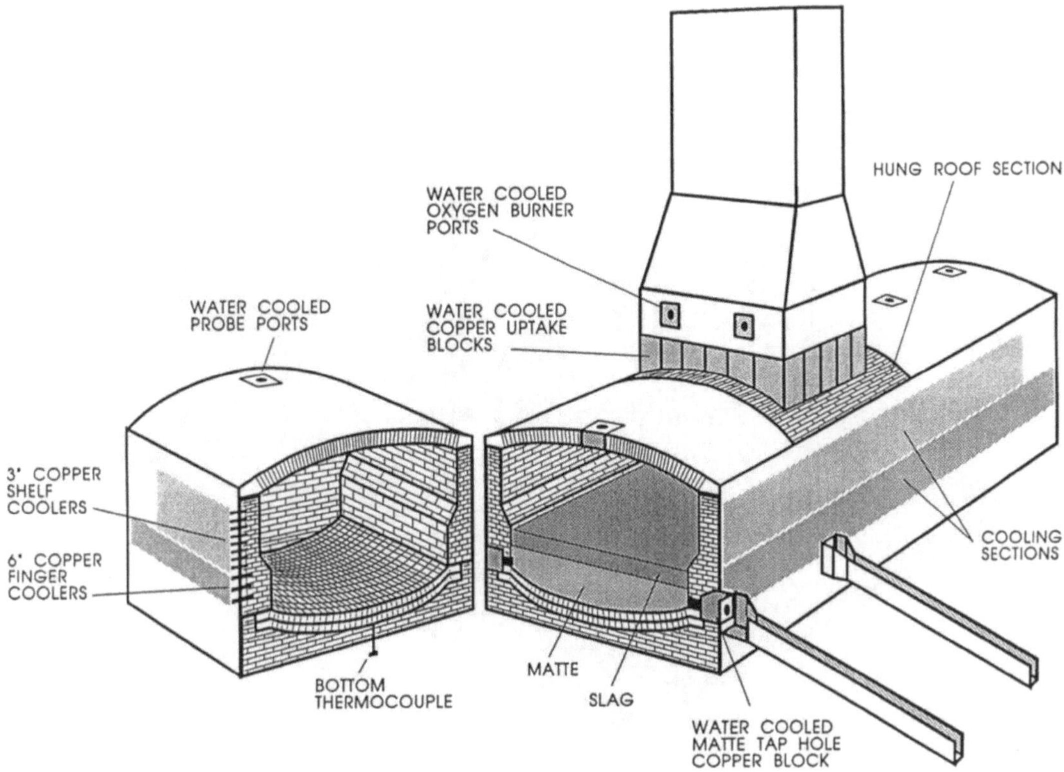

FIGURE 21.3. Inco Copper Cliff smelter flash furnace—internal view of refractory detail and cooling equipment.

capital expenditures for the construction of the new facilities and operating procedures ($600 million) has been estimated to be approximately 10 years, based on projected energy and worker productivity gains and reductions in maintenance costs at the new operation (Table 21.1).

Environmental Monitoring

In the early 1970s, a voluntary strategy to control ground-level concentrations of sulfur dioxide was initiated. A network of Ontario Ministry of Environment and Energy-owned and -operated fixed monitoring stations was installed to continuously monitor ground-level sulfur dioxide concentrations. In addition to the government stations, company-owned and -operated sulfur dioxide emission monitoring vehicles patrol and measure ground-level sulfur dioxide concentrations.

The company also operates a meterological station. Data gathered at this station are input to computer models. The computer models calculate a maximum allowable sulfur dioxide emission rate for the Copper Cliff smelter complex, which would theoretically keep ground-level sulfur dioxide concentrations below the government limits. These data are transferred to operators at the smelter who adjust the processes so that the average sulfur dioxide emission rates do not exceed the given value.

Air quality data from the governments fixed stations and the company's mobile monitors along with continuously updated meteorological data are used to adjust the smelter's sulfur dioxide emission rates to meet required ground-level concentrations. This combination of company and government monitoring for emissions control is unique to Sudbury.

In 1978, the government passed legislation that changed the formerly voluntary program into a legally binding abatement program and

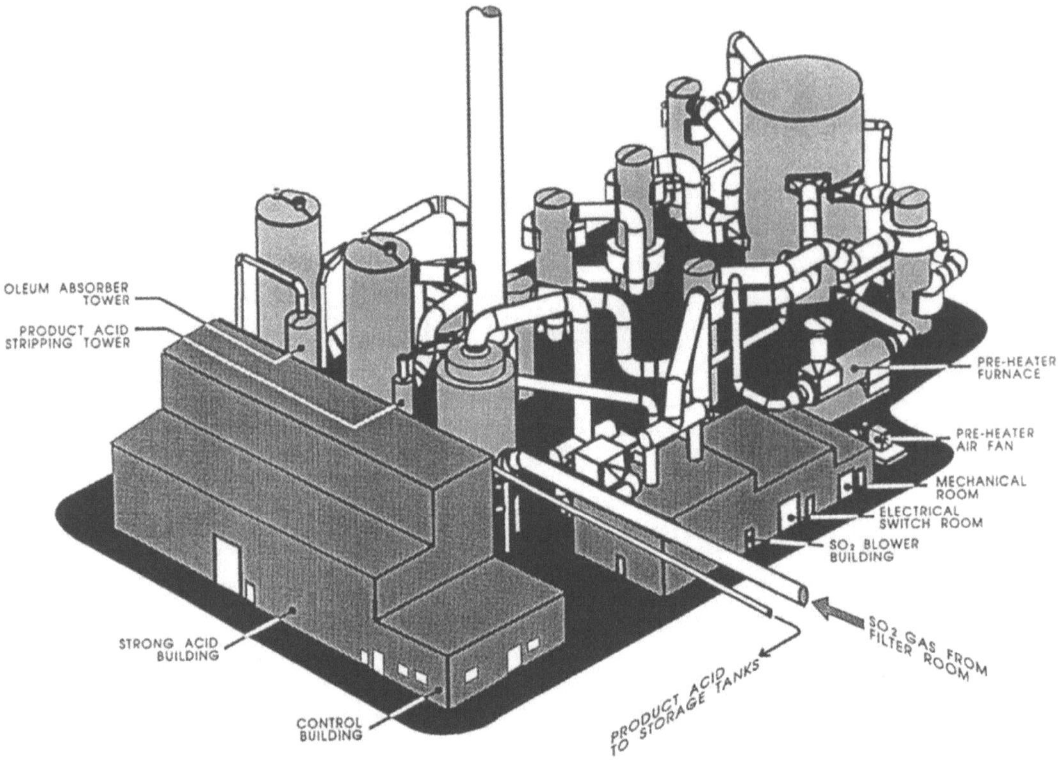

FIGURE 21.4. Inco Copper Cliff smelter—isometric view of acid plant.

imposed limits on sulfur dioxide emissions. To achieve these requirements, ground-level sulfur dioxide concentrations are measured every 5 minutes. Running averages of these readings are used to determine compliance with the government regulations. The government has imposed two different limits on emissions from the Copper Cliff smelter:

1. **The first limit is on emissions from the 381-m stack, either alone or in combination with other company stack emissions. The running hourly average of consecutive 5-min ground-level sulfur dioxide concentration readings caused by these emissions must not exceed 0.50 ppm.**
2. **The second limit concerns fugitive emissions from the complex. The running half-hour average of consecutive 5-min ground-level sulfur dioxide concentration readings caused by these emissions must not exceed 0.30 ppm.**

In addition to ground-level concentration limits, annual limits on sulfur dioxide emissions from all the company's Copper Cliff facilities were imposed by the government starting in the early 1970s. These limits have decreased in stages over the years, as described in Chapter 4. An annual sulfur dioxide emission limit of 265,000 tonnes from all of Inco's Sudbury operations was imposed effective for 1994. This represents a 90% reduction from the peak 1960 emission values.

Discussion

Given current technology and economics, the smelting of sulfide ore will inevitably result in the release of sulfur dioxide. Fixation of sulfur dioxide as liquid sulfur dioxide or sulfuric acid is a viable environmental protection strategy and can be economically favorable. It is argued that the sulfur content of these chemicals

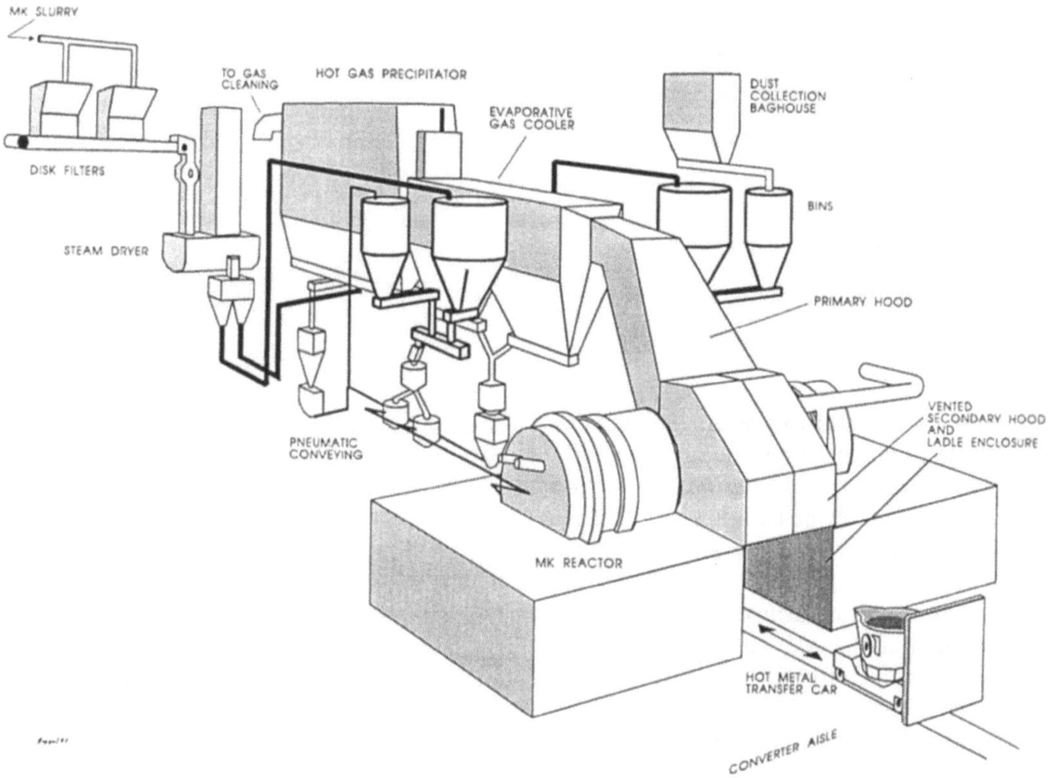

FIGURE 21.5. Inco Copper Cliff smelter—general arrangement of copper sulfide (MK) melter.

eventually ends up being discharged into the natural environment by consumers. However, the opportunity for pollution prevention is much greater with these solid forms than is possible with gaseous release as sulfur dioxide. One of the larger markets for sulfuric acid is for making phosphate fertilizer. Through the manufacture of fertilizers, the sulfur that previously became a pollutant, sulfur dioxide, can be used to produce a product that stimulates, rather than inhibits, plant growth.

Most other treatment methods involve some form of "scrubbing," which generates solid and/or liquid wastes. The amount of the secondary waste(s) that would be generated from a scrubbing creates a different set of pollution problems. Disposal of the wastes, such as a low-quality gypsum ($CaSO_4$), unsuitable for

TABLE 21.1. Energy and Productivity Gains from Implementation of the Sulfur Dioxide Emissions Abatement Program at Inco Limited Copper Cliff Smelter[a]

	Energy consumption (BTU/lb Ni+Cu)	Productivity lbs Ni+Cu per manshift	Acid Production (100% sulfuric acid basis) (tonnes per annum)
1980	15,500	1440	480,000
1989	10,000	2600	580,000
1994	4500	3030	720,000

[a]Assumes that liquid sulfur dioxide production remains at 80,000 tonnes/year.

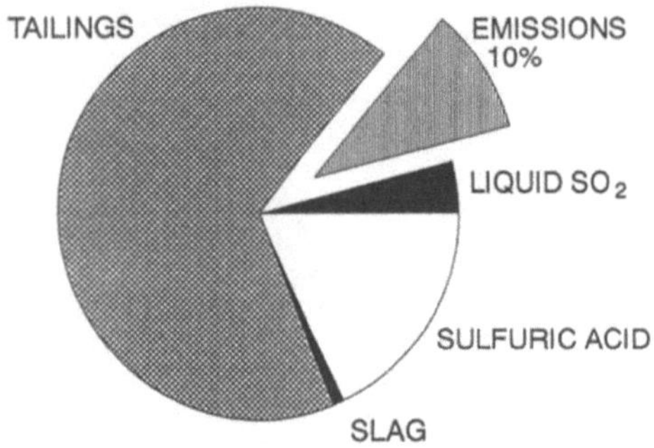

FIGURE 21.6. Inco Limited sulfur distribution—1994.

marketing, would add to present tailings disposal problems and result in the need for additional expensive disposal area. Rehabilitation of these additional disposal areas would then be required under current government legislation. While reducing the atmospheric problem, increased sulfur disposal to tailings can pose a storage and water treatment problem under current technology. Progress and remaining challenges in this field are the subjects of the next chapter (Chapter 22).

Recycling Metals: Converting a Problem into a Profit

Economic incentives can readily convert an environmental pollutant into a valuable resource. Inco Limited has become the world's largest recycler of automobile catalytic converters for the platinum and palladium content. This converts a potential environmental problem for one manufacturer, the automobile industry, into a valuable product for another company, the metal refining industry.

Most metallurgical smelters and refineries have implemented some type of recycling of scrap or waste materials produced both within the company's operations and from other producers. Substantial savings through reduced energy and processing costs can be realized by processing recycled materials while at the same time diverting waste materials from landfill sites and other such dumping locations. The recycling of aluminum cans is a prime example in which considerable economic benefits can be achieved in comparison with continued use of only the original raw materials.

Case Study Summary

Driven originally by government legislation and realizing the potential economic benefits to be derived, Inco Limited, the company in this case history, devised a program of emission control technologies and strategies designed to meet a wide variety of social, economic, and environmental needs. No single technology or strategy would have been effective in meeting these diverse needs. Only a well-integrated, multifaceted approach to the development of environmental control technologies and strategies could have achieved its high degree of success.

Inco Limited now contains 90% of the sulfur in mined ore, which is the reverse of the 1960s when 90% of the ore's sulfur was emitted to the atmosphere as sulfur dioxide. At yearend 1994, only 10% of the ore's sulfur content was emitted to the atmosphere as sulfur dioxide, a reduction of 1,735,000 tonnes/year in sulfur dioxide emissions (Fig. 21.6).

The company is now deriving the benefits of marketing its experience and technology internationally. Most applications of these technologies have taken place in the industrialized nations. Some applications of improved technology are occurring in developing nations such as Chile, where the national copper company has installed Inco's flash furnace technology.

Global Comparisons

The output of the nickel industry in Russia is similar to that of Inco Limited and Falconbridge Limited combined. Based on the referenced literature and the experiences of visiting scientists and managers, Russian facilities are at the stage of development in emission control technologies/strategies that Inco was in the early 1960s. The Norilsk Mining and Metallurgical Combine nickel facility has now become the world's largest point source of sulfur dioxide, with 1987 emissions estimated to be 2.4 million tonnes (Saunders 1990; Peterson 1993). The current economic uncertainty in Russia diminishes the likelihood that effective abatement technologies and strategies will soon be implemented. Economic and political problems in many of the former East Bloc countries also limit their ability to deal effectively with widespread environmental problems. Similar problems and fears exist in Asian countries, particularly China, where industrial development is occurring rapidly through the use of high-sulfur coal (Smil 1993). In China, approximately 900 million tonnes of coal per year is burnt to supply 75% of its energy needs. Only a small number of the most modern, large, coal-fired plants are equipped with electrostatic precipitators. Most smaller boilers have only mechanical dust separators. Millions of small sources are completely uncontrolled (Smil 1993).

Technology transfer to these needy nations is imperative if we are to meet the goals of sustainable development while reducing global environmental degradation from anthropogenic sources. Control systems, the norm in industrialized nations, must be implemented throughout the world to prevent further occurrences of the Sudbury experience.

Acknowledgments. I wish to thank Inco Ltd. Audiovisual Department, especially Aurel Courville, Charlie Hebert, Mike Barrette, and Gerald Sauve.

References

Air and Waste Management Association (AWMA). 1992. Air Pollution Control—Equipment, Inspection, and Maintenance, Fuels, Management. Papers from 85th Annual Meeting and Exhibition, Pittsburgh, PA.

International Union of Air and Pollution and Prevention Associations (IUAPPA). 1991. Clean Air around the World—National and International Approaches to Air Pollution. 2nd Ed. Brighton, England.

Lee, D.R. 1993. An economist's perspective on air pollution. Environ. Sci. Technol. 27(10):1980–1982.

Peterson, D.J. 1993. Troubled Lands—The Legacy of Soviet Environmental Destruction. A Land Research Study. Westview Press, Boulder, CO.

Saunders, A. 1990. Poisoning the arctic skies. Arctic Circle 1(2):22–31.

Smil, V. 1993. China's Environmental Crisis. M.E. Sharpe Inc., Armonk, NY.

Watson, S. 1992. Pollution prevention: policies and approaches, IU-28.01. *In* Air and Waste Management Association (ed.). Risk Assessment, Strategies and Pollution Prevention. Papers from the 9th World Clean Air Congress, Pittsburgh, PA.

22

Integrated Management and Progressive Rehabilitation of Industrial Lands

Ellen L. Heale

Member companies of the Mining Association of Canada and the Ontario Mining Association are committed to the concept of sustainable development. "Sustainable development requires balancing good project stewardship in the protection of human health and the natural environment with the need for economic growth" (Mining Association of Canada 1990; Ontario Mining Association 1993).

To meet the challenge of sustainable development and improve the level of environmental protection, it is necessary for companies to adopt several operating principles throughout exploration, mining, processing, and decommissioning activities. These basic principles include compliance with applicable legislation; applying cost-effective best management practices to minimize environmental risks; maintaining self-monitoring programs; supporting research to improve treatment technologies; expanding scientific knowledge of mining industry's impact on the environment; and assisting in the development of equitable, cost-effective, and realistic laws for environmental protection.

An integrated approach to decision making and management is essential to implement environmentally sustainable economic development (Lecuyer and Aitken 1987). Environmental decision making involves

1. the wise use of air, water, land, and energy
2. mitigating adverse environmental impacts arising from mining-related activities
3. safeguarding the health of people and the natural environment
4. recycling and reducing wastes
5. disposing of non-recyclable wastes in an environmentally sound manner
6. rehabilitating disturbed land to a safe, stable, and productive condition

In previous chapters, authors describe programs for the rehabilitation of tailings areas (see Chapters 9 and 10) and technological developments for the reduction of atmospheric emissions (see Chapter 21). In this chapter, integrated planning approaches, progress in achieving environmental improvements in air and water quality, and site remediations are presented. Most of the examples are drawn from work at Inco Limited's Sudbury area operations.

Improvements in Air and Water Quality

In compliance with provincial government regulations, Sudbury companies have significantly reduced sulfur dioxide emissions. Inco's sulfur dioxide abatement program (Inco Limited 1992) and Falconbridge Limited's smelter environmental improvement project (Falconbridge Limited 1992) represent the largest environmental projects ever undertaken by any mining companies. Since 1970, Falconbridge has reduced their sulfur dioxide output by more than 85% (Falconbridge Limited 1992). Inco's emissions reduction program will achieve an 87% reduction in sulfur dioxide emissions from 1972 to 1994. Of the sulfur in the ore, 90% will be contained and not emitted into the atmosphere (Inco Limited 1991).

As described in the previous chapter, the air pollution prevention approaches adopted in Sudbury have both environmental and economic benefits. Immediate improvements in air quality are some of the obvious environmental benefits (Dobrin and Potvin 1992), but as shown in other chapters in this book, the reduced emissions have also allowed for vegetation establishment (Allum and Dreisinger 1986), restoration of land (see Chapter 8), and biological recovery of lakes (Keller et al. 1992).

Water Management and Treatment

In addition to air quality improvements, both Inco and Falconbridge have developed extensive water management and treatment programs aimed at reducing the downstream impact of mining effluents, as well as reducing costs. Water conservation and recycling initiatives reduce the amount of fresh water that must be used for processing and, ultimately, the amount of process waste water requiring treatment before discharge. To reduce the amount of surface drainage that must be treated, considerable amounts of stormwater are diverted away from areas of potential contamination. Research is ongoing to determine the impact of revegetation on run-off water quantity and quality within industrially disturbed watersheds. With favorable research results, substantial long-term environmental and economic benefits are possible.

Inco's ore processing facilities at Copper Cliff operate entirely on recycled water decanted from the tailings area (Van Cruyningen and Puro 1987). Water levels in storage ponds and lakes, upstream of Inco's waste water treatment plants, are controlled to ensure that sufficient water is available for processing, yet capacity is available in the system to retain water that enters during storm events and spring freshet. This prevents process interruptions and avoids spills to the environment. Since 1992, the retention capacity upstream of the Copper Cliff waste water treatment plant has been increased to 5.26 billion liters.

Not all water can be retained on site, primarily due to the large influx of precipitation, which far exceeds the evaporation rate. Large volumes must be treated to satisfy government regulations before being released to the environment. Inco operates two waste water treatment plants in the Copper Cliff area, having a combined treatment capacity of 216,000 m^3/day. Reactor-clarifier technology is used, with hydrated lime and polymer as the reagents. According to an Ontario Ministry of the Environment and Energy report, this method has been identified as among the best available treatment technologies for base metal mine effluents. At smaller more-remote sites, batch lime addition is used, with clarification occurring in large retention ponds. Although much simpler, this method is equally effective.

Lime treatment is relatively inexpensive compared with other treatment methods. Lime ($Ca(OH)_2$) or calcium oxide (CaO) is added, at a controlled rate, to raise the pH of waste water for the purpose of precipitating dissolved metals. Precipitation of dissolved iron occurs at pH 7, whereas other metals such as zinc and nickel require pHs between 9 and 10.6. Sludges generated by this process tend to be voluminous and are not stable when exposed to rain and oxygen. Generally, recovery of metals from these sludges is currently not feasible, and disposal is expensive. At Copper Cliff, sludges are continually pumped from the waste water treatment plant to the tailings area. Remote batch lime facilities are periodically dredged and the sludge transported by truck to the Copper Cliff tailings area.

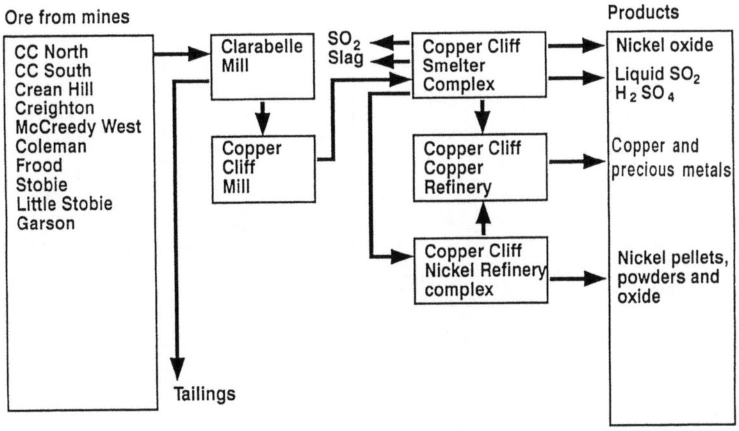

FIGURE 22.1. Inco Limited, Sudbury District operations flowsheet.

Many companies are required to reduce the alkalinity of their effluent after lime treatment and before discharge to the natural environment. Recently, carbon dioxide, rather than strong acids, has been used because it is less hazardous to handle and does not add chemicals to the treated water. There is less risk of overdosing with carbon dioxide, and a minimum pH of 5.6 is achievable. Falconbridge's Moose Lake facility incorporates carbon dioxide into the treatment process for neutralization of the final effluent to pH 7 (M. Wiseman, *personal communication*).

Although the requirement for pH adjustment is intended to reduce the toxicity of effluents, the desired effect is not always achieved. In tests performed on effluents of member companies, the Ontario Mining Association found that in some cases, toxicity actually increased after adjustment to pH 7.

Industrial Lands

Mining, milling, smelting, and refining operations are very large and visible operations in the Sudbury region. Privately owned industrial lands include tailings disposal areas, mine and plant sites, adjacent barren rocky outcrops, waste rock and slag piles, and sand pits. Integrated management of these lands is required due to the history, size, and scope of the impacted areas and because of the diversity of processes (Fig. 22.1), the variety of land uses, and the many interactions between environmental stresses.

Inco, a leading producer of nickel, copper, precious metals, and cobalt, has been operating since 1902. Falconbridge, a major producer of nickel, copper, and cobalt, has been in operation in Sudbury since 1928. Industrial operations are often complex, large, or spread out over a variety of terrains (Fig. 22.2). Other sites may be remote, with limited access or abandoned. Sites do not represent single confined processes or problems (Moore and Luoma 1990), and therefore management of these sites must reflect a wide variety of factors.

In addition to the plants and mines, byproducts of these operations are stored on industrial lands. For every 100 tonnes of ore that Inco currently mines, 90 tonnes is rejected as tailings waste. Inco produces 7.7–8 million tonnes of tailings each year. Approximately 25% of this material is used to fill mined-out areas underground. Falconbridge produces 1 million tonnes of tailings annually, and two-thirds of the tailings is pumped back underground. The balance of the tailings is disposed of in large tailings disposal areas (Fig. 22.3). Inco's Copper Cliff tailings area has a total storage capacity of 700 million tonnes and covers an area of 2225 h (Van Cruyningen and Puro 1987). Falconbridge stores 45 million tonnes of tailings in the Sudbury area.

Smelting produces 9 tonnes of slag (see Plate 13 following page 182) for every 100 tonnes of ore that is mined. Slag is a glasslike residue, primarily an iron silicate material. Falconbridge stores approximately 10 million tonnes of slag. More than

FIGURE 22.2. Inco Limited's Copper Cliff smelter complex. (Photo by Bob Chambers.)

FIGURE 22.3. Aerial view of tailings disposal area at Falconbridge. Treated waste water from the site has developed into a productive marsh area supporting various fish and wildlife species. (Photo by Ed Snucins.)

FIGURE 22.4. Framed by Inco's Frood-Stobie Mine complex, the 33.4 million-m^3 open pit was mined from 1938 to 1961. This is one of several pits that could be used for the disposal of acid-generating waste rock and tailings. (Photo by Bob Chambers.)

119 million tonnes of slag are stored on Inco property. The largest of four stockpiles is located at Copper Cliff, where more than 100 million tonnes of slag are stored in an area that covers 240 ha (see lower part of Fig. 22.2). Slag is slow-cooled on dumps and crushed or granulated. Some Inco slag is sold for road and rail ballast and domestic fill. Inco's 1993 rate of slag production was approximately 1.3 million tonnes/year.

Industrial sites also include areas for the storage of waste rock. Large quantities of waste rock are generated with the development of open pits, sinking of shafts, and mining of drifts (Fig. 22.4). It is estimated that Inco has 31 million tonnes of waste rock located in more than 65 piles. The largest storage pile contains 11 million tonnes of waste rock, spread to a maximum depth of 21 m. Falconbridge's total waste rock storage is in the order of 20 million tonnes.

Acid Rock Drainage

The large amounts of sulfur in Sudbury ores present unique environmental challenges, because for every 1 part of nickel, there are 8 parts of sulfur. Tailings and waste rock contain these sulfide minerals, and as the sulfides oxidize, in the presence of water and oxygen, acidic drainage is generated (Steffen Robertson and Kirsten 1989a,b). Traditional revegetation of tailings or waste rock piles will reduce the infiltration of precipitation or surface water. However, vegetation will not stop the acid generation process.

$$\underset{\text{metal sulfide}}{MS} + \underset{\text{oxygen}}{3O_2} + \underset{\text{water}}{2H_2O} \rightleftarrows \underset{\text{sulfuric acid}}{2H_2SO_4} + \underset{\text{dissolved metals}}{M^+}$$

Acidic drainage or seepage waters are characterized by low pH, elevated levels of dissolved solids, total acidity, trace elements, and inorganic compounds. Acidic seepage is also responsible for impacts on aquatic biota (Huckabee et al. 1975). Acid rock drainage is one of the most significant environmental and economic challenges facing the mining industry.

Both mining companies are active participants in Mine Environment Neutral Drainage, a cooperative research program sponsored by the Canadian mining industry and government agencies. Research and investigations are underway to develop reliable and affordable methods to prevent, monitor, control, and treat acid rock drainage (Skousen et al. 1987; Itzkovitch 1993) (see Chapter 10). From 1988 to 1992, contributions (from all participants) to the research program totalled $Can 8.1 million.

One solution to reducing the amount of sulfur released into the atmosphere during the smelting process is to ensure that much of the pyrrhotite in the ore (up to 30% sulfur) is rejected before the ore is smelted. However, storing pyrrhotite in the tailings area creates an additional source of acid generation.

To stop the oxidation process that leads to acid formation, pyrrhotite tailings are stored under water, or under a 30–100-cm cover of less reactive tailings slimes (Michelutti 1987), or at sites where the seepage can be collected and treated. Falconbridge is involved in research to develop oxygen barriers over tailings to control acid seepage effectively. Covers of partially decomposed domestic waste, digested sewage sludge combined with alkaline materials, or shallow water covers are being examined. Another approach that Falconbridge is investigating is the construction of a porous envelope around tailings disposal areas. An impervious cap on top can prevent ground- and surface water from entering the tailings whereas groundwater will flow around the tailings in the more porous envelope, minimizing its contact with tailings (Falconbridge Limited 1991). Treatment of acid seepage using wetlands to facilitate the development of natural alkaline-generating, sulfate-reducing microbial processes is being tested at Inco and Falconbridge (see Chapter 10).

To demonstrate more economically feasible methods for detecting acid rock drainage and identify areas requiring remediation, major projects were conducted in and around Inco property in 1993. It was determined that non-intrusive ground conductivity survey methods, using standard exploration equipment, were a useful tool in the environmental assessment of groundwater to detect acid rock drainage. Another project used an underwater probe, dragged by a boat, to detect sources of acid rock drainage discharging into surface waters. An investigation in the Copper Cliff tailings area involved piezocone technology. Measurements from an electronic piezocone, typically used to characterize soil stratigraphy and geotechnical parameters, will be correlated with chemical constituents from corresponding drilled samples to determine if a relationship exists (Itzkovitch 1993).

The production of a low-sulfur tailings is also currently being investigated by Inco. A new process in the milling circuit will produce a low-sulfur (0.4%) tailings to encapsulate and reduce the oxidation potential of the high-sulfur (15–20% sulfur) and pyrrhotite tailings stored in the center of Inco's current tailings disposal area. Low-sulfur tailings are a non-acid-generating waste.

Waste rock, containing sulfide minerals, is also a source of acidic drainage. As is practical, waste rock remains underground. It is used as backfill and for road bed construction. On surface, acid-generating waste rock is placed underwater in pits to stop the oxidation process. Inco has 23 pits, with a total volume capacity of 145 million tonnes. These pits are potential sites for the disposal of acid-generating waste rock (Fig. 22.4). When these alternatives are not feasible, seepage from waste rock piles is collected for treatment before to its release to the environment. Waste rock piles are currently being inventoried and assessed to determine acid-generating potential.

Progressive Rehabilitation

Mining companies in Ontario face additional environmental and economic challenges. Recent amendments to the Mining Act of Ontario have introduced legislation to set standards for the closure of mines and the rehabilitation of lands used for mining activities (Ontario Ministry of Northern Development and Mines 1992). Progressive rehabilitation and the development of closure plans are legislated to protect public health and safety, alleviate or eliminate environmental damage, and allow productive use of land in its original condition or an acceptable alternative.

Engineering and environmental studies required for the preparation of closure plans are underway by both companies. Inco is monitoring groundwater, conducting site inventories at active and abandoned mines, properties, and quarries, and sampling soil and vegetation. Closure plans and the estimated costs to implement the plans will be submitted to the Ontario Ministry of Northern Development and Mines for acceptance. Once approved, financial assurance will be negotiated to ensure economic resources for future work. Annual reports updating progressive rehabilitation and progress on studies will be required.

Progressive rehabilitation is essential to reduce the company's financial liability. Filling pits with acid-generating waste rock for underwater disposal, demolition of abandoned buildings, revegetation and restoration are current progressive rehabilitation activities. Perpetual monitoring and treatment of acidic seepage, for example, are not economically feasible or desirable. Companies may need to modify current operating practices. Also, industries cannot afford to wait until a facility closes to initiate rehabilitation activities.

The challenges for restoration of industrial lands include identifying and assessing environmental impacts (Bridges 1991), treatment, and rehabilitation costs. These factors must not be considered in isolation. Areas of impact, such as watersheds and neighboring lands, must be managed as an ecosystem unit.

At locations where mining operations are in close proximity to one another, an integrated approach to rehabilitation and closure is required. Inco's Levack Complex and Falconbridge's Onaping Complex, northwest of Sudbury, are within the same watershed management area. Joint technical planning groups have been established to implement a joint closure plan for this area. Integration of planning efforts within a company is also required. For example, Inco is currently developing an integrated watershed management plan for nine sites in the Copper Cliff area. For a detailed discussion of watershed or catchment planning, see Chapter 24.

Aerial Treatment Program

Surface run-off from the Nolin Creek watershed (850 ha), northeast of Copper Cliff, is currently being treated at one of Inco's waste water treatment plants. However, as shown in other parts of this book, the barren rocky outcrops in this area can be treated with surface applications of ground limestone to detoxify soil and allow plant growth (Winterhalder 1988) (see Chapter 8). Such treatments offer considerable promise for reducing containment export from metal-contaminated sites.

In 1980, Inco began a labor-intensive program on company-owned property to manually apply agricultural limestone, fertilizer, and grass seed to treat a few hectares of land in

Figure 22.5. Airplane being loaded with agricultural limestone. Between 1990 and 1994, Inco treated more than 650 ha of barren rocky outcrops with its innovative aerial treatment program. (Photo by Ellen Heale.)

Copper Cliff. These areas were virtually treeless, with sparse vegetation or shallow pockets or crevices of bare soil. By the mid-1980s, four-wheel-drive all-terrain vehicles, specially fitted with spreaders, treated 8–12 ha/year. In 1990, an innovative program to reclaim barren rocky outcrops began. Suitably equipped aircraft (Fig. 22.5), capable of carrying 1.5 tonnes of material, dropped agricultural limestone, followed by applications of a fertilizer and grass seed mix, over the stressed lands.

In 1990, an initial 50-ha site in the Nolin watershed was treated. With the establishment of successful vegetation covers (Fig. 22.6), the aerial treatment program was expanded. From 1990 to 1994, 650 ha of rocky outcrops have been treated. Productivity was increased in 1992 with the construction of a bulk lime-loading system for the aircraft (see Fig. 22.5). This aerial technique allows the company to treat relatively inaccessible areas in a safe, efficient, and cost-effective manner.

It is expected that revegetation will have a significant impact on the quantity and quality of surface run-off water and thus reduce the amount requiring treatment. With the application of agricultural limestone, a critical factor in the success of the revegetation program, natural colonizing species quickly become established. The grassed sites are subsequently planted with coniferous forestry seedlings.

Self-Sustaining Ecosystem Development

Restoration of mining land is necessary for many reasons, including responsible corporate business practice, surface stabilization, improving run-off water quality, aesthetic en-

FIGURE 22.6. Seven months after an aerial application of limestone, fertilizer, and grass seed, a vegetation cover has been established over rocky outcrops on Inco Limited property. Volunteer native tree seedlings were 15 cm high 3 months later. (Photo by Ellen Heale.)

hancement of the area, and compliance with legislation. These successful rehabilitation efforts lead to diversification of vegetation species, promotion of wildlife habitats, and overall ecosystem development (Australian Mining Industry Council 1987; Green and Slater 1987). This, in turn, leads to a reduction in further soil erosion and surface water run-off and the need for mitigation of the effects of metal contaminants in run-off waters. A variety of land use options is available (Powell 1992). Both Inco and Falconbridge have focused their efforts on the development of self-sustaining ecosystems with the promotion of wildlife habitats.

Inco's Copper Cliff tailings area has been designated as a wildlife management area (see Chapter 9). Efforts continue to diversify the number and types of vegetation species to supply suitable habitats and food sources for a wide variety of birds and wildlife. More than 86 different native and introduced species of vegetation have been identified on the tailings, including trees, shrubs, grasses, wetland species, legumes, field weeds, and mosses (Heale 1991). Over an 8-year period, 92 species of native and migratory birds have been identified on tailings (Peters 1984).

In less than 10 years, a 160-ha lifeless bog, adjacent to the Falconbridge smelter, was successfully converted into a productive wetland with a transient population of hundreds of birds, small fish, and mammals (Fig. 22.3). Treated alkaline waste water was directed into the bog to neutralize the existing acidity and reduce metal toxicity (Michelutti 1987). Cattails (*Typha latifolia*) and reedgrass (*Phragmites australis*) cover 90% of the area. Falconbridge has turned this productive marsh site into a wildlife sanctuary and conservation area. Fish and waterfowl thrive in a natural wildlife area in the settling pond of the Falconbridge tailings area. Red Pine Lake, 1 km from the smelter complex, provides a stopover point for migratory waterfowl and is a popular fishing spot. The lake also supports thousands of speckled trout (*Salvelinus fontinalis*). Falconbridge was a partner in the release of 45 Peregrine falcons in the Sudbury area from 1990 to 1993 (see Chapter 12).

Inco has initiated studies to identify potential sites for the development of aquatic ecosystems. Three abandoned sand pits, at various stages of rehabilitation, are being investigated for water chemistry, physical features, aquatic invertebrate species, and vegetation diversity. Work is ongoing, in conjunction with the Sudbury Game and Fish Protection Association, to develop walleye (*Stizostedion vitreum vitreum*) fingerling rearing ponds on company property. Habitat development is also being assisted with the construction and placement of artificial nesting islands for loons (*Gavia immer*) in area lakes and the installation of osprey (*Pandion haliaetus*) nesting platforms.

FIGURE 22.7. Environmental sampling is being conducted to characterize seepage from Inco's slag piles. (Photo by Inco Limited.)

Future Challenges

There have been significant improvements in water quality in and around mining operations (Ontario Mining Association 1993). However, water quality remains a focus for concern. The impacts of mining activities on water quality continue to be assessed. Methods to minimize risk, mitigate those impacts, or economically treat water must also be examined. The environmental impacts and economic realities of preventing, controlling, or treating acid rock drainage are only one set of challenges (see Chapter 10).

Sampling to assess the environmental impacts of slag is ongoing (Fig. 22.7). This includes identification and characterization of seepage from slag piles and leachate tests on different types of slag. A joint study between Laurentian University and Inco is examining the microbial leaching of slag.

Studies to assess the impacts of reclamation on surface run-off water quantity and quality are underway. Inco is in the first year (1993) of an intensive long-term study to establish a baseline by which to gauge future conditions in revegetated watersheds, compared with unreclaimed sites. Integrated management of watersheds involves multidisciplinary expertise (see Chapter 24). Work will include mapping watershed boundaries, water sampling, acute toxicity testing, and soil and vegetation identification, mapping, sampling, and analyses.

Research is also needed to develop a cost-effective method of removing ammonia from effluent streams. Groundwater investigations will continue to examine the hydrogeology and geochemistry of Inco's tailings area. This involves the installation of boreholes and monitoring wells and surface electromagnetic and borehole conductivity mapping. Additional research, technology development, and monitoring needs for the integrated management and progressive rehabilitation of industrial lands are outlined (Box 22.1).

The cooperative efforts of government, industry, academic institutions, and private citizens have resulted in significant progress and success in the restoration of the Sudbury region. Future environmental improvements depend on continuing cooperation, innovative technology and techniques, and multidisciplinary research (Moore and Luoma 1990). Inco and Falconbridge have formalized their commitment to sustainable development and decommissioning within corporate environmental policies.

To meet future challenges, environmental and economic decision making and integrated management will ensure continued progressive rehabilitation of industrial lands and ecosystem development in the Sudbury region.

Acknowledgments. I thank Mark Wiseman of Falconbridge Limited and Dr. Tom Peters, re-

Box 22.1.

Research, technology development, and monitoring programs are key requirements for environmental protection and the ongoing rehabilitation of industrial lands and watersheds in the Sudbury area. Although not prioritized, the following list represents some of the environmental challenges and research needs facing the mining industry, not only in Sudbury but elsewhere in Canada.

To ensure sustainable development of the mining industry, proposed solutions must be reliable, practical, and affordable.

1. **methods to prevent and control acid rock drainage**
2. **characterization of large waste rock piles**
3. **rehabilitation techniques for waste materials, including tailings disposal areas, waste rock, and slag piles**
4. **water use, re-use, treatment, and conservation management, including the impact of naturally occurring wetlands on industrial effluent**
5. **impact of industrial activities on groundwater movement, quality, and quantity and methods for mitigation**
6. **decommissioning of abandoned sites**
7. **impact and fate of trace metals in the environment**
8. **potential for biotechnology and genetic adaptation to enhance rehabilitation and restoration techniques**
9. **further initiatives for emission reductions and metals recovery**
10. **recycling and waste reduction**
11. **potential for ecosystem recovery with no direct treatment**
12. **maintenance and monitoring of perpetual treatment systems**

tired from Inco Limited, for providing information for this chapter. The assistance of Inco personnel Dr. Larry Banbury, Brian Bell, Carolyn Hunt, Marty Puro, and Paul Yearwood is also appreciated.

References

Allum, J.A.E., and B.R. Dreisinger. 1986. Remote sensing of vegetation change near Inco's Sudbury mining complexes. Int. J. Remote Sensing 8:399–416.

Australian Mining Industry Council. 1987. Mining and the Return of the Living Environment. Canberra, Australia.

Bridges, E.M. 1991. An evaluation of surveys of soil contamination in the city of Swansea, South Wales, pp. 40–49. *In* M.C.R. Davies (ed.). Land Reclamation: An End to Dereliction? Elsevier Applied Science, Oxford.

Dobrin, D.J., and R. Potvin. 1992. Air Quality Monitoring Studies in the Sudbury Area 1978 to 1988. Technical Report. Ontario Ministry of Environment, Toronto, Canada.

Falconbridge Limited. 1991. Falconbridge Limited and the Environment. Falcon Nov./Dec. 1991 Special Edition. Sudbury, Canada.

Falconbridge Limited. 1992. Falconbridge Limited Report on the Environment. Toronto, Canada.

Green, J.E., and R.E. Slater. 1987. Methods for Reclamation of Wildlife Habitat in the Canadian Prairie Provinces. Prepared for Environment Canada and Alberta Recreation, Parks and Wildlife Foundation by the Delta Environmental Management Group Ltd., Edmonton, Canada.

Heale, E. L. 1991. Reclamation of tailings and stressed lands at the Sudbury, Ontario operations of Inco Limited, pp. 529–541. *In* Proceedings of the Second International Conference on the Abatement of Acidic Drainage, Montreal, Canada. Vol. 2. MEND, Ottawa.

Huckabee, J.W., C.P. Goodyear, and R.D. Jones. 1975. Acid rock in the Great Smokies: unanticipated impact on aquatic biota of road construction in the regions of sulphide mineralization. Trans. Am. Fish. Soc. 104:677–684.

Inco Limited. 1991. Inco Limited Ontario Division Publication. Copper Cliff, Canada.

Inco Limited. 1992. Inco Limited 1992 Annual Report. Toronto, Canada.

Itzkovitch, I.J. 1993. Mine Environment Neutral Drainage Program 1992 Annual Report. Toronto, Canada.

Keller, W., J.M. Gunn, and N.D. Yan. 1992. Evidence of biological recovery in acid-stressed lakes near Sudbury, Canada. Environ. Pollut. 78: 79–85.

Lecuyer, G., and W.R.O. Aitken. 1987. Report of the National Task Force on Environment and Economy. Submitted to the Canadian Council of

Resource and Environment Ministers. Downsview, Canada.

Michelutti, R.E. 1987. Reclamation programs and research at Falconbridge Limited's Sudbury operations, pp. 1–10. *In* P.J. Beckett (ed.). Proceedings of the 12th Annual Meeting of the Canadian Land Reclamation Association, Sudbury, Canada. CLRA, Guelph, Ontario.

Mining Association of Canada. 1990. Guide for Environmental Practice. Ottawa, Canada.

Moore, J.N., and S.N. Luoma. 1990. Hazardous wastes from large-scale metal extraction. Environ. Sci. Technol. 24:1278–1285.

Ontario Mining Association. 1993. Sustainable Mining in Ontario. Environment Committee Report. Ontario Mining Association, Toronto, Canada.

Ontario Ministry of Northern Development and Mines. 1992. Rehabilitation of Mines Guidelines for Proponents. Version 1.2. Sudbury, Canada.

Peters, T.H. 1984. Rehabilitation of mine tailings: a case of complete ecosystem reconstruction and revegetation of industrially stressed lands in the Sudbury area, Ontario, Canada, pp. 403–421. *In* P.J. Sheehan et al. (eds.). Effects of Pollutants at the Ecosystem Level. Wiley, New York.

Powell, J.L. 1992. Revegetation options, pp. 49–91. *In* L.R. Hossner (ed.). Reclamation of Surface-Mined Lands. Vol. 2. CRC Press, Boca Raton, FL.

Skousen, J.G., J.C. Sencindiver, and R.M. Smith. 1987. A Review of Procedures for Surface Mining and Reclamation in Areas with Acid-Producing Materials. West Virginia Surface Mine Drainage Task Force, University Energy and Water Research Center and Mining and Reclamation Association, Morgantown, WV.

Steffen Robertson and Kirsten (B.C.) Inc. 1989a. Draft Acid Rock Drainage Technical Guide. Vol. 1. British Columbia Acid Mine Drainage Task Force, Vancouver, Canada.

Steffen Robertson and Kirsten (B.C.) Inc. 1989b. Acid Rock Drainage Draft Technical Guide. Vol. 2—Summary Guide. British Columbia Acid Mine Drainage Task Force, Vancouver, Canada.

Van Cruyningen, J.P., and M.J. Puro. 1987. Tailings disposal area development at Inco Sudbury operations. *In* Proceedings of the Conference of Canadian Mineral Processors, Ottawa, Canada.

Winterhalder, K. 1988. Trigger factors initiating natural revegetation processes on barren, acid, metal-toxic soils near Sudbury, Ontario smelters, pp. 118–124. *In* U.S. Department of the Interior Circular 9184. Mine Drainage and Surface Mine Reclamation Conference, Pittsburgh, PA.

23

Remote Sensing and Geographic Information Systems: Technologies for Mapping and Monitoring Environmental Health

J. Roger Pitblado and E. Ann Gallie

> The face of our land looks to the sky. To see its many features, we must get above it and look down.
>
> *(Dill 1958)*

Henry Dill's statement was made in the context of using air photographs to evaluate changes in agricultural land use. Three and a half decades later, his "bird's-eye view" is even more relevant as specialized cameras, electronic scanners, satellites, and computer technology have been added to the arsenal of tools that help us gather, map, and monitor the characteristics of earth resources (Rudd 1974; Harper 1976; Richards 1986; Star 1991). In this chapter, we illustrate how some of these tools, *remote sensing* and *geographic information systems* (GIS), are being used to assist in the rehabilitation of the Sudbury region.

Environmental monitoring is fundamental to the complex task of natural resource management. A well-designed monitoring program should enable (Yan and Keller 1991)

- measurement of natural variation in the structure and dynamics of ecosystems
- identification of sites with special attributes
- exploration of patterns at various scales
- detection of changes that are gradual or cyclic in nature
- recognition and evaluation of unusual events
- improved capability to predict the ecological outcome of future events

But monitoring is neither a simple nor an inexpensive endeavor, especially of processes such as the acidification and recovery of the Sudbury region, which affect large areas over long periods of time. It is here that remote sensing and GIS have a unique and important role to play.

Remote sensors, especially satellite sensors, produce images that cover large areas for a few cents per square kilometer, and most important, the same coverage is repeated month after month. If information relevant to the monitoring program can be extracted from these images, then remote sensing can offer a full two-dimensional view that cannot be matched by point sampling. GIS offers other advantages. Developed from computer-assisted cartography, GIS has many tools designed to produce maps from point data, to analyze existing maps, and to explore the spatial patterns within and between maps. Moreover, GIS enables us to combine and transform useful but static maps into dynamic scenarios of our resources. In combination, GIS and remote sensing complement and enrich ground-based sampling programs, contributing to the cost-effective achievement of monitoring goals.

Box 23.1. Landsat 6 Disaster!!

In the fall of 1993, remote sensing scientists throughout the world were shocked to hear the news that the $200 million (US) Landsat 6 had failed to reach orbit after launch and probably plunged to the ocean. Many had been counting on this satellite, which included new instrumentation, to supply applications data until the year 2000. Fortunately, satellite data for resources monitoring continues to be received from Landsat 5 as well as French, joint European, Russian, Indian, and Japanese satellite programs. As well, Canada's RADARSAT program is expected to be fully operational by 1995 and promises to provide all-weather coverage of exceptional value for geology, oceanography, sea-ice monitoring, flood monitoring, agriculture, and forestry.

Remote Sensing

Remote sensing is the measurement or acquisition of information by a recording device that is not in physical contact with the object under study. This includes computer or digital images taken from satellites, photographs taken from airplanes, and color measurements taken with hand-held sensors. In the Sudbury region, several remote sensing studies have been carried out, most of which have used digital images taken from sensors mounted on the Landsat series of satellites.

Since the Landsat program began in 1972, there have been five Landsat satellites (Box 23.1). All five carried a recording instrument called the multispectral scanner (MSS). An improved scanner, the thematic mapper (TM), was added to the last two satellites (Landsat 4 1982; Landsat 5 1984). These instruments measure the earth's reflected radiation in specific and relatively narrow bands of the electromagnetic spectrum (Table 23.1). Also, TM is capable of sensing emitted or thermal radiation from the earth.

The Landsat sensors collect continuous readings over a 185-km-wide swath along the orbital track of the satellites. The smallest area on the ground for which measurements can be made is called a pixel, short for "picture element." An MSS pixel covers an area that is 79 m by 79 m, slightly larger than 0.6 h. For TM, the pixel size is 30 m by 30 m (0.1 ha). Each image is made up of a set of four (MSS) or seven (TM) bands, with each band composed of a grid of several million pixels.

In theory, every spot on earth is imaged at 16–18-day intervals, although this varies depending on the satellites in operation and cloud cover. The repetitive coverage is considered to be one of the great advantages of satellite resource monitoring.

The key to interpreting remotely sensed data is held in the concept that earth materials reflect the sun's radiation differently, making

Table 23.1. Landsat Spectral Bandwidths for the Multispectral and Thematic Mapper Scanners

Multispectral scanner (Landsat 1, 2, 3, 4, 5)		Thematic mapper (Landsat 4, 5)	
Band no.	Channel wavelengths[a]	Band no.	Channel wavelengths[a]
1	0.5–0.6 (green/yellow)	1	0.45–0.52 (blue)
2	0.6–0.7 (red)	2	0.52–0.60 (green/yellow)
3	0.7–0.8 (near-infrared)	3	0.63–0.69 (red)
4	0.8–1.1 (near-infrared)	4	0.76–0.90 (near infrared)
		5	1.55–1.75 (mid-infrared)
		6	10.4–12.5 (thermal IR)
		7	2.08–2.35 (mid-infrared)

[a]Wavelength bounds expressed in μm.

each appear in unique but typical colors. Thus, materials are said to have a characteristic spectral signature or spectral response pattern (Swain and Davis 1978; Richards 1986; Campbell 1987). In reality, different materials may sometimes look alike, or the same material may display a range of spectral responses. However, the spectral response measured by a satellite sensor usually can be interpreted in terms of the materials being viewed. For example, deep clear water is most reflective in the blue wavelengths, but as sediment is added to the water, the peak response shifts to the green. Vegetation, which humans see as green, in fact is most reflective in the near-infrared (NIR), and any decrease at this wavelength is usually interpreted as a sign of plant stress.

Remote Sensing Applications

Sudbury Vegetation Surveys

The damaged area surrounding Sudbury is of sufficient size that satellite imagery provides the only convenient and affordable means of monitoring it. One of the earliest studies was undertaken using MSS imagery, with the goal of mapping zones of anthropogenic influence in the Sudbury region (Pitblado and Amiro 1982). This project accompanied an extensive field program (Amiro and Courtin 1981) in which many vegetation plots were described to provide baseline data for future vegetation monitoring.

For this study, a map of the vegetation index (*VI*) was prepared from a late-summer MSS image using the second MSS NIR band (band 4) and the red (band 2).

$$VI = \frac{NIR - Red}{NIR + Red}$$

Vegetation indices (of which VI is but one version) have long been reported in the remote sensing literature (Tucker 1979). These indices are highly correlated with the density of vegetation canopy cover, biomass, leaf area, and certain seasonal vegetation characteristics. Dense vigorous vegetation gives high VI values because of its strong NIR reflection. Water, bare soil or rock, clouds, and shadows give low VI values because these materials reflect almost equally at red and NIR wavelengths.

For the image analyzed, the lowest VI values were centered around the smelter sites where the ground was barren or only a few stunted trees survived, and progressively higher VI values were found with distance from the smelters. Two such maps are shown in Figure 23.1 (1973) and Figure 23.2 (1986). The unvegetated runways of the Sudbury airport show as a dark gray cross just south of Wanapitei Lake. Dark gray barren areas surround the Falconbridge smelting complex near the airport, the Coniston smelter (closed in 1972) east of Ramsey Lake, and the Copper Cliff smelting complex west of Ramsey Lake along the left edge of the image.

A simple visual comparison of these two maps shows that the vegetation has increased dramatically over the 1973–1986 period (i.e., far less-dark, low biomass areas in 1986). Research is currently in progress to quantify the changes by revisiting the original field sites (Winterhalder and Sinclair, *personal communication*) and relating actual biomass to the MSS-derived VI (Courtin and Beckett, *personal communication*). These efforts will enable us to provide quantitative estimates of change over the entire region and to monitor the spatial patterns of change. Without remote sensing, such analyses would be impossible at the scale required.

Vegetation change mapping has also been undertaken by Inco Limited to monitor the effects of emission reductions and reclamation efforts (Allum and Dreisinger 1986). The method used allows only qualitative estimates of vegetation change. In this example, change was recorded for periods as short as 3 years. The company concluded that Landsat imagery provides a cost-effective and reliable means of monitoring vegetation change over large areas.

Lake Water Quality Surveys

Lake water sampling is expensive, especially of many lakes scattered over a large area with limited accessibility. Remote sensing has the capability to measure at least some surface water parameters and thus can be used to extend and complement traditional sampling

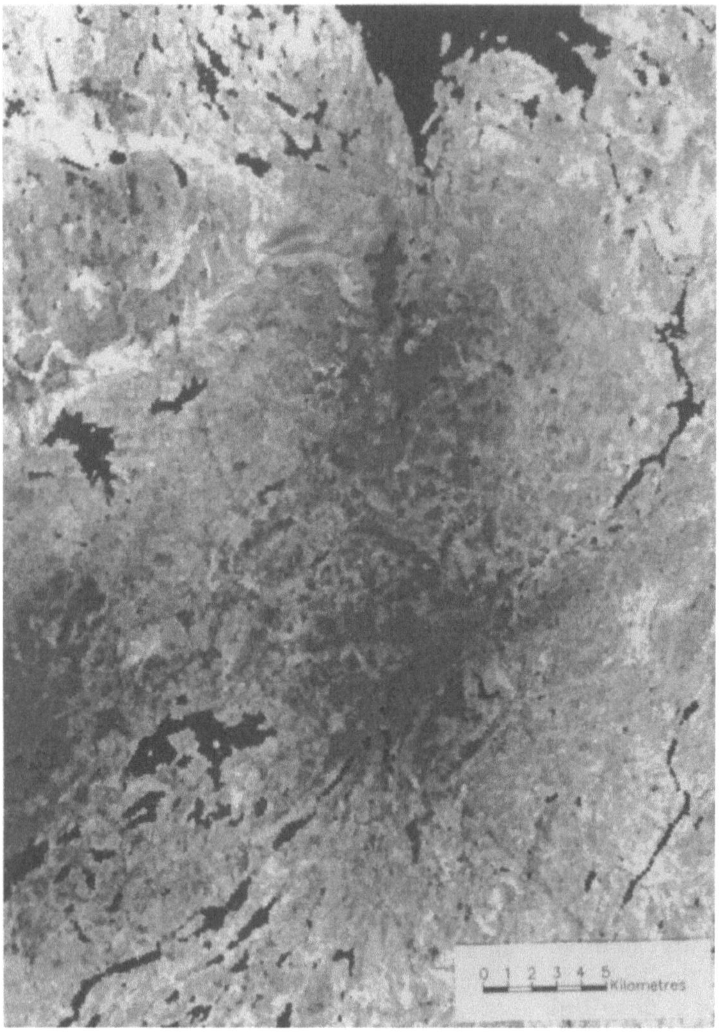

FIGURE 23.1. Relative biomass (vegetation index) image derived from a Landsat 1 multispectral scanner scene taken July 13, 1973. Wanapitei Lake is located at the top center of the image. Ramsey Lake is one-third up from the bottom and one-quarter in from the left edge of the image. Lighter tones represent higher biomass values.

programs. A few studies have focused on the feasibility of remotely measuring acidity (Hardy and Jefferies 1981; Moniteq 1982). However, only one study has been undertaken in the Sudbury region to map acidic and non-acidic lakes (Pitblado 1992a,b).

In this study, 227 lakes throughout northeastern Ontario (Fig. 23.3) were used to calibrate relationships between Landsat TM radiance measurements and concentrations of dissolved organic carbon (DOC) collected from field surveys. With the exception of bog or marsh waters, acidic lakes have very low DOC concentrations (Yan 1983), possibly due to aluminum, which may cause DOC to precipitate out of the water column (Effler et al. 1985). Because DOC is the dominant coloring agent in northern lakes, acidic lakes are very clear or transparent. In turn, this affects the radiance viewed by satellite sensors, especially blue radiance viewed by TM band 1.

For the 227 lakes, the DOC of 70% of the lakes was predicted within ±1 mg/L, whereas 90% were within ±2 mg/L (Fig. 23.4). This is relevant to mapping pH because of the strong link between pH and DOC for anthropogenically acidified lakes. The pH-DOC relationship is similar, although weaker, when a wider selection of lakes is included. Thus, DOC can be used as a surrogate for mapping pH, especially

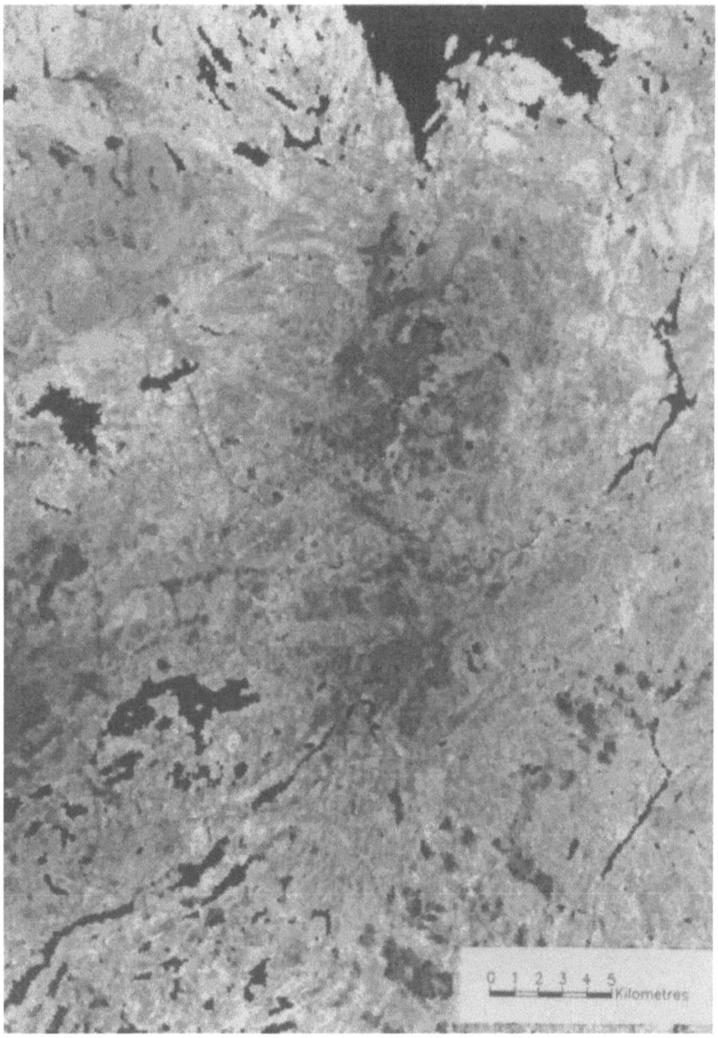

Figure 23.2. Relative biomass (vegetation index) image derived from a Landsat 5 multispectral scanner scene taken August 13, 1986. Vegetation recovery is suggested here by the increase in areas of high biomass values (lighter tones) compared with the 1973 image. Dark patches that look like lakes in the lower right section but that do not appear on the 1973 image are clouds and cloud shadows.

in the greater Sudbury area. When this approach was applied to the lakes north of Wanapitei Lake, every one of the known acidic lakes (pH <5.6) was identified, and it seems reasonable to expect that the other lakes so identified have also been mapped correctly.

To meet the goals of monitoring, however, one must be able to measure whether water quality is changing with time. This is a more challenging problem for remote sensing because differences in the atmosphere between images may mask changes in a lake. Nine images of Bowland Lake (located 70 km north of Sudbury) from 1973 to 1986 were used to look for small changes in DOC due to liming (Pitblado 1992a,b). The images provided tantalizing evidence of an increase, but the data were not statistically significant. The actual change in DOC over the short time period, 1982–1985, was only 0.5 mg/L (Molot et al. 1990), and it would appear that larger differences are required before MSS or TM can detect them.

Geographic Information Systems

Multifaceted and multifunctional, a GIS is a computerized database management system for the capture, storage, retrieval, analysis, and

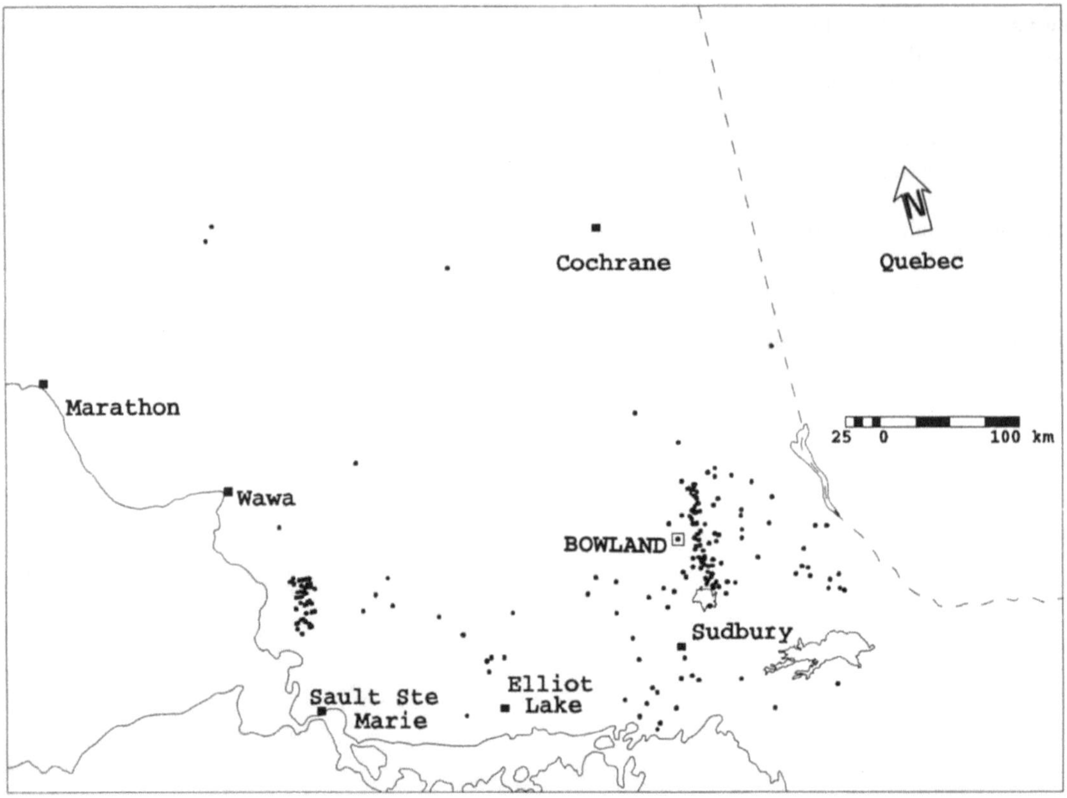

FIGURE 23.3. Location of 227 lakes used to calibrate Landsat thematic mapper radiance measurements and concentrations of dissolved organic carbon. As well, Bowland Lake was used to assess thematic mapper capability of detecting temporal changes in water quality (pre- and postneutralization with lime).

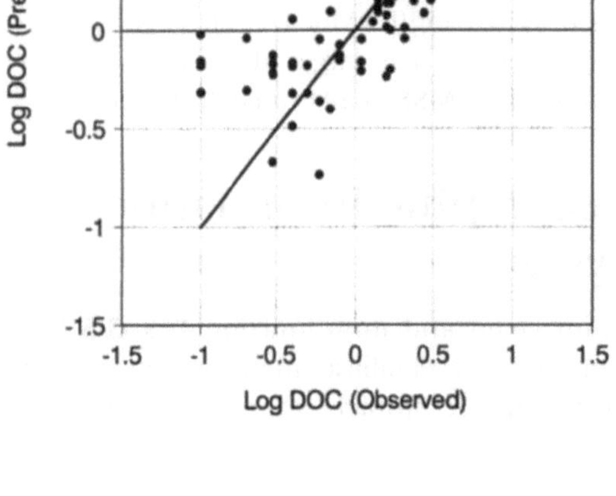

FIGURE 23.4. Highly correlated dissolved organic carbon measurements. Observed measurements (transformed here to logarithms of the original mg/L values) were obtained using traditional field survey techniques, and the predicted measurements were computed using Landsat thematic mapper data.

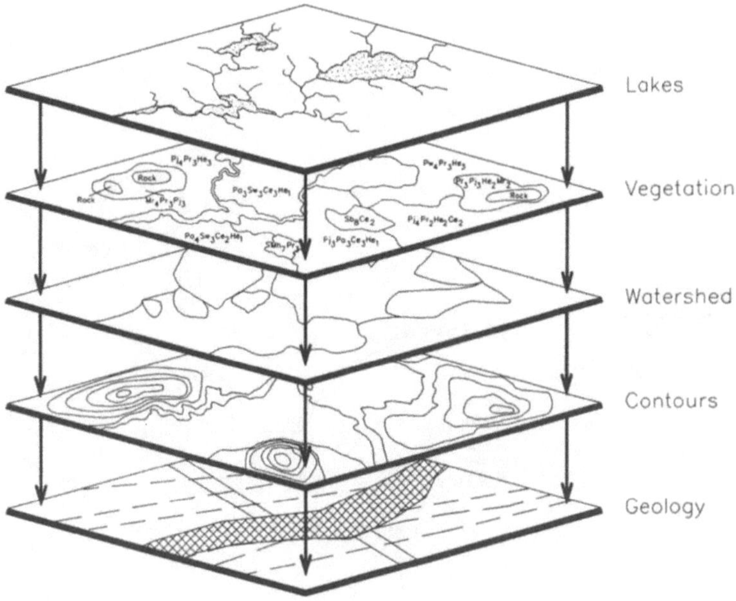

FIGURE 23.5. Schematic representation of thematic layers used in an overlay analysis using a geographic information system.

display of spatial or mapped data (Burrough 1986; Aronoff 1989). Evolving from computer mapping programs written in the 1960s, GIS programs transform paper maps into a digital or computer form, calculate areas, lengths, and perimeters, and facilitate the process of comparing two or more maps using overlay analyses. Thus, the integration of information from several single-discipline maps can be performed more quickly and reproducibly (Fig. 23.5).

The capabilities of GIS programs rapidly progressed beyond simple measurement and overlay techniques, however. Users soon realized that entirely new map layers could be developed from existing information. For instance, the elevation contour data found on topographic maps can be used to create maps of slope and aspect. In turn, these and other GIS parameters can be used in ecological or physical models that lead to a better understanding of ecosystem function or the consequences of land use planning decisions.

GIS Applications

Sudbury Lake Acidification Zone

In many of the previous chapters, a map of the Sudbury lake acidification zone has appeared. This map was created using GIS and illustrates overlay analysis and the capability to create maps from point data stored in a database (Neary et al. 1990; *personal communication*).

Between 1979 and 1988, water chemistry was sampled from more than 2000 lakes in central Ontario by the Ministry of the Environment and the Ministry of Natural Resources. Data were stored in an extensive lakes database and used to prepare two maps. The first is a map of sulfate concentration (Fig. 23.6). The isolines (lines joining points of equal value) were found by applying the value of sulfate concentration from each lake sample station to a circular area (30-km radius) surrounding the station. Where the circles from two or more stations overlapped, the value was based on a weighted average, with the weighting of adjacent points declining exponentially with distance.

The second map was prepared showing the ratio of sulfate concentration to sulfate plus alkalinity. This ratio approaches zero when sulfate concentrations are low or when carbonate-rich rocks maintain high alkalinity and approaches 1.0 when acidification has reduced the alkalinity of lakes (Fig. 23.7).

The zone adversely affected by the Sudbury smelters was defined as the area where sulfate

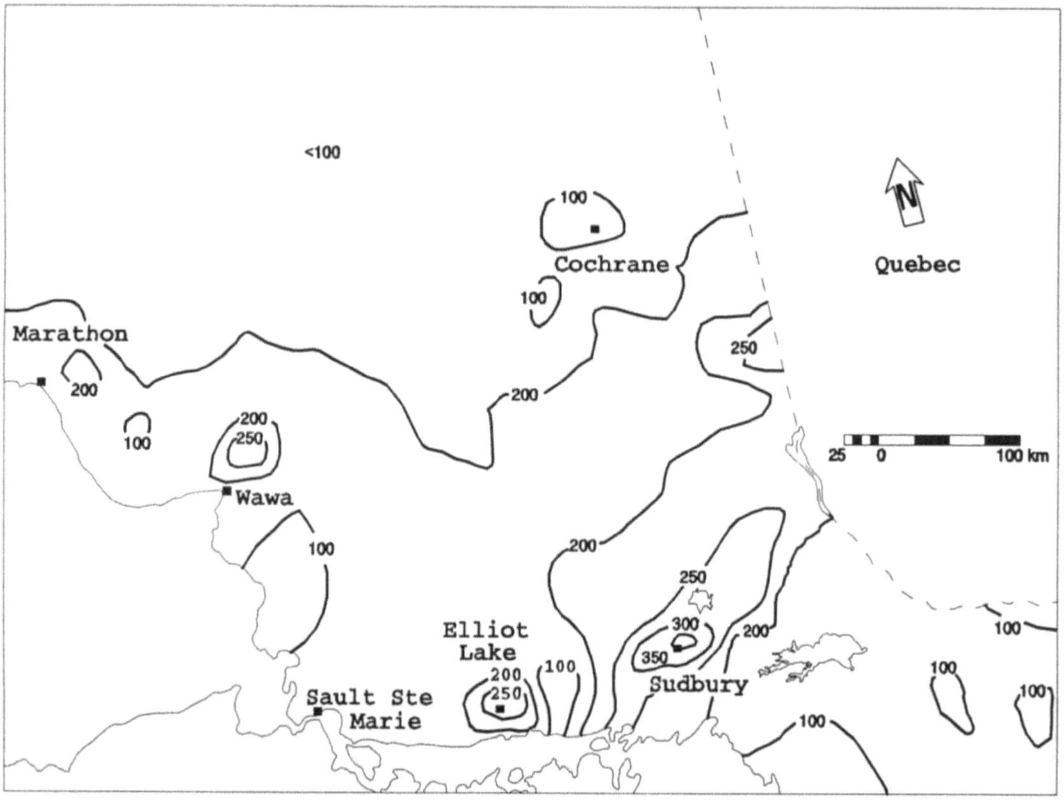

FIGURE 23.6. Lake surface water sulfate concentration. All contour values are in µeq/L SO_4 (from Neary et al. 1990).

concentration exceeded 200 µg/L and the sulfate/(sulfate + alkalinity) ratio exceeded 0.7. The intersection identifies zones affected by local sulfur emissions as opposed to long-range transport of sulfur. The Sudbury Lake Acidification Zone stands out, calculated to cover approximately 17,000 km^2 (Fig. 23.8).

GIS was essential to this study. Traditionally, detailed data have been stored in files or, more recently, in databases. To see how the information varied locally or regionally, either many maps had to be prepared or the information had to be generalized. Because the process of plotting hundreds of points and then contouring them was costly, the spatial variation of data was seldom studied. GIS allows isoline maps to be prepared rapidly and provides new and advantageous techniques for doing so. Also, specialized maps can be produced as required using many combinations of variables in the database.

Daisy Lake Watershed Liming Study

Daisy Lake is a metal-contaminated acidified lake (pH 4.9) about 3 km from the former Coniston smelting complex. Two small subwatersheds have been intensively studied for several years in preparation for a watershed liming experiment. GIS has helped with the visualization and understanding of the data. For example, soil pH, mapped from point samples, shows a distinct spatial pattern (see Plate 14 following page 182). Low pH values in the northeast are easy to explain because this area is closest to the former smelter. The reason for the remaining pattern becomes clear when the pH map is draped over the topography. Low pH is found on exposed slopes facing the Coniston complex, slopes that

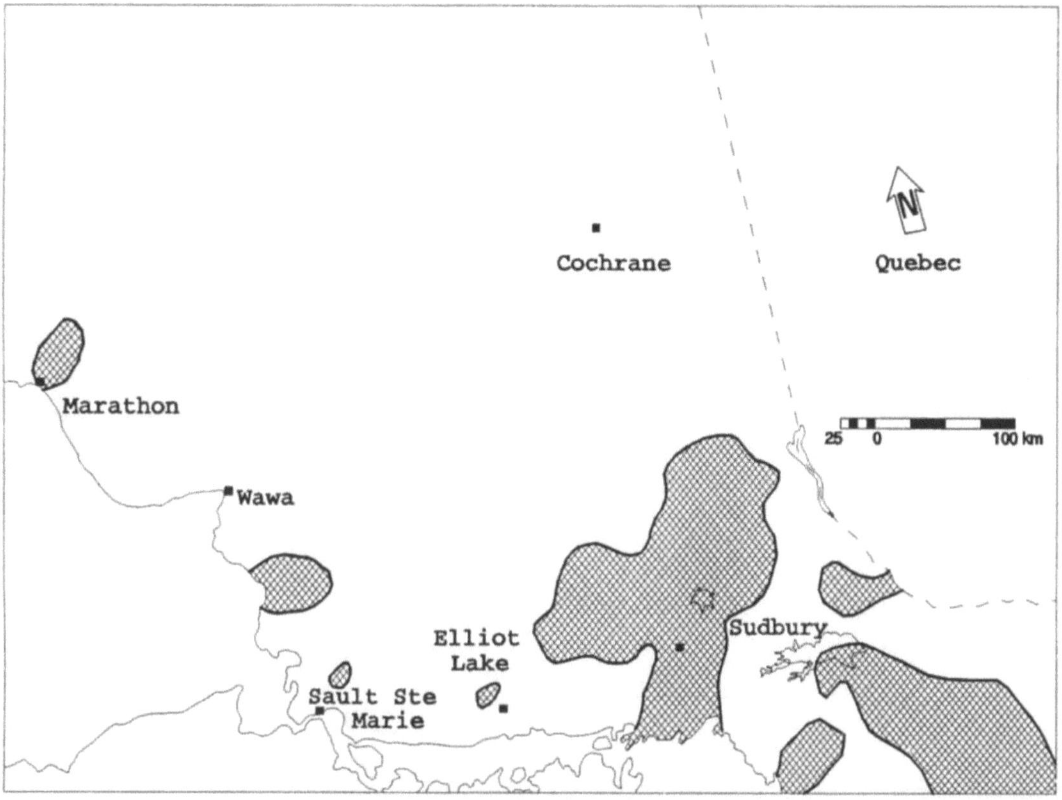

FIGURE 23.7. Zones where lakes have sulfate/(sulfate+alkalinity) ratios greater than 0.7. The zone around Sudbury is primarily due to local smelter effects. The other zones are probably due to long-range transport of sulfur compounds (from Neary et al. 1990).

were fumigated by clouds of sulphur gases drifting down the Daisy Lake valley. Highest pH is found in areas most protected from northeast winds (see Plate 17 following page 182).

Remote Sensing and GIS Integration

In the remote sensing examples discussed earlier, analyses were undertaken without the help of GIS. Similarly, in the GIS applications, all the information was derived from ground surveys. As the two fields mature, the common spatial view of the world that they share has lead to increasing fusion of the technologies.

Counts and Measures

For small-scale (large area) investigations, an excellent example of remote sensing/GIS integration is the work recently completed by Hélie et al. (1993). Using 129 Landsat TM scenes for most of eastern Canada (south of 52° north latitude from the Manitoba-Ontario border to the Atlantic provinces), 881,634 water bodies were counted and measured in terms of size. GIS facilities enabled the researchers to associate these water bodies with major drainage basins and with digitized maps of ecodistricts interpreted as areas having a low, moderate, or high potential to reduce acidity (see Chapter 1). Seventy-three percent of these water bodies were found to lie in areas having low potential.

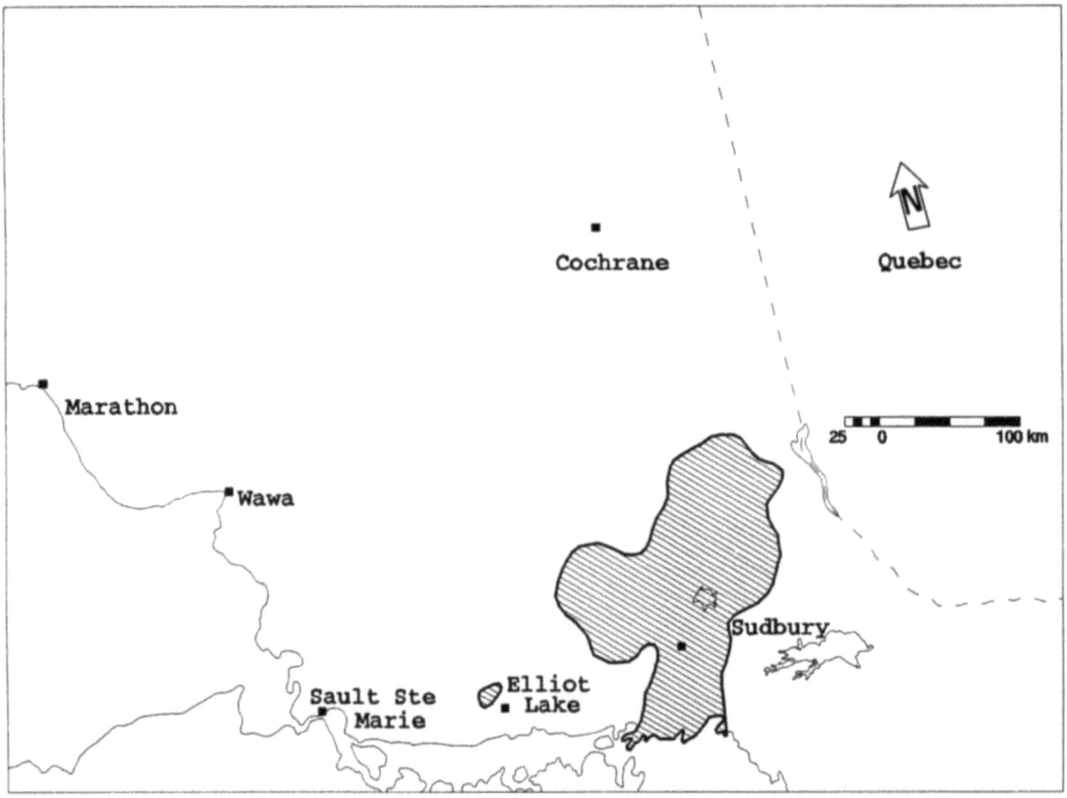

FIGURE 23.8. Lake acidification zones. The small area near Elliot Lake is based on data from very few lakes and may be altered with more lake data (from Neary et al. 1990).

Hélie et al. (1993) correctly argued that "counts and measures data are an integral part of ongoing modelling efforts to describe and predict the effects of anthropogenically induced acidification on surface water." The integration of remote sensing and GIS tools permitted them to address three critical characteristics of the surface water bodies in eastern Canada: extent, size, and the numbers deemed to be at risk due to acidification.

Wanapitei/Algoma Dissolved Organic Carbon Study

At a much larger scale, similar integration of remote sensing/GIS tools is being performed in northeastern Ontario. The remote sensing lakes monitoring study described earlier in this chapter found a correlation between water color as measured by TM and lake DOC concentration and pH. This study is being expanded to include GIS with two goals in mind: to develop our understanding of watershed-water quality processes; and to improve the ability to interpret water color and hence water quality monitoring capabilities.

There are two study sites: the Wanapitei area north of Sudbury, which has been affected by acid rain, and the Algoma area north of Sault Ste. Marie, which is relatively pristine. GIS databases have been developed for both sites that include information from many different sources. Field surveys have provided water quality information for about 150 lakes in each area. Existing map data have been digitized, giving coarse descriptions of geology, surficial deposits, and soils. Topographic maps have been used to develop digital terrain models of the landscapes, and from these, secondary layers have been developed including

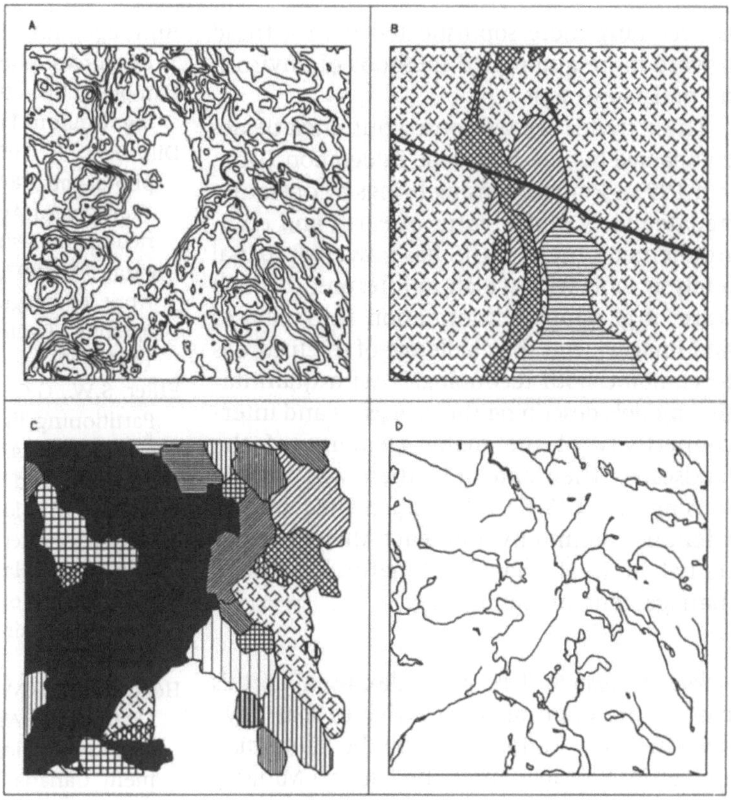

Figure 23.9. Selected thematic maps from the geographic information system layers of the Wanapitei area Dissolved Organic Carbon Study. (**A**) elevation contours; (**B**) geology; (**C**) watersheds; (**D**) lakes and streams.

slope, aspect, and watershed boundaries. Vegetation cover has been classified using TM imagery, which also provides water color measurements for all lakes. Some of the GIS layers that are being used in the Wanapitei study area are illustrated as separate maps in Figure 23.9.

The next step is to address the interactions between these landscape components. Most of the colored DOC in a lake comes from the surrounding watershed and wetlands. Anthropogenic acidity removes this DOC from the water, making the lakes clear. But some lakes are naturally clear, and their transparency does not mean that they are acidic. The intent is to develop GIS models describing how DOC moves from a watershed into a lake. By combining such models with TM measures of color, it should be possible to monitor pH and DOC even more accurately.

The Wanapitei/Algoma study represents a significant step of combining remote sensing and GIS through the development of an integrated view of the landscape. This is an important move in the progress of these complementary technologies for monitoring environmental health.

Future Developments and Applications

The examples of remote sensing and GIS for mapping and monitoring environmental health that have been outlined in this chapter represent a minute fraction of the capabilities of these technologies. We have only dealt with one sensor program, Landsat. But in this and many other space and airborne scanning programs, rapid progress is being made to enhance our ability to characterize and discriminate earth features, particularly by increasing the spatial, spectral, temporal, and radiometric resolution of the scanners. At the same time, image processing and GIS software is becoming more sophisticated with a trend toward more automated methods of environmental assessment.

These trends are welcomed but must be accompanied by true landscape integration (Dobson 1993). Remote sensing brings a constantly updated view of the status of earth and environmental processes. GIS has powerful spatial analysis tools with which to interpret the vast store of geographic information in databases and imagery. But the challenge of the future is to combine both technologies with quantitative models describing the processes and interconnectivity of the many elements of the landscape. Then and only then will remote sensing and GIS be fully appreciated by the scientific community and reach their true potential in local and global natural resource management and monitoring.

Acknowledgments. The work discussed in this chapter draws on research activities that have been funded at various times by the Laurentian University Research Fund, the Ontario Ministry of the Environment, the Ontario Ministry of Northern Development and Mines (Environmental Youth Corps), the Northern Ontario Heritage Fund, and the Natural Science and Engineering Research Council. We thank Léo Larivière (Department of Geography, Laurentian University) and Michael Courtin (Remote Sensing/GIS Laboratory, Laurentian University Elliot Lake Field Station) for their assistance with the illustrations and John Fortesque (Ontario Geological Surveys) for reviewing the manuscript.

References

Allum, J.A.E., and B.R. Dreisinger. 1986. Remote sensing of vegetation change near Inco's Sudbury mining complexes. Int. J. Remote Sens. 8: 399–416.

Amiro, B.D., and G.M. Courtin. 1981. Patterns of vegetation in the vicinity of an industrially disturbed ecosystem, Sudbury, Ontario. Can. J. Bot. 59:1623–1639.

Aronoff, S. 1989. Geographic Information Systems: A Management Perspective. WDL Publications, Ottawa.

Burrough, P.A. 1986. Principles of GIS for Land Resources Assessment. Clarendon Press, Oxford.

Campbell, J.B. 1987. Introduction to Remote Sensing. Guilford Press, New York.

Dill, H.W., Jr. 1958. Information on land from airphotos, pp. 381–384. *In* A. Stefferud (ed.). Land, Yearbook of Agriculture. U.S. Department of Agriculture, Washington, DC.

Dobson, J.E. 1993. Commentary: a conceptual framework for integrating remote sensing, GIS, and geography. Photogramm. Eng. Remote Sens. 59: 1491–1496.

Effler, S.W., G.C. Schafran, and C.T. Driscoll. 1985. Partitioning light attenuation in an acidic lake. Can. J. Fish. Aquat. Sci. 42:1707–1711.

Hardy, N.E., and W.C. Jefferies. 1981. Chromaticity analysis of color aerial photography and its application to detection of water quality changes in acid-stressed lakes. Can. J. Remote Sens. 7:4–23.

Harper, D. 1976. Eye in the Sky Introduction to Remote Sensing. Multiscience Publications Ltd., Quebec.

Hélie, R.G., G.M. Wickware, and M. Sioh. 1993. Quantitative Assessment of Surface Water at Risk Due to Acidification in Eastern Canada. Environment Canada. Canada Communication Group, Ottawa.

Molot, L.A., P.J. Dillon, and G.M. Booth. 1990. Whole–lake and nearshore water chemistry in Bowland Lake before and after treatment with $CaCO_3$. Can. J. Fish. Aquat. Sci. 47:412–421.

Moniteq. 1982. Evaluation of Historical Landsat Data on Water Reflectance for Acidic Precipitation Applications in the Sudbury Area. Monitoring Environmental Quality, Ltd., Concord, Ontario.

Neary, B.P., P.J. Dillon, J.R. Munro, and B.J. Clark. 1990. The Acidification of Ontario Lakes An Assessment of Their Sensitivity and Current Status with Respect to Biological Damage. Technical Report. Ontario Ministry of the Environment, Dorset, Ontario.

Pitblado, J.R. 1992a. The Mapping of Lake Surface Water Characteristics of Northeastern Ontario Using Satellite Imagery. Research Advisory Committee Project 354G. Research and Technology Branch, Ontario Ministry of the Environment, Toronto.

Pitblado, J.R. 1992b. Landsat views of Sudbury (Canada) area acidic and non–acidic lakes. Can. J. Fish. Aquat. Sci. 49(Suppl. 1):33–39.

Pitblado, J.R., and B.D. Amiro. 1982. Landsat mapping of the industrially disturbed vegetation communities of Sudbury, Canada. Can. J. Remote Sens. 8:17–28.

Richards, J.A. 1986. Remote Sensing Digital Image Analysis. Springer-Verlag, New York.

Rudd, R.D. 1974. Remote Sensing A Better View. Duxbury Press, North Scituate, MA.

Star, J.L. (ed.). 1991. The Integration of Remote Sensing and Geographic Information Systems. American Society for Photogrammetry and Remote Sensing, Bethesda, MD.

Swain, P.H., and S.M. Davis (eds.). 1978. Remote Sensing: The Quantitative Approach. McGraw-Hill, New York.

Tucker, C.J. 1979. Red and photographic infra–red linear combinations for monitoring vegetation. Remote Sens. Environ. 8:127–150.

Yan, N.D. 1983. Effects of changes in pH on transparency and thermal regimes of Lohi Lake, near Sudbury, Ontario. Can. J. Fish. Aquat. Sci. 40: 621–626.

Yan, N.D., and W. Keller. 1991. The value of spatial and temporal reference sites: responses of crustacean zooplankton to changes in habitat acidity, pp. 82–86. *In* N.D. Yan (ed.). Natural Resources: Riches or Remnants. Proceedings of the Canadian Society of Environmental Biologists. CSEB North York, Ontario.

24

Catchment Management in the Industrial Landscape

Peter J. Dillon and Hayla E. Evans

A catchment can be defined as the area that encompasses a particular aquatic environment (e.g., a lake or a stream) including the land that drains into it. In Sudbury, a catchment can vary from a few hectares of area drained by a small temporary stream to the thousands of square kilometres drained by major rivers in the area.

Catchments can be considered as individual ecosystems with their own sets of biological, physical, and chemical inter-relationships; however, it must be recognized that catchments are linked to each other through processes including material transport that occur in the atmosphere and by hydrologic transport from one catchment to another. Catchments are therefore very important ecological units, but the concept of the catchment is also a very tangible and appealing one from the human perspective. For example, municipal planners have made increasing use of the catchment (or watershed) as the appropriate level for environmental planning (RCFTW 1992). The sense that the drainage water leaving a catchment integrates the inputs to the area, as well as the human activities within the area, creates a real sense of "place" for many people and thus a recognition of the need for responsible management planning. Some suggestions and challenges for providing effective management of catchment are the focus of this chapter.

Integrated Approach to Catchment Management

Managing catchments as ecosystems requires that the effort be integrated at various levels or through various processes. First, the efforts of government, industry, academia, and non-government organizations must be integrated effectively. This includes establishing agreed-on goals for the management of the ecosystem and requires the integration of responsibilities and expertise from a wide cross section of organizations and, in many cases, from within an organization (Fig. 24.1). For example, within government, integration must occur at all jurisdiction levels. That is, it is essential that the municipal, the provincial (or state), and the federal government be aware of what every other level is doing and not be working at cross purposes or duplicating efforts.

In the Sudbury region, an excellent example of a successful multidisciplinary group is the Vegetation Enhancement Technical Advisory Committee. This committee, established in 1974, consists of members from the local mining industries, various government agencies, elected politicians, the academic community, public interest groups, and individual citizens. It has successfully developed, initiated, and carried out many projects aimed at improving highly visible barren land in and

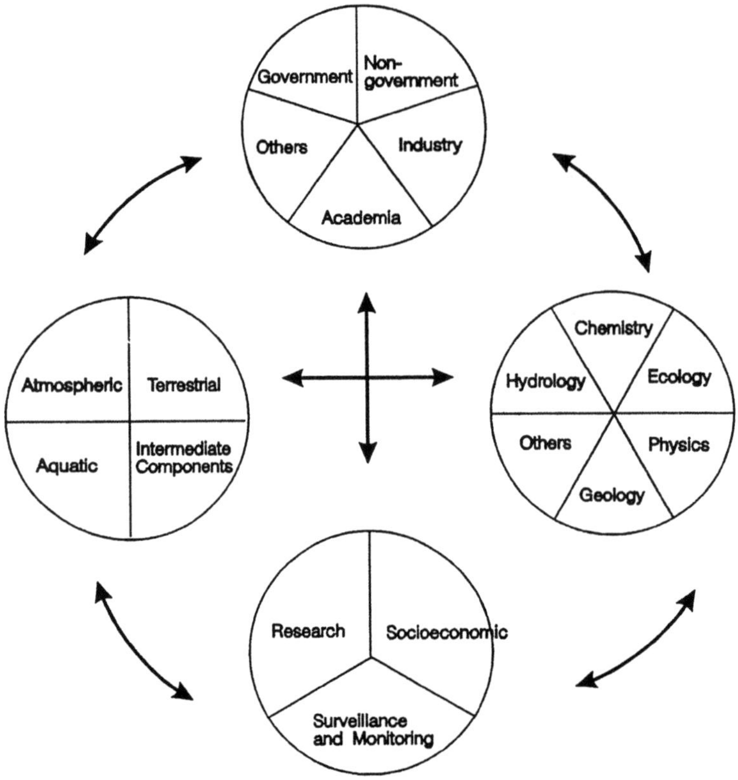

Figure 24.1. Levels of integration required for catchment management.

around the environs of Sudbury (see Chapter 8).

In a similar manner, scientists and managers must integrate the disciplines needed to examine the various components of the ecosystem (Fig. 24.1). Chemists, biologists, ecologists, hydrologists, atmospheric scientists, and others must work together. This means, too, that methodologies must be consistent among the various scientists and disciplines. Most important, the main subcomponents of the catchment ecosystem, the aquatic, terrestrial, and atmospheric components, must all be considered. Furthermore, it is necessary to study not only the terrestrial, the atmospheric, and the aquatic portions individually but also any interactions among them. For example, although sulfate is deposited from the atmosphere onto both the terrestrial and the aquatic portions of catchments, sulfate falling on forests or soils or wetland areas may be stored there for an extended period and ultimately enter a lake or stream (i.e., as groundwater or run-off) as a consequence of changes in other environmental factors. The questions then become how, when, and in what form.

It is also important when integrating the measurements on the physical and biological components of the catchment not to overlook areas of the ecosystem. In the Sudbury region, for example, very little work has been done on streams. Wetlands and the littoral zone of lakes are other areas that are sometimes under-represented in sampling and management strategies, but they are important because they link the aquatic and terrestrial environments. This lack of attention implies that these areas are less important as sources or sinks of pollutants or as refuges or dispersal routes for organisms. However, quite the contrary is true (Dillon and LaZerte 1992; Devito and Dillon 1993).

Surveillance and monitoring (i.e., assessment of the status of the ecosystem) and the research efforts of scientists also must be integrated, not only with respect to each other but also with respect to socioeconomic considerations. The establishment of surveillance and monitoring programs is critical to establish "baseline" conditions and to determine natural variation in the ecosystem. Only then can anthropogenic disturbances be readily detected. Although one does not always need these scientific data to demonstrate that an environment is degraded (i.e., it may be intuitive, as is the case in the Sudbury region), survey and monitoring data are necessary to identify less severely damaged areas at a stage at which abatement programs may be more effective. For example, it was only by routine monitoring of precipitation in the Muskoka/Haliburton area of Ontario (250 km south of Sudbury) in the mid-1970s, that the effects of long-range transport of sulfur dioxide on remote ecosystems became apparent (Dillon et al. 1978).

The data from monitoring studies also provide the foundation on which scientific research, including the important core component—predictive models, are built. These research activities allow scientists to predict the effects on the ecosystem of a reduction or an increase in the level of the stressor. The accuracy of these predictions can be tested only if monitoring studies are continued, even after the stress is entirely eliminated. Thus, the integration of monitoring and surveillance programs with scientific research is not only logical but essential. Unfortunately, monitoring is an activity that governments and academia are reluctant to undertake because of the amount of time and money involved. Furthermore, for many scientists, simply "describing the extent of the damage" (or lack thereof) is considered to be boring (Haukioja 1993). In other words, there is little interest in characterizing pristine ecosystems or in monitoring what is often a very gradual response to long-term changes in stressor levels. Thus, monitoring and surveillance programs often may not be initiated early enough to establish the initial conditions of the ecosystem(s).

The integration of scientific endeavors (monitoring, surveillance, and research) with socioeconomic considerations is perhaps the most important area of integration in terms of overall catchment management. This marriage of science and society is essential to produce sound management policies a priori and/or, in situations in which the stress has already occurred (such as in Sudbury), sound rehabilitation or remediation practices. The return of highly damaged areas to conditions that even closely resemble "historical" conditions (i.e., restoration) is often not possible (Moore and Luoma 1990). Thus, it is important for society to be able to assess the benefits (e.g., economic, health, environmental) to be achieved through reducing the magnitude of the stressor before deciding how much stress they are willing to let an ecosystem tolerate.

Environmental Protection Model for Managing Catchments

Recently, Somers et al. (1994) developed a conceptual model of the environmental protection process (Fig. 24.2) that shows the inter-relationship between monitoring/surveillance activities (collectively termed *assessment* in their model), scientific research, policy development, and reporting. This model is directly applicable to management of industrial areas such as the Sudbury region and emphasizes the need for integration at many different levels. Step 1 involves reporting on the state of the ecosystem. This requires sound and extensive data that have been collected through monitoring activities (temporal patterns) and surveys (spatial patterns). The response of the public, non-government agencies, governments, industry, etc., (step 2) can then be used to set or revise the ecosystem goals or objectives (step 3) (i.e., what society views as acceptable conditions). These are then used to judge if the current state of the ecosystem is acceptable or unacceptable (step 4). If the ecosystem is acceptable, then the "assessment" or monitoring/surveillance loop is followed. Ongoing characterization of the ecosystem through monitoring programs allows for the establishment of baseline conditions (step 5). However, because it is impossible to monitor all components of the ecosystem, chemical, physical, and/or biological indicators of ecosystem health are selected (step 6). This usually requires some sort

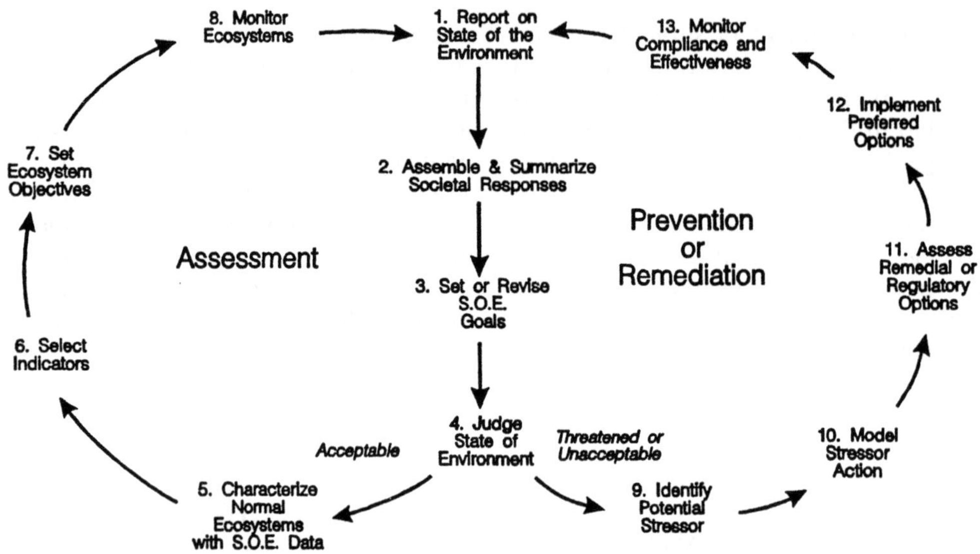

FIGURE 24.2. Essential activities inherent in the environmental protection process (from Somers et al. 1994).

of scientific research. Then, by using these indicators together with an idea of what the normal ecosystem should look like, scientifically based objectives that satisfy the goals established in step 3 can be developed (step 7). However, as mentioned earlier, continued monitoring and surveillance is required (step 8) to ensure that the ecosystem goals and objectives are met.

However, if in step 4 the state of the environment is deemed to be unacceptable, then the prevention/remediation loop is followed. First, the stressor must be identified (step 9), usually through scientific research efforts. Once the stressor is identified, its origin, fate, and mode of action must be evaluated and modeled (step 10) so that the effect of the stressor on the ecosystem indicators can be predicted. This allows for the assessment of remedial or regulatory actions (step 11), followed by implementation of the preferred options (step 12), which usually includes a reduction in the levels of the stressors, although in the worst scenario, it may include only ecosystem manipulation designed to counter the effects of the stressor without reducing the magnitude of the stressor. Then, as in the assessment loop, monitoring programs must be pursued to assess the compliance and effectiveness of the prevention/rehabilitation strategy. The prevention loop is identical to the rehabilitation loop except for the fact that prevention is initiated before the ecosystem damage.

In summary, the integration of monitoring and research data with socioeconomic factors means that although scientists can collect data, test hypotheses, and make recommendations, it is the responsibility of many people, often, most important, the public, to determine the direction that management policies will take. An excellent example is provided by the city of Sudbury, which recently used an "ecosystem approach" to lake management to develop an award-winning community improvement plan for Ramsey Lake (see Chapter 25).

Specific Management Issues in the Sudbury Region

As shown in previous chapters in this book, management of the Sudbury region as a whole generally provides a test of the integrated approach to catchment management

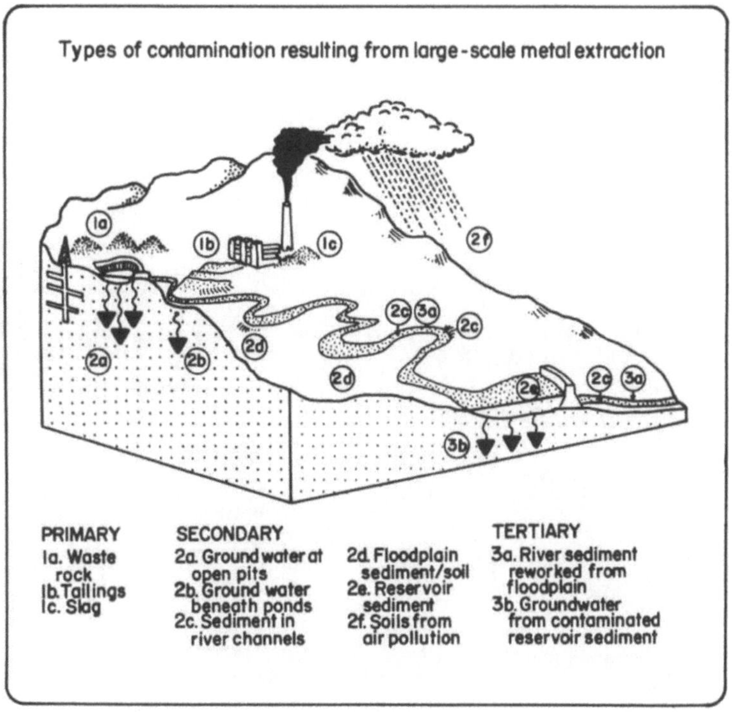

FIGURE 24.3. Contamination pathways resulting from large-scale metal extraction (from Moore and Luoma 1990).

(see Fig. 24.1). First, the ability to manage and the responsibility for management of the area has rested with and continues to rest with many organizations including government, industry, academia, and non-government organizations, but an informed public is increasingly becoming the force that ensures that these agencies honor these responsibilities. Second, a very significant body of scientific data, representing most compartments of the ecosystem, has been collected by hundreds of scientists over the past 30 years, beginning with the first published studies by Gorham and Gordon (1960). Finally, these data have provided the foundation for sound management practices and for the development of restoration techniques relevant to the regional problems.

Difficulties in Applying the Integrated Approach

Despite this potential for an integrated approach to catchment management, serious challenges have and still are faced by those attempting to set policies for the Sudbury region. The reasons for the difficulties stem from the fact that pollution often has no political, geographical, social, or temporal boundaries.

Secondary and Tertiary Pollution

First, as illustrated in Figure 24.3, there is the problem of secondary and tertiary pollution. This means that one cannot only look at one ecosystem (catchment) in isolation but also must study those downstream and downwind of that catchment. Most lakes in regions that have been glaciated have an outflow. Consequently, the presence, for example, of elevated metals in a lake means that potentially all the lakes downstream of it may become contaminated with metals (Dillon et al. 1982). This is illustrated by the fact that large areas of metal-contaminated sediments have been documented at the mouths of rivers that drain the Sudbury area (Fitchko 1978; Spanish Harbour Rap Team 1993), as is the case in other metal smelting areas of the world (Baumann 1984; Moore and Luoma 1990). Similarly, although the completion of the 341-m Superstack in 1972 alleviated the unacceptable air quality conditions in Sudbury, it increased the area

FIGURE 24.4. La Cloche Mountain lakes located approximately 40–60 km south of Sudbury were the first lakes in Canada where the damaging effects of acidification were identified (Ontario Ministry of Natural Resources photograph).

affected by high sulfate and metal (copper and nickel, primarily) deposition. It has been estimated that as much as 97% of the sulfur emitted from the Superstack is carried farther than 60 km from Copper Cliff and, thus, contributes to more widespread problems of acid deposition (Hutchinson 1982). Thus, what was perceived to be an acceptable solution to the local problem may have far-ranging consequences (Fig. 24.4).

Time Scale

Another consideration in managing areas such as Sudbury is one of time scale. From a management perspective, this is an issue because both short-term and long-term remediation efforts must be considered and the relative effort to be placed on each must be assessed. Thus, efforts spent on rehabilitation techniques such as lake liming (see Chapter 15) (Dillon et al. 1979; Molot et al. 1986, 1990) or fertilization (Yan and LaFrance 1984), fish stocking, and watershed liming and fertilization, which are only short-term solutions to the long-term problem (Havas et al. 1984), are not a substitute for reducing the levels of the stressors. These must be weighed against the efforts invested in establishing the acceptable sulfur and metal emission and deposition rates.

From a scientific perspective, time presents a problem with respect to determining long-term or sublethal effects of the stress on the ecosystem. Although the extirpation of fish species such as the aurora trout (see Chapter 11), the loss of entire fish populations (Beamish and Harvey 1972; Beamish 1974), and the denudation of the landscape (see Chapter 2) are obvious and direct lethal effects of high acid and metal deposition, long-term or chronic effects on biota are not so readily discernible. For example, the copper concentrations in many of the Sudbury area lakes typically exceed 2 μg/L (Keller et al. 1992) (i.e., are greater than the provincial water quality objective). Although the copper levels in the worst cases are high enough to be acutely toxic to some biota, in many cases the effects may be chronic and therefore less readily observable.

Time presents a complication when managing systems for another reason. It is known that the capacity for the terrestrial and wetlands portion of catchments to "store" strong acid (in the form of reduced sulfur compounds) and trace metals is high (Urban et al. 1989; Dillon and LaZerte 1992). However, evidence has shown (Bayley et al. 1987; RMCC 1990) that terrestrial areas may release these compounds for periods of years to decades. For example, in Plastic Lake, Ontario, the pH and acid neutralizing capacity (ANC) of the water decreased gradually from 1979 to 1985, despite the fact that atmospheric deposition of sulfate and strong acid decreased sharply during that same period (Dillon et al. 1987). The reason for this is that the wetlands in the catchment continued to supply sulfate to the water despite the reductions in the sulfate deposition. A similar situation has been documented in the Sudbury area (Keller et al. 1992). Thus, detrimental effects on biota may be observed and measured long after the stress (i.e., sulfate deposition) has been reduced or even eliminated. This presents a challenge to scientists who must try to predict the effect of the stress (see step 10 in Fig. 24.2) without knowing its future level in the ecosystem.

Interactions between Stresses

The interaction between stresses is another problem that became apparent during the integrated studies carried out in Sudbury. One example is forest denudation (resulting from both logging activities and acid metal deposition), which led to increased soil erosion, higher surface reflective temperatures, and enhanced frost action (see Chapter 18). A second example is the interaction between sulfate (and H^+ ion) deposition and trace metals. Because of smelting activities, copper and nickel concentrations in soil, water, and biota are elevated in the immediate vicinity of Sudbury. Thus, biota there must contend not only with high metal levels but also with low pH. It has been well documented that pH can have a synergistic (i.e., enhancing) effect on metal toxicity (e.g., Campbell and Stokes 1985; Cusimano et al. 1986; Hutchinson and Sprague 1986; Hickie et al. 1993). For example, Welsh et al. (1993) found that the toxicity (expressed as the 96-hour LC_{50}) of copper to fathead minnows (*Pimephales promelas*) increased fourfold as pH decreased from 7.2 to 5.6.

Acidic precipitation also can cause leaching and mobilization of other metal ions such as calcium, magnesium, aluminum, manganese, iron, and zinc from soils and bedrock (Mayer and Ulrich 1980; Jeffries et al. 1984). As a result, biota that are in ecosystems remote from industries that emit metals into the atmosphere may have to contend with the combination of high levels of both acid and metals. A discussion of the complexities of metal–H^+ or metal–metal interactions is beyond the scope of this chapter. However, depending on the organism in question as well as the mitigating circumstances in the lake/soil, the toxicity of chemical "mixtures" can be either (1) synergistic (i.e., the presence of one enhances the toxicity of the other), (2) additive (i.e., the presence of one adds to but does not have an effect on the toxicity of the other), or (3) antagonistic (i.e., the presence of one decreases the toxicity of the other, e.g., selenium/mercury, calcium/aluminum). When making policy decisions for a catchment, one must be aware of these potential interactions and, as discussed previously, conduct "integrated" scientific research.

External Factors

Although scientists can study, model, and predict to a certain extent the effects of these multiple stresses on the ecosystem, factors external to the principal stressors often can impinge on the situation and complicate the issue. A few examples follow.

Climate/Weather Change

As discussed above, wetlands, soils, and even the littoral zone of lakes are able to store strong acid in the form of reduced sulfur compounds. Catchment mass balances have shown that the storage of sulfate and/or the oxidation of stored reduced sulfur usually occurs during dry seasons (Dillon and LaZerte 1992) or in dry years when the water table becomes lowered. At the beginning of the next wet season, or in a wet year, the stored sulfate is then released back into the outflowing waters. Given that you cannot predict the weather, it is difficult to imagine managing a catchment if you are unable to predict when these pulses of acid might occur. However, long-term monitoring studies might provide insight into climatic trends and thus allow for modeling efforts.

Decrease in Atmospheric Ozone

The recent attention given by the scientific and non-scientific communities to the decrease in the ozone layer also has consequences with respect to biota in the Sudbury region. It is well documented that ultraviolet (UV) radiation can have detrimental effects on biota and that UV levels that lakes and other ecosystems are experiencing are increasing as a result of a decrease in ozone levels in the upper atmosphere. It is the dissolved organic matter (usually measured as dissolved organic carbon [DOC]) in lakes that absorbs strongly in the UV range. Unfortunately, acidification results in the reduction of DOC levels in streams and lake water (Dillon et al. 1987), and the soft water, low alkalinity, acidic lakes typical of the Sudbury region generally have low DOC concentrations. Thus, they are particularly vulnerable to a decrease in the ozone layer.

Nitrate Release from Catchments

After sulfate, deposition of nitrogen compounds is the next most important class of anthropogenic acids entering aquatic systems in Canada (RMCC 1990). However, emission of NO_x and deposition of related compounds should not be directly influenced significantly by mining/smelting activities but perhaps indirectly because most mining industries make massive uses of energy, which is often derived from the combustion of fossil fuels. Thus, although emissions of sulfur dioxide and declines in the atmospheric deposition of SO_4 may have occurred in Ontario since the mid-1970s, NO_x emissions and NO_3 deposition have not decreased (Dillon et al. 1988). In the past, this has not been thought to be a problem because terrestrial ecosystems are often limited by nitrogen, and thus, they have been a sink for nitrogen inputs (Driscoll and van Dreason 1993). However, declining vegetational demands coupled with increased deposition of nitrogen from the atmosphere may result in saturation of the terrestrial ecosystem.

Consequently, in Europe and in parts of the northeastern United States, there has been evidence of surface water acidification resulting from elevated NO_3 (Henriksen and Brakke 1988; Driscoll and van Dreason 1993). This means that even if large-scale reductions in sulfur dioxide emissions were implemented, recovery of lakes from acidic deposition may be delayed due to offsetting increases in NO_3 concentration.

Other Stresses

Kelly Lake in Sudbury, like several other lakes in the region that have calcareous deposits in their catchments, at present is not adversely affected by acidification, even though it lies at the center of the high sulfate deposition zone. However, it is the recipient of sewage from the city of Sudbury and, thus, suffers from all the problems typical of municipal lakes—it is eutrophic, bacterial counts often exceed guidelines established for human health protection, and macrophytes proliferate along its shoreline. It is important for those who are involved in management decisions for communities, such as Sudbury, that are affected by mining activities, not to overlook lakes such as Kelly in their eagerness to remediate only those lakes that have problems due to smelting activities. However, there may be indirect links between acid rain and eutrophication. Caraco et al. (1993) suggested that anthropogenically induced increases in sulfate concentrations in lakes can cause an increase in the magnitude of phosphorus released from the sediments, as well as an increase in the availability of that phosphorus to biota. Thus, two seemingly independent stressors may, in fact, behave in an inter-active way, again demonstrating it is apparent that the need for integrated research and management policies is essential to overall catchment management.

Summary

The management of catchments that have been damaged by anthropogenic activities or that may be potentially stressed in the future is a complex problem requiring (1) an informed public, (2) the cooperation of many individuals, government agencies, and non-government organizations, (3) the coordination of all scientific endeavors, and (4) the consideration of both scientific and socioeconomic factors. Consequently, in this chapter, we have described and emphasized the need for an integrated whole-ecosystem approach to catchment management (see Fig. 24.1).

Although the management in the Sudbury area generally has set an example of how the integrated approach to catchment management should work, serious challenges have been faced by those attempting to set policies for the region. The reasons for these difficulties arise from the fact that pollution transcends political, geographic, social, and temporal boundaries. Secondary and tertiary pollution (see Fig. 24.3) and multiple stresses such as interactions between sulfate and trace metals and forest denudation leading to increased soil erosion, higher surface reflective temperatures, and enhanced frost action have complicated scientific findings. Sulfur storage in the terrestrial and wetland portions of catchments and its potential release long after sulfate deposition has been reduced also presents a challenge to scientists who must try to predict the effect of the sulfate without knowing its future concentration in the ecosystem. Also, the element of time confuses the issue, not only for the scientist, who must determine both long-term (chronic) and short-term (acute) effects of the stress on the ecosystem but also for the manager, who must weigh the advantages/disadvantages of short-term versus long-term remedial actions.

Other factors, external to the principal stressors (high sulfate and metal deposition) can affect the Sudbury region and complicate matters even further. These include climate or weather changes, decreasing ozone levels in the atmosphere, nitrate release from the terrestrial portion of the catchment, and phosphorus inputs from sewage.

In conclusion, we emphasize that only when scientific research activities are integrated with responsive and responsible management policies can overall catchment management be successful.

References

Baumann, A. 1984. Extreme heavy metal concentrations in sediments of the Oker—a river draining an old mining and smelting area in the Harz Mountains, Germany, pp. 579–591. *In* J. Nriagu (ed.). Environmental Impact of Smelters. Advances in Environmental Science Series. John Wiley and Sons, New York.

Bayley, S.E., D.H. Vitt, R.W. Newbury, K.G. Beaty, R. Behr, and C. Miller. 1987. Experimental acidification of a *Sphagnum*-dominated peatland: first year results. Can. J. Fish. Aquat. Sci. 44:194–204.

Beamish, R.J. 1974. Loss of fish populations from unexploited remote lakes in Ontario, Canada as a consequence of atmospheric fallout of acid. Water Res. 8:85–95.

Beamish, R.J., and H.H. Harvey. 1972. Acidification of the La Cloche Mountain lakes, Ontario and resulting fish mortalities. J. Fish. Res. Board Can. 29:1131–1143.

Campbell, P.C.G., and P.M. Stokes. 1985. Acidification and toxicity of metals to aquatic biota. Can. J. Fish. Aquat. Sci. 42:2034–2049.

Caraco, N.F., J.J. Cole, and G.E. Likens. 1993. Sulfate control of phosphorus availability in lakes: a test and re-evaluation of Hasler and Einsele's model. Hydrobiology 253:275–280.

Cusimano, R.F., D.F. Brakke, and G.A. Chapman. 1986. Effects of pH on the toxicities of cadmium, copper and zinc to steelhead trout (*Salmo gairdneri*). Can. J. Fish. Aquat. Sci. 43:1497–1503.

Devito, K.J., and P.J. Dillon. 1993. The influence of hydrologic conditions and peat oxidation on the phosphorus and nitrogen dynamics of a conifer swamp. Water Resource Res. 29:2675–2685.

Dillon, P.J., D.S. Jeffries, and W.A. Scheider. 1982. The use of calibrated lakes and watersheds for estimating atmospheric deposition near a large point source. Water Air Soil Pollut. 18:241–258.

Dillon, P.J., D.S. Jeffries, W. Snyder, R. Reid, N.D. Yan, D. Evans, J. Moss, and W.A. Scheider. 1978. Acidic precipitation in south-central Ontario: recent observations. J. Fish. Res. Board Can. 35: 809–815.

Dillon, P.J., and B.D. LaZerte. 1992. Response of Plastic Lake catchment, Ontario, to reduced sulphur deposition. Environ. Pollut. 77:211–217.

Dillon, P.J., M. Lusis, R.A. Reid, and D. Yap. 1988. Ten-year trends in sulphate, nitrate and hydrogen deposition in central Ontario. Atmos. Environ. 22:901–905.

Dillon, P.J., R.A. Reid, and E. de Grosbois. 1987. The rate of acidification of aquatic ecosystems in Ontario, Canada. Nature 329:45–48.

Dillon, P.J., N.D. Yan, W.A. Scheider, and N. Conroy. 1979. Acidic lakes in Ontario, Canada: characterization, extent and responses to base and nutrient additions. Arch. Hydrobiol. Beih. 13: 317–336.

Driscoll, C.T., and R. van Dreason. 1993. Seasonal and long-term temporal patterns in the chemistry of Adirondack lakes. Water Air Soil Pollut. 67:319–344.

Fitchko, Y. 1978. The distribution, mobility and accumulation of nickel, copper and zinc in a river system draining the eastern part of the metal-polluted Sudbury smelting area. Ph.D. thesis, University of Toronto, Toronto, Ontario.

Gorham, E., and A.G. Gordon. 1960. The influence of smelter fumes upon the chemical composition of lake waters near Sudbury, Ontario and upon the surrounding vegetation. Can. J. Bot. 30:477–487.

Haukioja, E. 1993. Research on ecological effects of aerial pollution: a Finnish perspective, pp. 67–69. *In* M.V. Kozlov, E. Haukioja, and V. Yarmishko (eds.). Aerial Pollution in Kola Peninsula. Proceedings of International Workshop, April 14–16, 1992, St. Petersburg, Russia. Kola Scientific Center, Apatity, Russia.

Havas, M., T.C. Hutchinson, and G.E. Likens. 1984. Red herrings in acid rain research. Environ. Sci. Technol. 18:176–186.

Henriksen, A., and D.F. Brakke. 1988. Sulfate deposition to surface waters. Environ. Sci. Technol. 22:8–18.

Hickie, B.E., N.J. Hutchinson, D.G. Dixon, and P.V. Hodson. 1993. Toxicity of trace metal mixtures to alevin rainbow trout (*Oncorhynchus mykiss*) and larval fathead minnow (*Pimephales promolas*) in soft, acidic water. Can. J. Fish. Aquat. Sci. 50: 1348–1355.

Hutchinson, N.J., and J.B. Sprague. 1986. Toxicity of trace metal mixtures to American flagfish (*Jordanella floridae*) in soft, acidic water and implications for cultural acidification. Can. J. Fish. Aquat. Sci. 43:647–655.

Hutchinson, T.C. 1982. The ecological consequences of acid discharges from industrial smelters, pp. 105–122. *In* F.M. D'Itri (ed.). Acid Precipitation: Effects on Ecological Systems. Ann Arbor Science.

Jeffries, D.S., W.A. Scheider, and W.R. Snyder. 1984. Geochemical interactions of watersheds with precipitation in areas affected by smelter emissions near Sudbury, Ontario, pp. 195–241. *In* J. Nriagu (ed.). Environmental Impacts of Smelters. Advances in Environmental Science Series. John Wiley and Sons, New York.

Keller, W., J.R. Pitblado, and J. Carbone. 1992. Chemical responses of acidic lakes in the Sudbury, Ontario area to reduced smelter emissions, 1981–89. J. Fish. Res. Board Can. 49:25–32.

Mayer, R., and B. Ulrich. 1980. Input to soil, especially the influence of vegetation in intercepting and modifying inputs—a review, pp. 173–182. *In* T.C. Hutchinson and M. Havas (eds.). Effects of Acid Precipitation on Terrestrial Ecosystems. Plenum Press, New York.

Molot, L.A., P.J. Dillon, and G.M. Booth. 1990. Whole-lake and nearshore chemistry in Bowland Lake, before and after treatment with $CaCO_3$. Can. J. Fish. Aquat. Sci. 47:412–421.

Molot, L.A., J.G. Hamilton, and G.M. Booth. 1986. Neutralization of acidic lakes: short-term dissolution of dry and slurried calcite. Water Res. 20: 757–761.

Moore, J.N., and S.M. Luoma. 1990. Hazardous wastes from large-scale metal extraction. Environ. Sci. Technol. 24:1278–1285.

Research and Monitoring Coordinating Committee (RMCC). 1990. The 1990 Canadian Long-Range Transport of Air Pollutants and Acid Deposition Assessment Report. Part 4. Aquatic Effects Studies, Toronto, Ontario.

Royal Commission on the Future of the Toronto Waterfront (RCFTW). 1992. Regeneration: Toronto's Waterfront and the Sustainable City (Final Report). Toronto, Canada.

Somers, K., G. Mierle, and N.D. Yan. 1994. An Environmental Protection Model for the Management of Ontario's Water Resources. Technical report. Ontario Ministry of Environment and Energy, Ontario.

Spanish Harbour Rap Team. 1993. The Spanish Harbour Area of Concern, Environmental Conditions and Problem Definition, Remedial Action Plan, Stage 1 Report. Intergovernmental report. Ontario Ministry of Environment and Energy, Sudbury, Ontario.

Urban, N.R., S.E. Bayley, and S.J. Eisenreich. 1989. Export of dissolved organic carbon and acidity from peatlands. Water Resource Res. 25:1619–1629.

Welsh, P.G., J.F. Skidmore, D.J. Spry, D.G. Dixon, P.V. Hodson, N.J. Hutchinson, and B.E. Hickie. 1993. Effect of pH and dissolved organic carbon on the toxicity of copper to larval fathead minnow (*Pimephales promelas*) in natural lake waters of low alkalinity. Can. J. Fish. Aquat. Sci. 50: 1356–1362.

Yan, N.D., and R. LaFrance. 1984. Responses of acidic neutralized lakes near Sudbury, Ontario to nutrient enrichment, pp. 243–282. *In* J. Nriagu (ed.). Environmental Impacts of Smelters. Advances in Environmental Science Series. John Wiley and Sons, New York.

25

Planning for the Environmentally Friendly City

Tin-Chee Wu and William E. Lautenbach

The growth of cities is a trend that has major impacts on both the global and local environment. In the 35 years since 1950, the number of people living in cities almost tripled, increasing by 1.25 billion. In the more developed regions, the urban population nearly doubled, from 447 million to 838 million. In the less-developed regions, it quadrupled, growing from 286 million to 1.14 billion (World Commission on Environment and Development 1987).

Girardet (1992) described these growing cities as parasites—they are energy drains whose survival depends on cheap food and energy produced elsewhere. Cities are also thirsty entities and produce large quantities of wastes. To qualify as environmentally friendly, cities must be able to break away from these trends, yet also must be able to sustain themselves.

To be sustainable, all human settlements must first meet the physical needs of its residents—needs that include air, water, food, shelter, energy, and raw and finished materials. Human settlements must also provide a healthy physical environment to support a healthy population. To meet these physical needs, a modern city provides clean water and waste disposal services; physical space and opportunities for employment and shelter; transportation and communications facilities and services; local energy distribution systems; community space for social interaction; healthy physical environments that include public and private open space for leisure and recreation; education facilities and services; and recreational facilities and services.

Urban planning is the integrating process by which these services and physical development are coordinated in cities. Some conceptual frameworks for ecosystem planning of environmentally friendly cities are highlighted in this chapter. Many challenges and problems are unique to a mining city (Fig. 25.1) in a north temperate climate, but instead of focusing only on planning needs for a city such as Sudbury, we discuss municipal planning principles that have wide application.

Ecosystem Planning

Ecosystem planning considers the natural environment more than just a medium on which urban development takes place. It sees humans living in the biosphere as a home rather than the planet being the house of humans. The word *home* evokes a much richer concept than does the word *house*—it involves a group of people who live together and jointly take care of and relate to their home; and it can be used to refer to a house, a home town, a country, or the planet all at the same time (Allen et al. 1993).

To achieve the harmony of humans at home, ecosystem planning has to begin with careful

Figure 25.1. Aerial view of downtown Sudbury with a portion of Ramsey Lake in the lower right corner.

consideration of the constraints as well as the opportunities offered by the biosphere. The built environment is then designed to fit into the home. Some useful operating principles for ecosystem planning are provided in the preface to the Royal Commission Report for the Future of the Toronto Waterfront (RCFTW 1992):

1. includes the whole system, not just parts of it
2. focuses on the inter-relationships among elements
3. understands that humans are part of nature, not separate from it
4. recognizes the dynamic nature of the ecosystem
5. incorporates the concepts of carrying capacity, resilience, and sustainability—suggesting that there are limits to human activity
6. uses a broad definition of environments—natural, physical, economic, social, and cultural
7. encompasses both urban and rural activities
8. is based on natural geographic units such as watersheds rather than on political boundaries
9. embraces all levels of activity—local, regional, national, and international
10. emphasizes the importance of species other than humans and of generations other than the present

11. **is based on an ethic in which progress is measured by the quality, well-being, integrity, and dignity it accords natural, social, and economic systems**

The philosophy behind 100-year plans is the creation of an overall vision for the long-term future, yet allowing detail planning and implementation to evolve in the future by adaptation to changes in circumstances over decades.

Long-Term Planning

Tonn (1986) proposed the concept and coined the term *500-year planning* to describe the goals, practice, and methodology of carrying out the design, development, and implementation of plans, programs, and laws to eliminate or substantially mitigate very long-term environmental problems—problems such as climate change, species extinction, soil erosion, salination and conversion of agricultural land, deforestation, groundwater contamination, nuclear waste, and chlorofluorocarbon pollution. These problems often take centuries either to develop or to show their full effects and take centuries to solve. The goals of 500-year planning are therefore designed to safeguard the physical/natural system for future generations and to protect present and future populations from health and safety risks caused by misuse of the environment as well as from environmental catastrophes that would restrict the future of the human species. As such, 500-year planning does not center on setting long-term goals for humankind nor to predict the future states of human society but to maximize the ways in which human socioeconomic evolution can occur (Tonn 1986).

Unlike 500-year planning, which would be more appropriate for planning at the national or international scale due to the nature and scope of the environmental problems that it addresses, 100-year planning is adaptable to local planning. In fact, the three 100-year plans existing in Canada today are all plans prepared at the local or regional scale (Moriyama and Teshima 1979, 1988, 1991). The latest is the Ramsey Lake Community Improvement Plan prepared for the City of Sudbury and the Region of Sudbury, Ontario (Moriyama and Teshima 1991; Regional Municipality of Sudbury 1992) (Box 25.1).

Techniques for Physically Planning an Environmentally Friendly City

Many planning techniques are available to plan for the environmentally friendly city. None of these techniques are new, although some have been used more often than others in the past. When used in combination, however, these techniques represent excellent starting points in achieving the desired goal.

Design with Nature

The design with nature principles and methods pioneered by McHarg (1971) still remains a powerful tool in physical design. It is both a philosophy and a technique. As a philosophy, it begins with an examination of values—both natural values and historical, cultural, and social values are considered. As a technique, it is a planning process that begins with an inventory of the individual components of the biophysical subsystem—factors such as bedrock geology, surficial geology, hydrology, physiographic features, soil drainage, vegetation, and wildlife habitats. Physical constraints to development are mapped individually and then put together as a set of map overlays. The resulting composite map helps indicate where constraints are most severe, less severe, or light. Development, whether in the form of a linear corridor such as a new road, a transmission line, or a trail or non-linear structures such as buildings, may then be located in areas where physical constraints are least severe. Alternatives can also be evaluated with the same technique, and choices can be made based on identifiable environmental factors.

Mixed Land Uses

Land use policies adopted in North America in the past few decades have been based on the

Box 25.1. Ramsey Lake Community Improvement Plan

Ramsey Lake is an 870-h urban lake located within the municipal boundaries of the City of Sudbury and within walking distance to downtown Sudbury (see Plate 18 following page 182). It is also the main source of drinking water for the city. During periods of peak demand, Ramsey Lake supplies up to 60% of the total quantity of drinking water for the city. The lake is surrounded by established residential neighborhoods and several activity centers and institutions—a university, three hospitals, a science center (see Fig. 8.12), two major city parks, and a 970-ha open space conservation area. The lake is also an important recreational area supporting swimming, sailing, rowing, power boating, and fishing. Several areas also support more passive activities such as walking and wildlife viewing, and the entire lake is valued as a scenic resource.

The Region and the City of Sudbury jointly prepared and adopted a Ramsey Lake Community Improvement Plan in 1992. The plan examines this vital resource of the city and begins with a 100-year vision of the lake and its watershed. It views Ramsey Lake and its watershed as a hydrogeological and ecological region that is shared by all people of the city. The plan identifies the highest and best use of the lake as "the green and natural heart of the City, a public domain where resources of City-wide importance can be gathered in a magnificent setting and made accessible, a place of enjoyment, discovery and recreation for all the people."

It is on the basis of this long-term vision that specific policies, programs, and projects are proposed in the Ramsey Lake Plan. Policies proposed include the preservation of the water quality of Ramsey Lake; the conservation of green space around the lake; the retention of these green spaces in public ownership over the long term and the acquisition of key open space properties by the public; and the protection of natural and environmentally sensitive areas such as wetlands, marshes, wildlife corridors, and fish spawning areas. Programs proposed include the development of the Ramsey Lake Interpretive and Recreational Trail around the lake and further development of the public park and conservation area properties for recreational uses. For the implementation of these policies and programs, the plan proposes the creation of a Ramsey Lake Trust to serve as the watchdog and guardian for the lake and to ensure long-term stewardship of the lake. Many of these programs and projects can be planned, budgeted, and carried out in 5-year time blocks, even though the completion of all the development projects for the Ramsey Lake area will probably take 20 years or longer to be accomplished.

---

premise of segregating incompatible land uses. Historically, segregation of land uses was needed to protect the health of residents by separating polluting industries from residences. Although this premise is still valid today for certain types of industries, it is not a necessity in most cases. This is due not only to the decline of many smokestack industries and the rise of service industries in the North American economy but also because many of the traditional industries have modernized and become much less polluting than their predecessors. Moreover, many of the emerging high-technology industries operate in industrial buildings that are physically indistinguishable from office buildings and carry out operations that have no higher environmental impacts than office work.

Furthermore, segregation of land uses has environmental costs. It requires daily transportation of workers between their homes and their workplaces. Means of transportation, whether in the form of public transit or private automobiles, are needed. It requires the construction of more roads and vehicles, the use of more energy for operating these vehicles, and the generation of more pollution from such operations. However, mixing of land uses could reduce some of these transportation needs and their associated environmental

Box 25.2. Neo-Traditional Neighborhood

Originally promoted by the husband and wife architectural team of Duany and Plater-Zyberk as neo-traditional development, this urban design concept has since evolved to become known as "New Urbanism." A primary objective of this design concept is to design urban space for people, not cars. This is in direct contradiction with the (usually unstated) urban design objective of the postwar period in North America, which is to design urban space for the convenience of the automobile.

Although the first community designed and built according to this principle was the resort community of Seaside, Florida, many other communities have been designed or built in many other locations in both the United States and Canada. In Canada, the latest examples can be found in the town of Markham, Ontario. For example, the Markham plan includes the following features on a 625-ha site that will accommodate 16,000 jobs and 10,000 housing units (Duany and Plater-Zyberk 1992; Gabor 1994; Wood-Brunet 1994):

1. **seven neighborhoods positioned no more than a 5-minute walk (400 m) from the central core**
2. **houses are set to narrow (15.5-m right of way) streets, planned on a modified grid pattern**
3. **extensive street parking and rear lanes for car access**
4. **a range and mix of housing in close proximity**
5. **space above rear lane double garages for living or working**
6. **strong regulation of the built form to ensure a human scale (but without regulation of architectural style)**
7. **a central core featuring main street shopping and the highest densities for housing and employment uses**
8. **higher net densities and mixed uses**
9. **a dedicated transit line built into a 4.5-km central corridor linking the neighborhoods**
10. **linked open space at the edge of the neighborhoods that contain schools, parks, and remnant woodlots**
11. **interesting building features, civic sites, or architectural follies used to terminate the views down streets**

costs. Similar benefits would also be realized by mixing residential and commercial land uses and thus reducing the number and length of shopping trips. Recently, there has been a small but growing number of new communities that are being designed and built according to what is generally known as new urbanism or neo-traditional principles that are being championed by Duany and Platter-Zyberk (1992) (Box 25.2).

Land use patterns take a long time to establish or change. Any resulting positive or negative environmental impacts will also remain for a long time. Planning therefore has to take a proactive role in influencing positive changes in the land use pattern at the earliest opportunity—the community design stage (Fowler 1991).

Green Space, Greenways, and Urban Forestry

Greenways are natural or landscaped linear open spaces for pedestrian or bicycle passage or for linking parks, nature reserves, cultural features, or historical sites with each other and with populated areas (Brown 1993). These linear corridors may include elaborate trail systems that permit walking, hiking, biking, riding, and skiing or simply a stretch of open space left in its natural state. Although recreation is a common use for greenways, they often serve the more important ecological function of protecting wildlife migration corridors and habitats. For these reasons, they are referred to by Wisconsin landscape architect Phil Lewis as "E-ways"—for environment, ecology, education, and exercise (Grove 1990). The Junction Creek Waterway Park that traverses the city of Sudbury is one example of such a linear open space system in an urban environment (Regional Municipality of Sudbury 1991).

Whether viewed as an extension of or a component of greenways, urban forestry is an effective means to bring the natural environment into the daily urban life and the consciousness of the urban population. The educational value of urban forestry is as important as its "practical" values such as providing shade and modifying the microclimate; absorbing carbon dioxide and improving air quality; or beautifying the urban landscape. The regreening of Sudbury is an excellent example of a community effort in large-scale urban forestry (see Chapter 8).

Bioclimatic Design and Winter Cities

Bioclimatic design relates the biological requirements of human comfort with the climate of the natural environment. From an energy efficiency perspective as well as a quality of life perspective, planning, designing, building, and retrofitting cities located in colder climatic zones according to the design principles developed under the umbrella concept of "winter cities" is an application of bioclimatic design (e.g., Manty and Pressman [1988] represents one of many publications on winter cities, and Matus [1988] illustrates solar design techniques). Although this will be a slow process—as cities have to develop and redevelop over time—the benefits will also be long-lasting for any cold climate community that is developed according to energy efficiency and winter quality of life criteria.

Community Energy Efficiency

Importing fuel to meet the energy needs of a community is a constant drain on the economic wealth that could have been retained within the community. In economic development, the strategy of import substitution is one that attempts to use local resources to substitute for imported resources. Although no community can rely entirely on locally generated energy resources, strategies such as substituting imported non-renewable energy with locally produced renewable energy; improving the energy efficiency of its building stock through appropriate community design, better construction, and retrofitting of older buildings; and reducing the community's reliance on the automobile are all components of this import substitution strategy. Over time, this strategy will create a less energy-demanding urban settlement.

Transit-Friendly Communities

Employment and commuting go hand in hand in North American cities. According to a 1992 Statistics Canada survey, an estimated 9.1 million urban Canadians commuted to work each weekday, whereas only 770,000 or 8% of the employed population were non-commuters. Among commuters who used a single mode of transportation exclusively—the automobile, public transit, bicycling and walking represented 69%, 3%, 2%, and 4%, respectively. Including those who used multiple modes of transportation, the automobile carried 87% of all commuters while public transit carried only 10%. The highest percentages of public transit users were found in Toronto, Montreal, and Ottawa-Hull—at 20%, 18%, and 16%, respectively (Marshall 1994). It is worth noting that both Toronto and Montreal are served by rapid transit systems.

Mixed land uses connected by a properly designed road system in a community with sufficiently high residential densities are all preconditions that support the provision of transit services. There is a direct relationship between development density and the use of transit. Compact urban areas are more supportive of transit than low-density areas typical of suburban developments. Generally, viable bus service requires a residential density of at least 10 dwelling units per hectare. Rapid transit generally requires considerably higher densities—30–80+ dwelling units per hectare and larger catchment areas. Even at these densities, however, transit services will continue to require public subsidies (Irwin 1992).

The planning tools highlighted so far invariably focus on design—either at the community design or site/building design level. Once designed and constructed, environmental impacts of the built environment will be long-lasting, remedies will be costly, and changes will be slow. It is therefore crucial that environmentally friendly designs be adopted whenever new urban development occurs (Fowler 1991).

Evaluation Tools

Supplementing these planning tools are also analytic tools that may be used to evaluate human settlements' interaction with and human actions on the natural environment. A few examples are highlighted.

Environmental Impact Assessment

Environmental impact assessment has been used for more than two decades as a technique to evaluate the environmental impacts of a proposed project before the project is carried out. For example, in the Province of Ontario, environmental impact assessment was legislated in 1975 with the passing of the Environmental Assessment Act. In addition to the project as proposed, environmental impact assessment also assesses the alternatives to the project and the alternatives in carrying out the chosen project alternative. Finally, the process also proposes actions that will mitigate the effects of the identified environmental impacts.

Although environmental impact assessment can be an effective tool for identifying and evaluating impacts before a project is undertaken, it is still difficult to identify the cumulative impacts of multiple projects on the same natural system. This shortcoming may be somewhat overcome if environmental assessment is applied to policies or plans that precede projects. Even so, such assessments are bound to be general in nature.

To identify the cumulative impacts of multiple projects on the same natural system, new technical evaluation tools and legal/regulatory frameworks must be developed.

State of the Environment Monitoring and Reporting

Reporting on the state of the environment is not new. In fact, by the time Canada produced its first state of the environment report in 1986, 16 of the 24 member countries of the Organization for Economic Co-operation and Development had produced a state of the environment report (Bird and Rapport 1986). The second Canadian report on the environment was produced in 1991 (Environment Canada 1991).

In recent years, municipalities have also adopted this tool—for example, the Edmonton Board of Health produced a report for the City of Edmonton in 1989, and the Ottawa-Carleton Health Department produced a report for the Region of Ottawa-Carleton in 1992. In Seattle, a coalition of local organizations and citizens has prepared a report on 20 indicators of sustainable community for the Seattle area. Twenty more indicators are in the process of being researched (Sustainable Seattle 1993). The Sudbury and District Health Unit is also in the process of preparing a report to cover the Region of Sudbury.

At the local level, a state of the environment report can be used (1) as the starting point for the creation of a library of baseline data for future environmental monitoring and environmental research in the community; (2) to identify the trends; (3) to identify actions required; (4) as the background information for evaluating the accomplishments and deficiencies of existing policies and programs; (5) as the starting point for the initiation of new policies and programs to fill identified gaps; and (6) as a tool for public education.

True Cost Accounting for Resource Uses

There are two main approaches to environmental protection: (1) the legal/regulatory approach in which legally enforceable standards are enforced by the regulatory agencies through the court system; and (2) the economic approach in which economic incentives or disincentives are used to bring forth compliance. There are advantages and disadvantages

Box 25.3. Green Housing

Modeled after its successful Energy Star Program, the City of Austin, Texas, offers a Green Builder Program to assist builders to produce and market environmentally friendly homes and to educate buyers to make informed choices. As part of the program, the city publishes the Eco-Home Guide that covers four main areas in which the built environment has an impact on natural resources and environmental quality—water, energy, building materials, and solid waste. Each of these four topics is examined from a life-cycle viewpoint (i.e., the product or system listed as an Eco-Option in the guide has been evaluated from an environmental standpoint through the stages of its existence from its source to its recycling or disposal). Each Eco-Option has a rating value assigned so that each component of a home and the home itself can be rated by either a builder or a buyer. This is believed to be the first program of its kind in the United States and has been awarded by the United Nations International Council for Local Environmental Initiatives Government Honours Programme.

in both of these approaches, but often, a combination of these approaches is used to deal with an environmental problem (e.g., the case of sulfur dioxide emissions reduction by the two Sudbury mining companies over the past decade) (see Chapters 4 and 21).

Currently, many public goods are delivered to the users at below their real costs of production. Examples include the provision of municipal drinking water, the collection and disposal of sewage and solid wastes, the construction and maintenance of roads to serve primarily private automobiles, and the provision of volume discounts for high-volume users of energy such as electricity and natural gas. By not paying the true economic costs of production, the consumers of these services and resources lack the market signal for conservation. In fact, these hidden subsidies often act as disincentives for the consumers to respond to administratively or legally imposed conservation measures. To change this, service providers must account for the true economic costs of their services and charge their customers at cost. Eventually, they must also include in the basic costs the environmental costs—the so-called externalities.

Given the proper market signal, consumers, be they individual households or high-volume private or public corporate users, will adjust their consumption accordingly over time. In practice, this user pay principle based on true cost accounting can be a powerful tool to meet both the fairness objective as well as the conservation objective. For example, in a municipal setting, the use of individual meters for water use and the adoption of a pricing structure that more closely reflects the true costs of municipal sewer and water services have been successful in achieving conservation objectives (Brooks et al. 1990).

A variation of this principle is the life-cycle costing technique, which takes into account the total energy and material resources used in manufacturing, distributing, operating, maintaining, and disposing/recycling of a manufactured product. For example, this technique may be applied to evaluate the different building materials, designs, and building techniques used in constructing buildings (Box 25.3). This evaluation technique may be applied to either public goods or private goods.

Conclusions

Kenneth Boulding (1973) likened the long process of transition in the relationship between humans and their environment as the transition from the "cowboy economy" of the past to the "spaceman economy" of the future. The bulk of this book describes the legacies of earlier cowboys and tells the story of how the descendants of those cowboys attempt to restore the damaged environment of the Sudbury Basin.

This chapter highlights some broad approaches and technical tools that should be used to build and retrofit an urban environment to become a healthy component of that future. Yet, strategies are only as effective as their users are willing to use them. Beyond these interim technical solutions and more sophisticated technical tools yet to be created, an environmental ethic must evolve to become the guiding principle for all human actions, including urban and regional planning. Leopold (1966) saw an ethic in the following way:

> An ethic, ecologically, is a limitation on freedom of action in the struggle for existence. An ethic, philosophically, is a differentiation between social and antisocial conduct. These are two definitions of one thing. The thing has its origin in the tendency of interdependent individuals or groups to evolve modes of co-operation. The ecologists call these symbioses.

Leopold further argued that a land ethic is "an evolutionary possibility and an ecological necessity."

As the Sudbury experience attests, environmental damages take a long time and great efforts to undo. Will an urban world environmental ethic evolve soon enough so that further damages will not be inflicted on our environment? In the final analysis, that is the great challenge facing all inhabitants of spaceship earth for decades to come.

References

Allen, T.F.H., B.L. Bandurski, and A.W. King. 1993. The Ecosystem Approach: Theory and Ecosystem Integrity. Unpublished report to the Great Lakes Science Advisory Board. International Joint Commission.

Bird, P.M., and D.J. Rapport. 1986. State of the Environment Report for Canada. Environment Canada, Ottawa.

Boulding, E.K. 1973. The economics of the coming spaceship earth, pp. 121–132. *In* H.E. Daly (ed.). Towards a Steady-State Economy. Freeman, San Francisco.

Brooks, D.B., R. Peters, and P. Robillard. 1990. Pricing: a neglected tool for managing water demand. Alternatives 17(3):40–48.

Brown, D.T. 1993. Reclaiming deserted corridors: rights of way as common property resources. Alternatives 19(3):24–28.

City of Austin. (no date). Eco–Home Guide. Austin, TX.

Duany, A., and E. Plater-Zyberk. 1992. The second coming of the American small town. Plan Canada (May 1992):6–13.

Environment Canada. 1991. The State of Canada's Environment. Government of Canada, Ottawa, Canada.

Fowler, E.P. 1991. Land use in the ecologically sensible city. Alternatives 18(1):26–35.

Gabor, P. 1994. Duany designs for people, not cars. Ont. Plan. J. 9(3):3–6.

Girardet, H. 1992. The Gaia Atlas of Cities. Doubleday, Toronto.

Grove, N. 1990. Greenways: paths to the future. Nat. Geog. 177(6):77–98.

Irwin, N.A. 1992. Transit and Land Use: Experiences in Ontario. Ont. Plan. J. 7(4):12–15.

Leopold, A. 1966. A Sand County Almanac. Ballantine Books, Inc., New York.

Manty, J., and N. Pressman (eds.). 1988. Cities Designed for Winter. Building Book Limited, Helsinki.

Marshall, K. 1994. Getting there. Perspectives (Summer 1994):17–22.

Matus, V. 1988. Design for Northern Climates: Cold–Climate Planning and Environmental Design. Van Nostrand Reinhold, New York.

McHarg, I.L. 1971. Design With Nature. Doubleday & Company, Garden City, NY.

Moriyama and Teshima Architects. 1979. The Meewasin Valley Project. Toronto.

Moriyama and Teshima Planners Limited. 1988. Ontario's Niagara Parks: Planning the Second Century. Niagara Parks Commission, Ontario.

Moriyama and Teshima Planners Limited. 1991. The Ramsey Lake and Watershed Community Improvement Plan: A 100 Year Vision. Toronto, Canada.

Regional Municipality of Sudbury. 1991. Junction Creek Waterway Park Community Improvement Plan. Sudbury, Ontario.

Regional Municipality of Sudbury. 1992. Ramsey Lake Community Improvement Plan. Sudbury, Ontario.

Royal Commission on the Future of the Toronto Waterfront (RCFTW). 1992. Regeneration: Toronto's Waterfront and the Sustainable City (Final Report). Toronto, Canada.

Sustainable Seattle. 1993. The Sustainable Seattle 1993: Indicators of Sustainable Community. Seattle, WA.

Tonn, B.E. 1986. 500-year planning: a speculative provocation. J. Am. Plan. Assoc. (Spring 1986): 185–193.

Wood-Brunet, E. 1994. The new urbanism in Markham. Ont. Plan. J. 9(3):7–8.

World Commission on Environment and Development. 1987. Our Common Future. Oxford University Press, Oxford.

26

From Restoration to Sustainable Ecosystems[1]

John M. Gunn, Nels Conroy, William E. Lautenbach, David A.B. Pearson, Marty J. Puro, Joseph D. Shorthouse, and Mark E. Wiseman

Vast areas of the world have been laid waste by destructive human activities: poor agricultural practices, industrial pollution, warfare, etc. (WCED 1987; Smil 1993; Edwards 1994). It has been estimated that at present there are approximately 2000 million ha of degraded land (approximately the combined size of Canada and the United States) and that this number increases by 5–7 million ha each year (Wali 1992). Mining and smelting have contributed heavily to these losses, being responsible for more than 20 million ha of some of the most severely damaged areas (Moore and Luoma 1990). These losses of productive lands and waters and the interferences with the health of natural systems on which they depend are continuing to occur at the same time that the need for ecological services (i.e., food, water, fibers, natural medicines, microbial decomposition of waste products, etc.) accelerates because of increasing individual demands for resources and a world human population that doubles approximately every 40 years (Ehrlich and Ehrlich 1991). For example, in China, a country expected to contain 1.25 billion people by year 2000, approximately 40 million ha of arable land (30–40% of the national total) has been lost in the past 50 years by soil erosion, urbanization, transportation, and industrial pollution (Smil 1993).

There is no consensus on what the ultimate carrying capacity of the earth will prove to be, but there is no doubt that accelerated growth cannot be sustained on a shrinking resource base (Gore 1992; Houghton 1994; Woodwell 1994). Losing vast areas of productive land is intolerable under such a strain on global resources.

Changing values in society create a new ethical environment in which companies must operate (Potter 1988; Dunlap and Scarce 1991). Society is now less willing to tolerate egotistical companies that suggest that what is good for the company is good for society because it generates wealth through jobs and useful products. Other values (essential ecological services, recreational activities, animal rights, protection of biodiversity, etc.) in land use choices have now begun to take precedence over the idea of "production at lowest possible cost." These values have become so significant that some industrial developments are stopped

[1]This paper was prepared by the synthesis group that consisted of representatives of government resource management and environmental agencies, the mining industry, municipal government staff, and the university. The group reviewed all previous chapters from the Sudbury case history before preparing this discussion paper.

by them. For example the construction of the Windy Craggy mine in British Columbia was recently stopped because of environmental concerns, even though more than $40 million had been spent by the company developing the site.

Sudbury Case History

The extensive destruction of land and water by industrial emissions from Sudbury represents one of the best-known environmental impacts in North America. These damages were not intentional; in fact, it can be argued that over the years "best technologies" were used to prevent their occurrence. But, unfortunately, severe damages did occur. A landscape that once supported a rich variety of natural resources—forests, fish, wildlife, etc.—was reduced to a barren wasteland in a few decades of mining and smelting.

There is no particular magic or uniqueness in the solutions to environmental damages in Sudbury. The same solutions apply everywhere: (1) reduce the contamination and (2) repair the damage. Progress on these so easily stated but difficult to implement solutions has been the subject of this book. In this final chapter, we ask ourselves what we have learned from the Sudbury experience that will maintain and encourage further restoration efforts in this area and whether there are some lessons of general application that others might take from Sudbury when working toward "sustainable ecosystems" elsewhere.

Irony of the Term *Sustainable*

It may seem ironic to use a case history of a hard-rock mining area to discuss ecosystem "sustainability." High quality ore deposits are a nonrenewable resource that can be rapidly depleted with current technologies. Mining has traditionally been a transient industry involving short-term use of land without regard to other future uses of that land. The presence of vast areas of derelict lands is the legacy of mining in most countries (Moore and Luoma 1990; Young 1992). However, in the same way that people cannot isolate themselves from the natural world that provides the elements of life, so too a mining industry or mining town cannot separate itself from the lives of its workers or the life of the planet. For example, stack emissions from Sudbury smelters make substantial contributions to the long-range transport of contaminants, but the Sudbury area also receives pollution from other areas. In a global context, it will do little good if the Sudbury area is cleaned up at great expense, but the rest of the world does not act in a similar manner.

It is important for the reader to remember that with the Sudbury case history we are discussing restorative change at a very early stage. As any recent visitor will attest, the Sudbury region is still badly damaged—only approximately 30% of the barren land has received remedial treatment (Fig. 26.1); natural recovery under emission controls is just beginning; and there is considerable uncertainty about the effectiveness of the remedial treatments or adequacy of the legislated control orders. It will be many decades, perhaps centuries (Croker and Major 1955), before these ecosystems can be healed (toxic metals return to background levels, stable podzolic soil structure re-established, insect epidemics under natural controls, diversity of biological communities re-established, etc.). However, the direction of environmental change is positive, and this will be our cautious focus for the remainder of this chapter.

Ingredients for Progress

Timing was essential to the environmental improvements that occurred in Sudbury. The recognition that change was needed was to a large degree simply part of a broad societal change in attitude that began in the late 1960s and early 1970s and rose rapidly through the 1980s. The enormous investments in time and money needed to initiate environmental

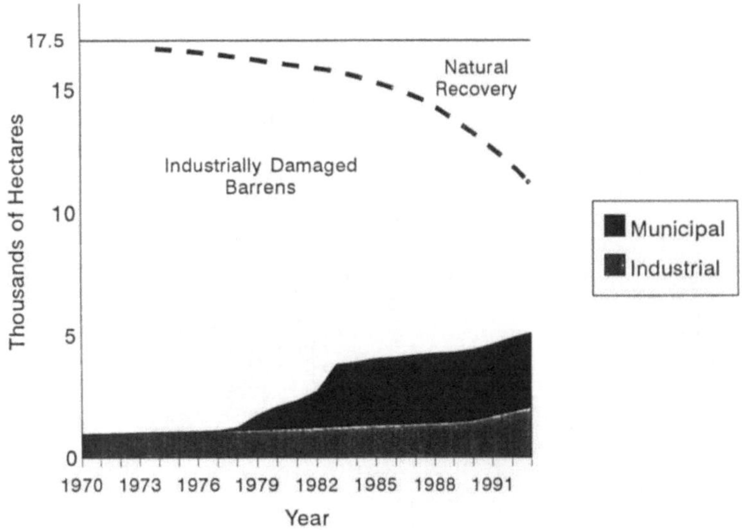

FIGURE 26.1. Changes in the extent of industrial barrens in the Sudbury area as a result of natural recolonization by plants after emission reductions in the early 1970s, and through re-vegetation efforts of municipal and industrial land reclamation programs. In neither the natural recovery areas nor the treated areas is restoration "complete"; the shrinkage in the barren area simply implies that a plant cover has been re-established.

cleanup in Sudbury would not have happened earlier.

Once initiated, what factors shaped or encouraged the particular directions that the Sudbury restoration efforts took?

1. Government pollution control legislation was the essential stimulus for industrial cleanup, but regulations were applied with "patience," giving the industries sufficient time and freedom to develop optimal strategies for pollution control.
2. The abatement programs occurred during a prosperous time for the mining industry (e.g., more than a billion dollars have been spent in Sudbury on emissions control and land and water treatments in the past 30 years), and government agencies were able to support expensive programs for environmental monitoring, land reclamation, research, etc. A great many other industries and countries would not have had these financial resources.
3. Economic benefits for the companies were obtained through the technology developments designed to meet environmental protection requirements (e.g., energy efficiency, worker productivity, marketable products from former waste).
4. Effective partnerships developed between industry, government, academia, and the public to design and implement restoration projects. Again, timing was important. People and groups were ready to work together and synergistic benefits from cooperation were soon obvious.
5. A minimal treatment approach for damaged lands proved effective, demonstrating that substantial gains could be made by assisting and working with nature—assisting the healing process—rather than striving for an overly designed and manipulated landscape (earlier engineered solutions). This cost-effective ecological approach emphasized the use and re-establishment of mainly native species.
6. Restoration projects did not wait for perfect solutions but focused on achievable goals by remaining flexible and making use of a variety of funding and staffing opportunities. Projects were supported and enriched through expert opinion and accumulated practical experience.
7. Efforts were made to involve the public in the restoration programs through direct participation and education. Volunteers are now recognized as essential to the continuation of the program. An informed and involved public is also needed to direct political actions through their elected representatives.

Figure 26.2. Environmental activist jumps off smokestack at Gavin power plant near Gallipolis, Ohio in October 1984 to protest acid-causing emissions. (Photo by Greenpeace.)

Public Pressure and Legislation

Public pressure was the incentive that forced governments to demand reductions in atmospheric emissions from industries such as the Sudbury smelters. The early advances in environmental control were largely driven by the concerns of local residents about poor air quality. However, by the 1980s widespread concern about "acid rain" created an enormous lobbying effort for environmental improvements (Fig. 26.2). Acid rain was probably the first regional, if not global, environmental problem around which people rallied and demanded government action. Public interest in environmental issues tends to vary with social and economic conditions in the country (Dunlap and Scarce 1991), but public pressure on governments will, no doubt, continue to force change. Indeed, Sudbury industries will probably be faced with more stringent regulations in the future, for although they have achieved nearly 90% reductions in sulfur dioxide, they remain as very large emissions sources (Table 26.1), and the worldwide problem of acidification has not gone away (Jeffries and Lam 1993; Galloway et al. 1994).

TABLE 26.1. Current and Historical (Highest Year) Sulfur Emissions from Inco and Falconbridge Smelters in Comparison with Current (1990s) Levels for the 25 Largest Emitters of Sulfur in Europe.[a] Estimated Emissions from Noril'sk in Eastern Russia Are Also Included[b]

Name of plant	Type	Location	Emission (kilotonnes sulfur)
Noril'sk (late 1980s)	Smelter	Russia	1,150,000
Inco (1960)	**Smelter**	**Canada**	**1,000,000**
1. Marista	Power station	Bulgaria	350,000
2. Puentes Garcia Rodriguez	Power station	Spain	271,000
3. Jänschwalde	Power station	Germany	215,000
4. Montsegorsk	Smelter	Russia	212,000
5. Nikel	Smelter	Russia	211,000
Falconbridge (1968)	**Smelter**	**Canada**	**190,000**
6. Reftinsk	Power station	Russia	184,000
7. Teruel	Power station	Spain	183,000
8. Turceni	Power station	Romania	183,000
9. Elbistan	Power station	Turkey	180,000
10. Belchatow	Power station	Poland	171,000
11. Afsin-Elbistan	Power station	Turkey	145,000
12. Soma	Power station	Turkey	143,000
13. Prunerov	Power station	Czech Republic	137,000
Inco (1994)[c]	**Smelter**	**Canada**	**133,000**
14. Drax	Power station	United Kingdom	132,000
15. Zaporzhe	Power station	Ukraine	129,000
16. Yatagan	Power station	Turkey	127,000
17. Boxberg	Power station	Germany	126,000
18. Moldavia	Power station	Moldavia	122,000
19. Kemerkoy	Power station	Turkey	118,000
20. Lukomyl	Power station	Belarus	116,000
21. Novocherkassk	Power station	Russia	116,000
22. Yenikoy	Power station	Turkey	113,000
23. Ryazan	Power station	Russia	110,000
24. Hagenwerder	Power station	Germany	110,000
25. Turow	Power station	Poland	103,000
Falconbridge (1993)	**Smelter**	**Canada**	**29,000**

[a]From report by M. Barrett and R. Protheroc, summarized by Agren (1994).
[b]Ehrensvard (1993).
[c]Maximum allowable emissions for 1994.

Public pressure also acts directly on polluting companies. A negative public image not only damages social and economic confidence in that company and industry, it runs the risk of inducing boycotts by consumers and investors (e.g., European attitudes toward North American harvest of "old growth" forests or fur bearers). Public attitudes and the reality of past-destructive activities also affects the ability of a company to attract and maintain high-quality staff. In the past, environmental concerns appeared to be less of an issue when workers sought places of employment. They simply went where the jobs were. However, now employees are often very concerned about the quality of their environment, and if given a choice would not raise a family in a "dirty" town or work for a "dirty" company. Under these socioeconomic pressures, it is therefore important for each mining company to not only actively participate in pollution prevention and restoration efforts but also to be perceived to be making progress in these areas, both as a company and as part of a larger industry.

Economic Benefits and Costs

There are now many economic incentives for pollution control. In this case study, the contaminants that were responsible for environmental damage (i.e., copper, nickel, sulfur) are actually the products (or had the potential to be, in the case of sulfur) that the companies market. Preventing their loss is of obvious economic benefit (see input management discussion by Odum [1989]). Once forced to invest in abatement programs and to conduct strategic planning (i.e., beginning when the ore is still in the ground, rather than trying to deal with a pollutant that is already in the smokestack), considerable economic efficiencies, particularly in energy consumption, were achieved. These were of direct benefit to the companies. The reduced fossil fuel use also lowers emissions of greenhouse gases, another global environmental problem. Interestingly, this energy efficiency gain was the same one that followed from technology development conducted in the 1970s by Japanese car companies to meet stringent U.S. automobile emission standards (Nishimura 1989). Not only did the Japanese meet the pollution standards, but they achieved energy efficiency in the new engines that greatly increased consumer demand for their product.

The technological developments that were needed to meet the reduced sulfur dioxide limits were part of an overall modernization program that was necessary even without the environmental legislation. The legislation may have speeded up the process, but improvements in the smelters and in the overall efficiency of the industrial operations had to occur if the local mining companies were to remain competitive and survive in the world market place. In fact, the companies in this case did not use standard pollution control technologies (e.g., scrubbers) to meet their emission abatement requirements. They achieved these requirements by reorganizing their entire operation as part of the larger modernization program.

Unfortunately, a frequent consequence of modernization of industries is the loss of jobs. In the past 20 years, the number of workers at the mines and mills of Sudbury has been reduced by more than 50%, while production of nickel has remained largely unaffected. The socioeconomic impacts on the community from such displacement of workers has at times been very severe. An aggressive community-based economic diversification and job creation program and considerable federal and provincial government support have cushioned many of the adverse effects of workforce reductions. However, many of the changes, both environmental and economic, have been very positive for the region (expanding tourism, government jobs, new industries, etc.). No adequate socioeconomic study has been done to access the net effect of these changes. Such studies are needed (Costanza 1991).

Although there were opportunities to gain economic benefits and improve the workplace environment within the industrial complex, solving the problem of a severely damaged external landscape is a different matter. Healing damaged ecosystems will no doubt prove to be a very expensive process, perhaps far more expensive than the costs of construction of facilities that caused the damage (Cairns 1993). Mining companies in North America and elsewhere are now forced through legislation to be financially responsible for environmental problems created by their operations and required to return mining and processing sites to a "rehabilitated state" before leaving an area. This responsibility greatly increases the cost of doing business (approaching true cost-accounting for the cost of extracting resources) and requires the development and application of much new science and technology. However, one of the many societal benefits from this new legislation is that restoration work will generate many jobs. These jobs may compensate for some of the jobs lost during modernization of industries (Renner 1992).

Role of Partnership in Restoration

If one had to choose a single reason for the achievements in Sudbury honored at the UN conference in Brazil, it would be the cooperation and partnerships that developed to assist restoration in this area. Partnerships did not develop through the policies of government agencies or industries. They occurred because government scientists, resource managers, university professors, municipal planners and staff, and a variety of industry personnel took the initiative and decided to collaborate and begin the restoration efforts. The enormity of the problem and the multidisciplinary nature of any potential solution demanded cooperation. The authority of government regulations still existed to drive the process, but time and expertise were not wasted on assigning blame.

It is difficult to assess why some partnerships succeed while others fail. Certainly, in our experience with cooperative projects, success is largely dependent on the quality and commitment of individual people involved. A few dedicated individuals can make a great deal of positive difference even when faced with severe environmental damages. Other guiding principles of successful partnerships were that they

1. begin with and continually enhance understanding of each of the partners' needs
2. consider that each discipline/stakeholder has something positive to contribute
3. require frequent and effective communication
4. function under the belief that cooperation can achieve more quickly and can attain larger goals than would result from the sum of individual efforts
5. remain flexible enough to seize opportunities (sources of funding, participation of volunteers, etc.)
6. measure even small progress
7. celebrate success (awards, certificates, media attention)

Role and Opportunities for Science—Restoration Ecology

This book began with a quote by A.D. Bradshaw that restoration research was the "acid test" of our understanding of how natural ecosystems function. Restoration projects represent unique and important opportunities to conduct research that will contribute in both basic and applied areas of ecology (Watson and Richardson 1972; Bradshaw 1983, 1993; Jordan et al. 1987). Such opportunities should not be missed. Authors in this volume have described some of the many research needs and questions that still exist in the Sudbury area. Several of these are very broad needs (e.g., rates and processes of biological recovery, chemistry and biology of degraded soils, socioeconomics of restoration), similar to some of the high-priority research items identified by the Ecological Society of America in support of its global sustainable biosphere initiative (Lubchenco et al. 1991).

There is a great need for more science in the field of restoration ecology. Far too little study has been conducted, and many industrially damaged areas could benefit from the published results of rigorous research programs in this field (Bradshaw 1993). Our experience indicates that it is easy to underestimate the need for proper scientific methods in this area of research.

Industrially damaged ecosystems such as Sudbury are the important "natural laboratories" where restoration ecology research must be conducted. An essential aspect of the design of any research work in this area is the need for controls and reference sites. Change in treated areas can only be realistically assessed against the standard of results from more pristine sites. In the same sense, some sites within the damaged area of Sudbury should be left untreated, both to illustrate to the public how far we have come but also, for purely scientific reasons, to study natural recovery. A damaged area "reserve" may also serve as a reminder about what is "just over the next hill," so that the buffer strips of trees along the highways will not prove to be a facade but simply a beginning.

A frustrating problem with ecological studies is that scientific understanding usually requires considerable time to develop. However, resource managers and administrators of large-scale restoration programs frequently cannot wait for perfect answers, and projects must often move ahead with the "best available information." Here, an experienced research scientist can make a substantial contribution by giving time and expert opinion to assist in restoration efforts, but time and funding must also be provided to at least measure change as a function of the applied restoration treatment. If we wish to improve treatment procedures, it is important that we carefully monitor environmental and ecological changes, and rigorously attempt to determine what caused these changes to occur.

Norton (1992) considered the participation of traditionally cautious and conservative research scientists in providing "expert opinion" as an essential part of a new paradigm in ecosystem management. Another aspect of this change, which we think deserves to be included in the use of the term *paradigm*, is the increasing involvement of industry in supporting environmental research. Industry must take a larger and more active direct role in science development in the field of restoration ecology, especially now in North America, when government support for research is dwindling under difficult financial constraints.

From "Environmental Policies" to "Environmental Ethics"

The assumption of responsibility by industry for environmental damages, the increased participation of industry in science development and in large-scale restoration programs, the open exchange of information by industry with government regulators and the public, and the importance of public presentation of environmental assessment findings before development of new sites (e.g., Thayer Lindsley Mine proposal by Falconbridge Limited in 1993) are some of the evidence that profound changes are occurring in how business is conducted by many companies. It is conceivable that in the future many industries will actively participate in policy and even legislative developments for environmental protection. In fact, some companies have already made commitments to achieve standards that exceed legislative requirements and to include environmental protection elements not considered by legislators (e.g., some recent initiatives by Shell Oil of Canada).

Certainly, public relations and economic pressures are important incentives for the development of corporate environmental policies. However, one should not dismiss the idea that a corporate "environmental ethic" (i.e., "that it is the right thing to do") is also emerging. The personal commitment of individual executives (Aitken 1991; see Foreword to Section E) and business practices such as the use of western environmental standards by companies setting up plants in developing countries without strong environmental regulations suggest that motives for change are not all profit-oriented. A cynic could easily dismiss these signals, but it is indisputable that such a change in attitude is needed to deal with the enormous environmental problems we face.

Steps toward Sustainable Ecosystems

We do not want to be labeled as naive "enthusiasts" who do not recognize the enormous global challenges we face (Hardin 1993), but this case history does provide many points for optimism and several suggestions for moving toward the goal of sustainable ecosystems. We got into this mess by adopting attitudes and actions that suggested that humans were not part of the global ecosystem, that resources were limitless, and that when ecosystems were damaged or soiled, we could simply move on. Now we know that these ideas were wrong. Humans are a part of nature, an increasingly large part (40% of net primary productivity of the land is in human enterprises [Ehrlich and Ehrlich 1991]), limits are rapidly being reached, and the "nomads" have nowhere else to go.

Environmental improvements can occur rapidly if people rethink and plan for the long-term future (NRC 1992). Hundreds of pieces of environmental legislation have been established within the past 25 years, and there are many dramatic cases of environmental improvement (e.g., Lake Erie, Thames River, Singapore, Sudbury). Environmental consciousness in is now well established in a great many people, including some important politicians and corporate executives. There are also encouraging signs that cooperation rather than confrontation is gaining momentum on the environmental front. However, time to begin restoration and recovery of damaged ecosystems cannot be wasted.

This book about Sudbury was originally designed to contribute knowledge to the global goal of being able to protect, restore (Bratton 1992), and maintain healthy sustainable ecosystems (WCED 1987; Turner 1988). However, as Sudbury residents, we have probably learned the most from this exercise. It is now clear that although substantial gains have been made, we still have a difficult job ahead of us for solving our own local environmental problems. In addition, we realize that few countries will ever have the same financial resources to conduct such a large-scale restoration program. However, this fact should not discourage others from beginning restoration work. Our land reclamation program began very modestly with very little money and mainly with the assistance of volunteers, demonstrating that great progress can be made from humble beginnings.

Finally, although people should not be discouraged by the challenge of restoration, the principal lesson from our work is that preventing such damage from occurring in the first place is surely the more sensible course of action.

Clever people know how to solve problems.
Wise people know how to avoid them.
Albert Einstein

Acknowledgments. We thank Michael Conlon for his assistance to the working group and for preparation of graphics. Harry Struik provided data through interpretation of aerial photography for the estimation of extent of natural recovery in Figure 26.1. David Balsillie, Bill Keller, Ed Snucins, and Frank Wilson reviewed earlier drafts of the manuscript.

References

Agren, C. 1994. The worst hundred sources. Acid News 3:1–3.

Aitken, W.R.O. 1991. Industry's response, pp. 283–284. *In* C. Mungall and D.J. McLaren (eds.). Planet under Stress. Oxford University Press, Toronto.

Bradshaw, A.D. 1983. The restoration of ecosystems. J. Applied Ecol. 20–1–17.

Bradshaw, A.D. 1993. Restoration ecology as a science. Restoration Ecol. 1:71–73.

Bratton, S.P. 1992. Alternate models of ecosystem restoration, pp. 170–189. *In* R. Costanza, B.G. Norton, and B.D. Haskell (eds.). Ecosystem Health. Island Press, Washington, DC.

Cairns, J., Jr. 1993. The Zen of ecological restoration: eight steps on the path towards enlightenment. Ann. Earth 9:13–15.

Costanza, R. (ed.). 1991. Ecological Economics: The Science and Management of Sustainability. Columbia Univ. Press, New York.

Crocker, R.L., and J. Major. 1955. Soil development in relation to vegetation and surface area at Glacier Bay, Alaska. J. Ecol. 43:427–448.

Dunlap, R.E., and R. Scarce. 1991. The polls: environmental problems and protection. Public Opinion Q. 55:651–672.

Edwards, M. 1994. Pollution in the former USSR: lethal legacy. Nat. Geog. 186:70–98.

Ehrensvard, A. 1993. A profitable polluter. Acid News 5:20.

Ehrlich, R.R., and A.H. Ehrlich. 1991. Healing the Planet. Addison-Wesley Publishing Co., New York.

Galloway, J.N., H. Levy II, and P.S. Kasibhatla. 1994. Year 2020: Consequences of population growth and development on deposition of oxidized nitrogen. Ambio 23:120–123.

Gore, A. 1992. Earth in the Balance. Houghton Mifflin, Boston.

Hardin, G. 1993. Living within Limits: Ecology, Economics and Populations Taboos. Oxford University Press, New York.

Houghton. R.A. 1994. The worldwide extent of land-use change. Bioscience 44:305–313.

Jeffries, D.S., and D.C.L. Lam. 1993. Assessment of the effect of acidic deposition on Canadian lakes: determination of critical loads for sulphate deposition. Wat. Sci. Tech. 28:183–187.

Jordan, W.R., M.E. Gilpin, and J.A. Aber (eds.). 1987. Restoration Ecology: A Synthetic Approach to Ecological Research. Cambridge University Press, Cambridge.

Lubchenco, J., and 15 other coauthors. 1991. The sustainable biosphere initiative: an ecological research agenda. Ecology 72(2):371–412.

Moore, J.N., and S.M. Luoma 1990. Hazardous waste from large-scale metal extraction. Envir. Sci. Technol. 24:1278–1289.

National Research Council (NRC). 1992. Restoration of Aquatic Ecosystems. National Academy Press, Washington, DC.

Nishimura, H. (ed.). 1989. How to Conquer Air Pollution—a Japanese Experience. Elsevier, Amsterdam.

Norton, B.G. 1992. A new paradigm for environmental management, pp. 23–41. *In* R. Costanza, B.G. Norton, and B.D. Haskell (eds.). Ecosystem Health. Island Press, Washington, DC.

Odum, E.P. 1989. Input management of production systems. Science 243:177–182.

Potter, V.R. 1988. Global Bioethics: Building on the Leopold Legacy. Michigan State University Press, East Lansing.

Renner, M.G. 1992. Saving the earth; creating jobs. World Watch 5(1):10–17.

Smil, V. 1993. China's Environmental Crisis. East Gate Book, M.E. Sharpe, Armonk, NY.

Turner, R.K. (ed.). 1988. Sustainable Environmental Management. Principles and Practices. Westview, Boulder, CO.

Wali, M.K. 1992. Ecosystem Rehabilitation. SPB Academic Publishing, The Hague, The Netherlands.

Watson, W.Y., and D.H.S. Richardson. 1972. Appreciating the potential of a devastated land. Forest Chron. 48:313–315.

Woodwell, G.M. 1994. Ecology: The restoration. Restoration Ecol. 2:1–3.

World Commission on Environment and Development (WCED). 1987. Our Common Future. Oxford University Press, Oxford.

Young, J.E. 1992. Mining the earth, pp. 99–118. *In* L.R. Brown et al. (eds.). State of the World 1992. W.W. Norton Ltd., New York.

Index

Entries occurring in figures are followed by an f; those occurring in tables, by a t.

MIX
Papier aus verantwortungsvollen Quellen
Paper from responsible sources
FSC® C105338

If you have any concerns about our products,
you can contact us on
ProductSafety@springernature.com

In case Publisher is established outside the EU,
the EU authorized representative is:
Springer Nature Customer Service Center GmbH
Europaplatz 3, 69115 Heidelberg, Germany

Printed by Libri Plureos GmbH
in Hamburg, Germany